BOTANICALS

BOTANICALS

Methods and Techniques for Quality & Authenticity

EDITED BY

KURT A. REYNERTSON
Johnson & Johnson, New Jersey, USA

KHALID MAHMOOD
Johnson & Johnson, New Jersey, USA

CRC Press
Taylor & Francis Group
Boca Raton London New York

CRC Press is an imprint of the
Taylor & Francis Group, an **informa** business

CRC Press
Taylor & Francis Group
6000 Broken Sound Parkway NW, Suite 300
Boca Raton, FL 33487-2742

First issued in paperback 2019

ISBN-13: 978-1-4665-9841-6 (hbk)
ISBN-13: 978-0-367-37784-7 (pbk)

This book contains information obtained from authentic and highly regarded sources. Reasonable efforts have been made to publish reliable data and information, but the author and publisher cannot assume responsibility for the validity of all materials or the consequences of their use. The authors and publishers have attempted to trace the copyright holders of all material reproduced in this publication and apologize to copyright holders if permission to publish in this form has not been obtained. If any copyright material has not been acknowledged please write and let us know so we may rectify in any future reprint.

Visit the Taylor & Francis Web site at
http://www.taylorandfrancis.com

and the CRC Press Web site at
http://www.crcpress.com

Contents

Preface...ix

Acknowledgments...xi

Editors.. xiii

Contributors xv

Chapter 1 The Importance of Quality and Authenticity for Botanical R&D 1

Kurt A. Reynertson and Khalid Mahmood

Chapter 2 The Importance of Proper Selection of Product Quality
Specifications and Methods of Analysis for Botanical Product
Evaluation 5

Paula N. Brown, Michael Chan, and Joseph M. Betz

Chapter 3 Cultivating Botanicals for Sensory Quality: From Good
Agricultural Practices to Taste Discernment by Smallholder
Tea Farmers 15

Selena Ahmed, John Richard Stepp, and Xue Dayuan

Chapter 4 Using Traditional Taxonomy and Vouchers in Authentication
and Quality Control.. 31

Wendy L. Applequist

Chapter 5 The DNA Toolkit: A Practical User's Guide to Genetic
Methods of Botanical Authentication 43

Danica T. Harbaugh Reynaud

Chapter 6 Metabolic Profiling and Proper Identification of Botanicals:
A Case Study of Black Cohosh 69

Tiffany Chan, Shi-Biao Wu, and Edward J. Kennelly

Chapter 7 A Model for Nontargeted Detection of Adulterants.......................... 91

James Harnly, Joe Jabolonski, and Jeff Moore

Chapter 8   The Promise of Class Prediction: How Multivariate Statistics
                Can Help Determine Botanical Quality and Authenticity 107

                Stephan Baumann

Chapter 9   Conformation of Botanical or Bio-Based Materials
                by Radiocarbon and Stable Isotope Ratio Analysis 125

                Randy Culp

Chapter 10 Botanical Isoscapes: Emerging Stable Isotope Tools
                for Authentication and Geographic Sourcing.................................. 143

                Jason B. West

Chapter 11 Nuclear Magnetic Resonance: A Revolutionary Tool for
                Nutraceutical Analysis 157

                Kimberly L. Colson, Jimmy Yuk, and Christian Fischer

Chapter 12 Quality and Authenticity of Complex Natural Mixtures:
                Analysis of Honey Using NMR Spectroscopy and Statistics........... 183

                István Pelczer

Chapter 13 Application of CRAFT to Data Reduction of NMR Spectra
                of Botanical Samples.. 199

                Krish Krishnamurthy and David J. Russell

Chapter 14 Practical Use of FT-NIR for Identification and Qualification
                of Botanicals: A Fit-for-Purpose Approach 215

                Cynthia Kradjel

Chapter 15 High-Performance Thin-Layer Chromatography for the
                Identification of Botanical Materials and the Detection
                of Adulteration... 241

                Eike Reich, Débora A. Frommenwiler, and Valeria Maire-Widmer

Chapter 16 Sensory and Chemical Fingerprinting Aids Quality
                and Authentication of Ingredients and Raw Materials
                of Vegetal Origin 261

                Marion Bonnefille

Chapter 17 The Hidden Face of Botanical Identity: An Industrial
Perspective on Challenges from Natural Variability
and Commercial Processes on Botanical Authenticity 279

*Leila D. Falcao, Camille Durand, Alexis Lavaud, Marc Roller,
and Antoine C. Bily*

Chapter 18 Aspects of Quality Issues Faced by Botanicals Used
as Cosmetic Ingredients 293

Jean-Marc Seigneuret

Index 305

Preface

Anyone who works with botanicals has at one time or another been confronted with concerns about the quality and authenticity of their materials. The news media fan the flames of these issues; propagating stories of contamination, adulteration, and poor quality; driving apprehension; and seeding doubt in the minds of consumers.

Whether working in academia, government, or industry, at some point in the life cycle of botanical materials, we relinquish some aspect of certainty to others. The concerns are universal and must be addressed by experts, researchers, sourcing managers, and consumers alike.

To that end, the purpose of this book is to compile methods and techniques that can be used to help guide quality and authenticity determinations. Assembling the most current information in one place provides guideposts for method applications and helps point to gaps in this important area of botanical quality assurance.

With the natural trend growing in popularity annually for food, dietary supplements, cosmetics, and personal care products, it was our hope that this volume would serve as a focal point and reference for identifying appropriate quality and authenticity methods specific to the need of the reader.

As new technologies are developed, software and computing power have advanced rapidly to assure that this is an area that will continue to modernize and adapt. The ability to statistically analyze big data sets is emerging as an important aspect, even when applied to more *traditional* chemometric analyses. This edition contains methods which are currently in use as well as methods which are still being developed, with the promise of becoming more routine soon. We deliberately chose to have both the new technologies and more traditional techniques. The authors represent industry, government, and academia. And while we realize that there are as many omissions as inclusions in the topics covered, we have tried to make this first edition as comprehensive as possible. Gaps remain, which future editions may rectify. For that reason, we hope that the readers and users of this book will reach out with their feedback and constructive suggestions.

Kurt A. Reynertson
Johnson & Johnson

Khalid Mahmood
Johnson & Johnson

Acknowledgments

I thank and acknowledge my brother Tariq Mahmood. Without him, I would not have been able to initiate or complete this project.

Khalid Mahmood

The original idea for this book sprang out of a conversation with Hilary Rowe following a scientific session on the topic at the International Congress on Natural Products Research in New York City in 2012. It has been a long journey to assemble, and I appreciate the support of my family and management in allowing me the time to dedicate to the project.

Kurt A. Reynertson

Acknowledgements

Editors

Kurt A. Reynertson has worked in natural products chemistry for over 15 years and published numerous research articles in international journals. He completed his PhD in phytochemistry at the City University of New York and worked as a postdoctoral researcher in cancer pharmacology at Weill Cornell Medical College before joining the Naturals Platform team at Johnson & Johnson Consumer Products, Inc. His work is primarily focused on early discovery R&D, which puts him on the front lines of botanical quality and authenticity issues.

Khalid Mahmood graduated with a PhD degree in organic synthesis chemistry from a Natural Products Institute in Pakistan. Pursuing a synthesis-based graduate program at the center of excellence for natural products chemistry provided an opportunity to learn both disciplines. This turned out to be the basis upon which he would build his professional career. Since his arrival to the United States as a postdoctoral scholar, Dr. Mahmood has accumulated significant exposure in the areas of brain receptor research at University of Pittsburgh and consumer products research working for small-sized companies in Illinois and California. For the past eight years, Dr. Mahmood has been the Naturals Platform leader at Johnson & Johnson Consumer Products, Inc. This is where combined experience from various sectors has helped him obtain inventorships on eight patents and multiple patent applications claiming natural technologies useful for personal care products. Dr. Mahmood has won multiple leadership awards for his initiatives at Johnson & Johnson to innovate and to manage discovery programs. He is recognized by Johnson & Johnson Consumer Products, Inc. for his research and outstanding record of collaborations earning him a Johnson & Johnson Consumer Fellow designation.

Contributors

Selena Ahmed
Department of Health and Human
  Development
Montana State University
Bozeman, Montana

and

Minzu University of China
Beijing, People's Republic of China

Wendy L. Applequist
William L. Brown Center
Missouri Botanical Garden
St. Louis, Missouri

Stephan Baumann
Agilent Technologies, Inc.
Santa Clara, California

Joseph M. Betz
Office of Dietary Supplements
U.S. National Institutes of Health
Bethesda, Maryland

Antoine C. Bily
Naturex SA
Avignon, France

Marion Bonnefille
Alpha MOS
Toulouse, France

Paula N. Brown
Natural Health & Food Products
  Research Group
Centre for Applied Research and
  Innovation
British Columbia Institute of
  Technology
Burnaby, British Columbia, Canada

Michael Chan
Natural Health & Food Products
  Research Group
Centre for Applied Research and
  Innovation
British Columbia Institute of
  Technology
Burnaby, British Columbia, Canada

Tiffany Chan
Department of Biological Sciences
Lehman College and The Graduate
  Center
The City University of New York
Bronx, New York

Kimberly L. Colson
Bruker BioSpin
Billerica, Massachusetts

Randy Culp
Center for Applied Isotope Studies
The University of Georgia
Athens, Georgia

Xue Dayuan
Minzu University of China
Beijing, People's Republic of China

Camille Durand
Naturex Inc
South Hackensack, New Jersey

Leila D. Falcao
Naturex SA
Avignon, France

Christian Fischer
Bruker BioSpin GmbH
Rheinstetten, Germany

Débora A. Frommenwiler
CAMAG Laboratory
Muttenz, Switzerland

Danica T. Harbaugh Reynaud
AuthenTechnologies
Richmond, California

and

The Center for Herbal Identity
University and Jepson Herbaria
Department of Integrative Biology
University of California
Berkeley, California

James Harnly
Food Composition and Methods
    Development Lab
Beltsville Human Nutrition Research
    Center
U.S. Department of Agriculture
Beltsville, Maryland

Joe Jabolonski
Institute for Food Safety and Health
Food and Drug Administration
Chicago, Illinois

Edward J. Kennelly
Department of Biological Sciences
Lehman College and The Graduate
    Center
The City University of New York
Bronx, New York

Cynthia Kradjel
Integrated Technical Solutions
New Rochelle, New York

Krish Krishnamurthy
Research Products Division
Agilent Technologies
Santa Clara, California

Alexis Lavaud
Naturex SA
Avignon, France

Valeria Maire-Widmer
CAMAG Laboratory
Muttenz, Switzerland

Jeff Moore
U.S. Pharmacopeial Convention
Rockville, Maryland

István Pelczer
Department of Chemistry
Frick Chemistry Laboratory
Princeton University
Princeton, New Jersey

Eike Reich
CAMAG Laboratory
Muttenz, Switzerland

Marc Roller
Naturex SA
Avignon, France

David J. Russell
Research Products Division
Agilent Technologies
Santa Clara, California

Jean-Marc Seigneuret
Alban Muller Group
Montreuil, France

John Richard Stepp
Minzu University of China
Beijing, People's Republic of China

and

Department of Anthropology
University of Florida
Gainesville, Florida

Jason B. West
Department of Ecosystem Science
  and Management
Texas A&M University
College Station, Texas

Shi-Biao Wu
Department of Biological Sciences
Lehman College and The Graduate
  Center
The City University of New York
Bronx, New York

Jimmy Yuk
Bruker BioSpin
Billerica, Massachusetts

1 The Importance of Quality and Authenticity for Botanical R&D

Kurt A. Reynertson and Khalid Mahmood

The international trade in plants is growing steadily as the worldwide demand for natural and botanical raw materials increases. In 2012, herbal dietary supplement sales were the highest ever, at approximately $5.6 billion [1]. However, this doesn't include cosmetics and teas. One recent estimate of the global market in herbal personal care and cosmetics sets the value at $12 billion in 2005 [2].

Increasingly, customers value natural products and botanicals as *green* alternatives; herbs and naturally derived ingredients are perceived as safer for both human use and environmentally and socially responsible choices. They are increasingly distrustful of what is perceived as *unnatural* chemical ingredients. Mainstream consumers expect that the companies who make these products are leaders in this respect.

Although it may be true that plants with a long history of traditional use are generally regarded as safe for human consumption, when considering the production of raw materials, the life of a plant from seed to shelf involves many complicated issues. Botanicals are different from traditional chemical raw materials and require a specific type of quality assurance and quality control. Botanical ingredients for herbal supplements and personal care products are partially manufactured as living organisms. In addition, they also typically undergo processing to become crude extracts or highly processed ingredients. Regardless of the level of final processing, quality considerations fall into three general categories: safety, authenticity, and sustainability.

There have been many well-publicized accidental and intentional contamination of botanicals with pharmaceuticals and/or incorrect plant species. In the absence of good manufacturing practices (GMPs) for herbals in 1997, several lots of plantain (*Plantago* spp.) were found to be adulterated with *Digitalis lanata* [3]. Lack of proper quality measures resulted in the mislabeling of over 3000 pounds of a relatively innocuous herb with one containing potent cardiac glycosides. Over 15 years later, we still face the need to develop new quality control methods for botanical materials. Unintentional contamination and economically motivated adulteration are issues that must be considered when dealing with plant materials or extracts that have been purchased from traders and other consolidators, or when the provenance of the raw materials is unclear or unknown. In order to insure that botanical raw materials are safe and effective, the supply chain must be tightly monitored and carefully considered. It is also necessary to have testing methods that are accurate, appropriate, and cost-effective in order to test and analyze materials.

Many countries have regulations governing the safety and labeling of botanicals for food, medicine, and supplements. In the United States, the Dietary Supplement Health and Education Act of 1994 (DSHEA) defines the regulatory environment for supplements; 21 CFR 111 states that a *scientifically valid* method of analysis must be used to ascertain the purity, authenticity, and quality of a dietary supplement. However, the onus of rigorously applying safety standards comes from peer review, consumer pressure, and industry alignment.

Generally, plants should be grown in accordance with good agricultural and collection practices and should be stored and processed in accordance with GMPs. In addition, consideration must be given to international treaties such as the Convention on International Trade in Endangered Species, the Convention on Biodiversity, and the Endangered Species Act. This insures a level of quality that affects all levels of safety, authenticity, and sustainability; issues of access and benefit-sharing related to ethnobotanically sourced material are also fundamental to ethical business practice.

Quality and authenticity assurance for natural materials may have its roots in classical taxonomic methods, but there are limitations. The shifting natural land-scapes calls for a need to develop appropriate technologies in addition to traditional tools. The development of new and sensitive methodologies that can answer specific questions is partly due to the increased popularity of natural ingredients and partly due to technological advances that have been applied in new ways to answer those questions. Most of the publications detailing the use of new methodologies for the authentication of natural materials tend to focus on high-value commodities like flavors and fragrances [4,5], spices [6,7], and honey [8,9]. The quality of botanical products, however, is more than an economic issue; regardless of whether the raw materials are used for food, supplements, or medicine; the identification and authen-tication issue is also a public health and safety concern. In a post-DSHEA United States, the quality issue as it relates to botanicals has been brought to the forefront of consumer awareness.

Today natural ingredients are used in cosmetics, dietary supplements, herbal medicines, personal care, and household products. These industries deploy nat-ural ingredients in slightly different ways requiring differing needs of quality assurance. We saw a need to consolidate various methodologies in one place for the user to consider how to apply appropriate methods to achieve authentication of new and existing natural ingredients, thereby enhancing quality and safety of the ingredients. This knowledge should also help to mitigate economically moti-vated adulteration or incidental contaminations. This volume is an attempt to bring together several different tools for botanical quality and authentication in one place, including traditional, taxonomic, and newer analytical tools. It is not perfect to the satisfaction of every need but a good start. We hope that future revisions of the volume, if taken up, continue to capture most recent advances of this area. This book may not provide how-to instructions, but should be able to point the reader in the right direction.

REFERENCES

1. Lindstrom, A., C. Ooyen, M.E. Lynch, and M. Blumenthal. Herb supplement sales increase 5.5% in 2012. *HerbalGram*, 2013, **99**: 60–64.
2. Bird, K. North American naturals market returns to growth. http://www.cosmeticsdesign.com/Market-Trends/North-American-naturals-market-returns-to-growth.
3. Blumenthal, M. Industry alert: Plantain adulterated with digitalis. *HerbalGram*, 1997, **409**: 28.
4. Mosandl, A. Capillary gas chromatography in quality assessment of flavors and fragrances. *Journal of Chromatography*, 1992, **624**(1/2): 267–292.
5. Mosandl, A., A. Dietrich, B. Faber, C.J. Frank, V. Karl, D. Lehmann, B. Maas, and B. Weber. Authenticity assessment of flavors and fragrances. in *Current Status and Future Trends in Analytical Food Chemistry, Proceedings of the European Conference on Food Chemistry*, 8th edn., Vienna, Austria, September 18–20, 1995.
6. Oberdieck, R., Cardamom. *Fleischwirtschaft*, 1992, **72**(12): 1657–1663.
7. Poole, S.K., W. Kiridena, K.G. Miller, and C.F. Poole. Planar chromatographic methods of determination of the quality of spices and flavors as exemplified by cinnamon and vanilla. *Journal of Planar Chromatography—Modern TLC*, 1995, **8**(4): 257–268.
8. Hawer, W.D., J.H. Ha, and Y.J. Nam. The quality assessment of honey by stable carbon isotope analysis. *Analytical Science & Technology*, 1992, **5**(2): 229–234.
9. Mazzoni, V., P. Bradesi, F. Tomi, and J. Casanova, Direct qualitative and quantitative analysis of carbohydrate mixtures using 13C NMR spectroscopy: Application to honey. *Magnetic Resonance in Chemistry*, 1997, **35**(Spec. Issue): S81–S90.

2 The Importance of Proper Selection of Product Quality Specifications and Methods of Analysis for Botanical Product Evaluation

Paula N. Brown, Michael Chan, and Joseph M. Betz

CONTENTS

Introduction 5
Limitations of Indirect Methods—Protein Products Adulterated with Melamine 8
Limitations of Indirect Methods—Amaranth Dye Adulteration in Bilberry 9
Method Precision, Accuracy, and Intended Purpose in Relation to Product
Specifications 10
Impact of Specification Selection on Accuracy and Precision—Determination
of Ginsenosides in Ginseng 11
Conclusion 12
References 12

INTRODUCTION

Recently, there has been a significant public outcry over the presence of adulterated food products in the marketplace. The most highly publicized of these incidents resulted in human deaths or injuries in China and was caused by adulteration of wheat gluten and other protein-based ingredients with melamine and cyanuric acid [1]. The public reaction focused on the quality of imported food and the inadequacy of import monitoring systems, but a deeper issue is the adequacy of test methods used for the evaluation of product and ingredient quality. The problem is not unique to conventional plant-derived food products. Many tests that are used for the evaluation and quality assurance of botanical products, regardless of whether they are considered drugs or dietary supplements, can give misleading answers if not selected and interpreted correctly. The proper selection and implementation of analytical methods is essential for the assurance of product quality, safety, and regulatory compliance.

In 2007, the US Food and Drug Administration (FDA) issued a final rule on current good manufacturing practice (CGMP) in manufacturing, packaging, labeling, or holding operations for dietary supplements [2]. This is only the most recent of many regulations that have been introduced in the United States to address product quality. The first such legislation was the Drug Importation Act of 1848 [3–5]. This particular legislation was introduced in response to adulterated botanical drug products entering the United States [3–5]. Among its provisions, the legislation required that imported drug products be inspected and tested prior to acceptance into the country [3]. The products had to conform to standards set in the pharmacopoeias and dispensatories of the United States and Europe [3].

Passage of the Act resulted in a reduction in the number of adulterated products entering the US market [4–6]. Large quantities of *greatly adulterated* botanical products, some *deteriorated by age and other causes* and others having had their *active properties removed*, were seized and condemned by customs authorities. Among the notable seizures were opium and scammony that already had their alkaloids and purgative glycosides, respectively, removed; jalap root of "spurious or bastard varieties, mixed with a small proportion of the genuine root"; and yellow bark and cinchona bark that was of a "bastard variet[y] that afford[ed] no quinine and very little if any cinchonine" and was considered "worthless for medicinal purposes" [6]. Yet, despite these successes, the Act could not solve all the adulteration problems associated with botanicals as several technical barriers that hindered its enforcement and limited its impact became readily apparent [4,5]. For instance, although the different pharmacopeias and dispensatories served as guides, they did not always agree on quality standards, leading to confusion among both regulators and suppliers over exactly what constituted a quality product [4,5]. Furthermore, there was a paucity of analytical methods for testing against many of the standards, and several of the methods that were available were complicated, were time consuming, and required specific training and expertise to perform [4,5].

Over 150 years later, the aforementioned final rule for CGMPs was published in 2007 [2]. The CGMP regulations require that manufacturers identify steps requiring control and establish specifications for identity, purity, strength, and composition, including limits on those types of contaminants that may adulterate or lead to adulteration of the finished product [2]. Furthermore, Section 111.75 of the CGMP final rule compels manufacturers to verify that these specifications have been met through appropriate, scientifically valid testing or examination [2]. In the century and a half between the two sets of requirements, there have been significant advances in technology and science, yet the issues faced in compliance and enforcement of the dietary supplement CGMP regulations are similar to those that plagued the Importation Act of 1848.

The current regulations specify that manufacturers "ensure that the test and examinations that you use to determine whether the specifications are met are appropriate, scientifically valid methods" and state that "a scientifically valid method is one that is accurate, precise, and specific for its intended purpose" [2]. These three requirements cannot be addressed independently, and it is not possible to say how precise and accurate a given measure is without first knowing the nature of the measurand.

The determination of what is measured and how it is measured is the core of the method's intended purpose. In turn the intended purpose of a method is very much dependent on the matrix in which measurements will be made, the quality standards it is meant to measure against, and what these standards ultimately reveal about the product and/or ingredient.

Many factors need to be considered when selecting and/or developing a method for evaluation against specifications and there exist numerous approaches to evaluating method performance characteristics [7–10]. This chapter will discuss some of these topics but will focus primarily on the importance of the measurand or property being analyzed when selecting and evaluating a method of analysis. These measurands or properties become the measureable components of the product specification through which product quality is judged. A thorough understanding of what each specification represents in terms of its characteristics, liabilities, and limitations is essential to ensure proper selection and evaluation of the methods underpinning the specifications.

Ideally, a product would be easily recognizable and would always conform to its specifications. Unfortunately, products can become adulterated through accidental, negligent, or even intentional means, and proper specifications will shield consumers against these mishaps.

Proper specifications that include scientifically valid methods suitable for the uses detailed therein are essential to product and ingredient characterization. When used as aids in development and maintenance of quality assurance programs, complying with regulations, and planning and interpreting scientific studies; this characterization is a critical element in assuring public safety and effectively documenting salubrious and adverse outcomes [2].

Quality specifications, including levels and profiles of desirable and undesirable constituents, are created for a variety of reasons. Specific chemical constituents, parameters, and/or profiles can be set to define product properties such as permissible natural toxin and pesticide levels and microbial load (and by proxy, safety), phytochemical marker levels (and by proxy, efficacy), ingredient identity, and provenance. It is important to keep in mind that the existence of a specification for an efficacy surrogate does not necessarily establish a causal relationship between the surrogate and the biological endpoint. Associations between markers and biological effects have been sometimes inferred by observing positive clinical or other biological endpoints and correlating these with the composition of the product used, but unless studies designed to establish causality have been performed, only associations can be made, and the identity of the specific active agent may be unknown. For that reason, measuring the definitive bioactive agent or property may be difficult or not possible. In fact, for most herbal products, the identity of these compounds is not known. In other cases, the identity of the putative active chemical(s) may be known, but they are unstable or not accessible to current analytical techniques, making them extremely difficult and/or costly to analyze directly. In these cases, it may be more feasible and economically attractive to measure other compounds or properties that are in some way associated with the causative agents or properties. When utilizing such surrogate methods however, it is important to be aware of and be able to account for their limitations. Failure to do so can have significant consequences, a notable example of which can be seen with the aforementioned protein adulteration.

LIMITATIONS OF INDIRECT METHODS—PROTEIN PRODUCTS ADULTERATED WITH MELAMINE

One of the most publicized series of incidents of product adulteration was the recent reports of adulterated protein products from China. Although the products associated with the incident are not strictly botanicals, the issues, limitations, and consequences associated with the analytical methods at the center of the incident are very similar to the ones that are faced by the botanical industry. Starting in early 2007, severe adverse event reports led to the discovery that *high protein* pet foods, wheat gluten, baby formula, and other dairy products imported from China were adulterated with melamine and cyanuric acid [1]. For these products, protein content is a measure of their quality [11]. The two common analytical methods used to measure protein content, the Kjeldahl method and the Dumas method, do not actually measure protein [12]. Instead, they measure total nitrogen content as a surrogate, assuming that the source of any nitrogen in the product is from protein. The methods cannot differentiate between protein and nonprotein nitrogen. Total nitrogen values obtained using the method(s) are converted to an estimated protein content by using a response factor. The addition of nonprotein nitrogen-containing compounds to products is thus one way to fool these tests and increase their apparent protein content.

The adulteration of these products meant product quality could no longer be assured simply through the determination of the product's nitrogen level. Specifications for these products were expanded to include a test for the detection of certain known adulterants. To ensure safety and quality, several regulatory agencies now require that certain protein-containing products be analyzed for melamine and other nitrogen compounds [1,12]. The FDA published two methods for detecting these substances in 2008, a gas chromatography-mass spectrometry (GC-MS) screening method for dry protein materials [13] and an liquid chromatography-tandem mass spectrometry (LC-MS/MS) method for infant formula [14].

Although these methods have been effective at detecting many adulterated products, it is still important to keep in mind that they too possess a very limited scope. They are intended to test against only one aspect of the product's specification, namely, the absence of melamine and certain other nitrogen-containing compounds. These methods do not provide any information on protein content, and there is no guarantee that they could detect other nitrogen-containing compounds.

The addition of nonprotein nitrogen compounds to these products appears to be predominantly motivated by a desire to increase the products' apparent protein level and mask other adulteration activities [1]. As such, perhaps the most effective method for product quality would be one that would provide a definitive protein level that does not rely on estimation via nitrogen content. Near-infrared methods that measure peptide bonds [15] and methods that measure amino acids [16] are available but were previously considered more expensive and difficult to perform than the nitrogen methods. Concerns about adulteration expressed by the regulatory, scientific, and industrial community have affected this view and driven further development of more specific methods that have shown promise in at least some matrices [12,15,16].

LIMITATIONS OF INDIRECT METHODS—AMARANTH DYE ADULTERATION IN BILBERRY

Another example of overreliance on a nonspecific quality test that has led to adulteration involved bilberry extracts. Bilberry, *Vaccinium myrtillus* L., is a popular dietary supplement and phytomedicine used for vascular and vision conditions [17]. Several clinical trials lend support to claims made about the fruit's therapeutic action [17]. The majority of these trials utilized bilberry fruit extracts standardized to a total anthocyanin content of 25% [17]. Although anthocyanins have not been definitively demonstrated to be responsible for the positive clinical outcomes, the association of 25% bilberry anthocyanin in products with positive outcomes has led to the adoption of this content as a quality specification for the dark red to purple bilberry extracts [17].

For a number of years, a spectrophotometric assay that measured absorbance at 528 nm (a wavelength characteristic of bilberry anthocyanins) was widely used to estimate anthocyanin content in bilberry extracts [18,19]. The method, published in the 2004 British Pharmacopeia, is simple, rapid, and effective, provided the material analyzed is genuine bilberry, an assumption that cannot always be made. Unfortunately, the method is not specific to bilberry extract, as it will respond to any substance that is more or less the same color as bilberry extract. In 2006, Penman et al. analyzed commercial bilberry extracts using two different analytical methods [20]. When the accepted spectrometric method was used, the measured anthocyanin content was 24%; however, when the investigators used a liquid chromatography (LC)/photodiode array detection method, the measured anthocyanin content was only 9% [20]. Further mass spectrometry (MS) and nuclear magnetic resonance (NMR) analyses revealed that the extract contained amaranth dye, a banned synthetic dark red to purple azo-dye commonly referred to as Red Dye No. 2 [20]. As in the protein case described above, an adulterant within the product was used to fool the indirect method used to measure one of the product specifications.

To address the adulteration issues, additional tests that could assess a quality standard other than absorbance at 528 nm had to be added to product specifications. The American Herbal Products Association published several tools and methods to detect the adulteration of bilberry powdered materials [21]. The first is a simple procedure that involves raising the pH of a dilute solution of bilberry extract and observing the presence or absence of a color change. At elevated pH, very dilute anthocyanin-containing solutions turn blue while dilute solutions containing amaranth dye do not. A high-performance thin-layer chromatography (HPTLC) method provides a visual image of separated individual anthocyanins from any dye present [21]. The latter method not only detects the presence of the dye but also provides for the visualization of an anthocyanin profile characteristic of bilberry providing a higher level of confidence that the extract has not been adulterated with other anthocyanin-containing berries [21].

Neither of these methods is intended to replace the original spectrometric method, which allows determination of whether or not the amount of anthocyanins present in the extract is equivalent to the amount found in the product used in the successful clinical trials. To achieve confidence in identity and compositional quality, both the HPTLC and spectrophotometric methods must be used.

To accomplish both qualitative and quantitative evaluation in a single test, more sophisticated methods such as the high performance liquid chromatography (HPLC) method utilized by Penman et al. are required [20]. In 2007, the botanical extract supplier Indena SpA published a validated liquid chromatographic method for bilberry extract, noting in their publication that if used as directed, only 6 of 40 marketed products they evaluated (15%) would deliver a quantity of anthocyanins that was equivalent to the standard of quality established in the previous clinical trials [22]. Furthermore, 10% of the products analyzed lacked anthocyanins and 25% of those analyzed exhibited an LC profile different from typical bilberry extract [22]. This method was subsequently adopted by the European Pharmacopoeia as the official analytical method for bilberry extract [23].

METHOD PRECISION, ACCURACY, AND INTENDED PURPOSE IN RELATION TO PRODUCT SPECIFICATIONS

Even though the use of indirect methods for quantitative determinations in both the melamine and bilberry examples led to significant problems, the use of these methods, or indirect methods in general, may in some cases be appropriate. As with all methods, the context in which they are used and the assumptions associated with their use must be considered when assessing their applicability. The UV method for bilberry extracts as described in the British Pharmacopeia is appropriate for the particular quality specification for which it was designed. Clinical trials had shown that bilberry extracts exhibiting a prescribed value of absorbance at 528 nm was of sufficient quality to be considered effective. The specific purpose of the method was to test this specification. The assumption underlying the utility of this test is that the extract being tested is a bilberry extract. An additional specification accompanied by a test or tests would be needed to determine that the extract is bilberry and does not contain other substances such as dyes or other anthocyanin-containing fruits. The single spectrophotometric specification is insufficient to ensure the quality of the product. Additional tests for the presence of dyes or nontarget anthocyanins are required. Once these specifications are determined, test methods whose purposes are specific for those specifications can be developed and selected. Note that a working knowledge of potential confounders (dyes, other berries) would be very useful in establishing meaningful specifications.

Additional considerations in the creation or selection of quality specifications involve the capabilities of the analytical methods required to test whether or not specifications are met. For dietary supplements, 21 CFR Part 111.75 requires manufacturers to "ensure that the tests and examinations that you use to determine whether the specifications are met are appropriate, scientifically valid methods," and notes that "a scientifically valid method is one that is accurate, precise," as previously discussed, the method must also be "specific for its intended purpose" [2]. Accuracy can be simply defined as the closeness of the value obtained from an analysis to the true value in the sample, whereas precision is defined as the measure of how close individual measurements are to each other [24]. Detailed definitions of these two parameters, as well as the procedures through which they are determined and assessed, are available in guidelines published by several organizations including the AOAC International, FDA, the International Union of Pure and Applied Chemistry, and the International Conference on Harmonisation of Technical Requirements for Registration of Pharmaceuticals for Human Use [7–10]. By following these guidelines, it is possible to determine if a method is accurate and/or precise when used to measure particular properties. However, it is still imperative that the properties being measured are the ones that give the desired information about quality. The Kjeldahl and Dumas methods demonstrated high accuracy and precision in the analysis of nitrogen [11], and the bilberry spectrophotometric method provides precise and accurate absorbance values [20]. Yet, despite their high accuracy and precision, these methods are not useful for the identification of ingredients or the detection of the adulterants associated with the respective products.

Typically, highly precise and accurate methodologies are desirable; however, these traits can be dictated by the quality specification being examined. The most direct example of this can be observed in a phenomenon referred to as the *Horwitz Trumpet* [25]. While reviewing collaborative study data, Dr. William Horwitz discovered a striking relationship between analyte concentration in a matrix and standard deviation [25]. Horwitz observed that in general, as concentration of analyte decreased across two orders of magnitude, the relative standard deviation of reproducibility ($rRSD_r$) increased by a factor of 2 [25]. This relationship was observed regardless of the nature of the analyte or the type of methodology used [25], and thus, the achievable accuracy and precision of a measurement of a selected quality parameter are related to the concentration of the analyte in the test article.

The relationship between $rRSD_r$ and analyte concentration has since been used as a means to evaluate analytical method performance [7,26]. Horwitz developed a formula through which a predicted relative standard deviation of reproducibility ($pRSD_r$) could be determined based on the concentration of analyte [26]. Methods that have $rRSD_r$ that is substantially different from the $pRSD_r$ can be considered flawed [26]. Such methods may benefit from further optimization or development. In certain cases, a reevaluation of the quality standard under scrutiny could be warranted.

IMPACT OF SPECIFICATION SELECTION ON ACCURACY AND PRECISION—DETERMINATION OF GINSENOSIDES IN GINSENG

The main constituents of interest in North American and Asian ginseng (*Panax quinquefolius* L. and *P. ginseng* C.A. Mey., respectively) are the triterpene saponin ginsenosides, which are present in both neutral and acidic malonyl forms [27]. The malonyl ginsenosides are more polar and water soluble than their corresponding neutral counterparts and are not measured by most ginsenoside assays [27]. They are, however, susceptible to hydrolysis and can be readily converted to their neutral counterparts during analytical extraction [28,29]. Uncontrolled hydrolysis during sample preparation can result in inconsistent values when quantifying the neutral ginsenosides [29,30]. A recent interlaboratory study evaluating a HPLC method for ginsenoside quantification demonstrated the poor precision that can result under uncontrolled circumstances [30]. The analysis of two separate ginseng finished product samples following a controlled alkaline hydrolysis step designed to quantitatively convert the acidic malonyl ginsenosides to their neutral counterparts prior to HPLC determination demonstrated greatly improved precision compared to analyses performed without forced hydrolysis [30]. Subsequent analyses included the controlled hydrolysis step and the improved method was found to be precise enough to meet AOAC International requirements [30].

An alternative approach to achieving precision in the ginsenoside analysis could have been to simply remove the malonyl ginsenosides by some other technique, such as solid-phase extraction, and measure only the native neutral compounds. However, although the malonyl ginsenosides are apparently not as biologically active as the neutral forms, they can exert a pharmacological effect in the body after they are converted to their neutral counterparts in the gut [27]. As in the bilberry example, a specification designed to assure the quality of a ginseng product in terms of its ability to elicit a biological effect *in vivo* must be expanded. In this case, the method is required to take the malonyl ginsenosides into account.

CONCLUSION

Concerns about product quality and the means through which it is assessed have existed since the earliest development of commerce and trade. Both Pliny the Elder's (23–79 CE) *Naturalis Historia* [31] and Dioscorides' (40–90 CE) *Materia Medica* [32] described herbal products based on their predominant characteristics as well as methods through which adulteration of these products could be detected. Although there have been significant advances in science and technology since the original publications of these books, the same fundamental questions and challenges remain, namely, what specifications should be selected to represent the quality of a product and how those specifications can be evaluated. A firm understanding of specifications and what they are intended to represent is essential to ensuring product quality and consistency. This understanding is incomplete unless the properties of the methods used to evaluate against specifications are also known. These properties include the nature of the measurand, the nature of the product matrix, a grasp of what the measurement results represent, the inherent limitations of the method, and any potential confounders. Even with this knowledge, specifications (and tests) should be continuously evolving as new confounders and technologies come into existence.

REFERENCES

1. World Health Organization (WHO). Toxicological and health aspects of melamine and cyanuric acid: Report of a WHO expert meeting in collaboration with FAO, Supported by Health Canada. Geneva, Switzerland, WHO, 2009. http://www.who.int/foodsafety/publications/chem/Melamine_report09.pdf (accessed on September 2, 2014).
2. Food and Drug Administration (FDA). Current good manufacturing practice in manufacturing, packaging, labeling, or holding operations for dietary supplements (Title 21 Code of Federal Regulations Part 111). Department of Health and Human Services, Washington, DC, 2010. http://www.accessdata.fda.gov/scripts/cdrh/cfdocs/cfcfr/CFRSearch.cfm?CFRPart=111 (accessed on November 22, 2013).
3. An act to prevent the importation of adulterated and spurious drugs and medicines. 30th Congress, 1st Session, chapter 70, 1848: 237–239.
4. D.B. Worthen. Pharmaceutical legislation: An historical perspective. *International Journal of Pharmaceutical Compounding* 1, 2006: 20–28.

5. S. Foster. A brief history of adulteration of herbs, spices, and botanical drugs. *HerbalGram* 92, 2011: 42–57.
6. M.J. Bailey, Report on the practical operation of the law relating to the importation of adulterated and spurious drugs, medicines, etc., D. Fanshaw, New York, 1849. https://archive.org/stream/101161381.nlm.nih.gov/101161381#page/n1/mode/2up (accessed on September 4, 2014).
7. AOAC International. Guidelines for single laboratory validation of chemical methods for dietary supplements and botanicals. AOAC International, Gaithersburg, MD, 2003.
8. Center for Drug Evaluation and Research, Reviewer guidance validation of chromatographic methods, Department of Health and Human Services, Washington DC, 1994.
9. M. Thompson, S.L.R. Ellison, and R. Wood. Harmonized guidelines for single laboratory validation of methods of analysis (IUPAC technical report). *Pure and Applied Chemistry* 74, 2002: 835–855.
10. ICH, International conference on harmonisation of technical requirements for registration of pharmaceuticals for human use, ICH harmonised tripartite guideline—Validation of analytical procedures: Text and methodology Q2(R1), ICH, Geneva, Switzerland, 1996.
11. J.L. Lewis. The regulation of protein content and quality in national and international food standards. *British Journal of Nutrition* S2, 2012: S212–S221.
12. M. Thompson, L. Owen, K. Wilkinson, R. Wood, and A. Damant. A comparison of the Kjeldahl and Dumas methods for the determination of protein in foods, using data from a proficiency testing scheme. *Analyst* 127, 2002: 1666–1668.
13. J.J. Litzau, G.E. Mercer, and K.J. Mulligan. GC-MS screen for the presence of melamine, ammeline, ammelide, and cyanuric acid. http://www.fda.gov/Food/FoodScienceResearch/LaboratoryMethods/ucm071759.htm (accessed on September 30, 2013).
14. S. Turnipseed, C. Casey, C. Nochetto, and D.N. Heller. Determination of melamine and cyanuric acid residues in infant formula. http://www.fda.gov/Food/FoodScienceResearch/LaboratoryMethods/ucm071637.htm (accessed September 30, 2013).
15. A. Melfsen, E. Hartung, and A. Haeussermann. Robustness of near-infrared calibration models for the prediction of milk constituents during the milking process. *Journal of Dairy Research* 80, 2013: 103–112.
16. P. Feng and S. Baugh. Determination of whey protein content in bovine milk-based infant formula finished products using amino acids calculation method: AOAC first action 2012.08. *Journal of AOAC International* 96, 2013: 795–797.
17. M. Barrett. Bilberry. In *The Encyclopedia of Dietary Supplements, Second Edition*, P.M. Coates, J.M. Betz, M.R. Blackman et al. (Eds.), Informa Healthcare, London, 2010, pp. 37–42.
18. British Pharmacopoeia Commission Office. Bilberry Fruit. In *British Pharmacopoeia*, The Stationary Office, London, 2004, vol. 1, p. 258.
19. S. Foster and M. Blumenthal. The adulteration of commercial bilberry extracts. *HerbalGram* 96, 2012: 64–73.
20. K.G. Penman, C.W. Halstead, A. Matthias, J.J. De Voss, J.M. Stuthe, K.M. Bone, and R.P. Lehmann. Bilberry adulteration using the food dye amaranth. *Journal of Agricultural and Food Chemistry* 54, 2006: 7378–7382.
21. American Herbal Products Association. Resource links bilberry fruit extract. http://www.ahpa.org/default.aspx?tabid=164 (accessed September 30, 2013).
22. C. Cassinese, E. Combarieu, M. Falzoni, N. Fuzzati, R. Pace, and N. Sardone. New liquid chromatography method with ultraviolet detection for analysis of anthocyanins and anthocyanidins in *Vaccinium myritillus* fruit dry extracts and commercial preparations. *Journal of AOAC International* 90, 2007: 911–919.
23. European Pharmacopoeia. Fresh bilberry fruit dry extract, refined and standardized, *European Pharmacopoeia* 7, 2010: 1869.

24. J.M. Betz, P.N. Brown, and M.C. Roman. Accuracy, precision, and reliability of chemical measurements in natural products research. *Fitoterapia* 82, 2011: 44–52.
25. W. Horwitz. Evaluation of analytical methods used for regulation of foods and drugs. *Analytical Chemistry* 54, 1982: 67A–76A.
26. W. Horwitz and R. Albert. The Horwitz ratio (HorRat): A useful index of method performance with respect to precision. *Journal of AOAC International* 89, 2006: 1095–1109.
27. D.V.C. Awang. The neglected ginsenosides of North American ginseng (*Panax quinquefolius* L.). *Journal of Herbs, Spices & Medicinal Plants* 7, 2000: 103–109.
28. W.A. Court, J. Hendel, J. Elmi, and G. Jama. Reversed-phased high-performance liquid chromatographic determination of ginsenosides of *Panax quinquefolius. Journal of Chromatography A* 755, 1996: 11–17.
29. P.N. Brown. Determination of ginsenoside content in Asian and North American ginseng raw materials and finished products by high-performance liquid chromatography: Single-laboratory validation. *Journal of AOAC International* 94, 2011: 1391–1399.
30. P.N. Brown and R. Yu. Determination of ginsenoside content in *Panax ginseng* C.A. Meyer and *Panax quinquefolius* L. root materials and finished products by high-performance liquid chromatography with ultraviolet absorbance detection: Interlaboratory study. *Journal of AOAC International* 96, 2013: 12–19.
31. H. Rackham (Transl.), *Pliny: Natural History, Volume VII, Books 24–27*, Loeb Classical Library, Cambridge, MA, 1956.
32. J.M. Riddle, *Dioscorides on Pharmacy and Medicine.* University of Texas Press, Austin, TX, 1985.

3 Cultivating Botanicals for Sensory Quality
From Good Agricultural Practices to Taste Discernment by Smallholder Tea Farmers

Selena Ahmed, John Richard Stepp, and Xue Dayuan

CONTENTS

Introduction 16
    Botanical Quality 16
    Agro-Ecosystem Management 17
Good Agricultural Practices 19
    Site Selection 19
    Crop Production Plans 20
    Seeds 20
    Soil 21
    Water Irrigation and Drainage 21
    Climate 21
    Harvest and On-Farm Processing and Storage 21
    Energy 21
    Waste Management 22
    Social Factors 22
Case Study of Smallholder Tea Farmers 22
    Tea Management Systems 22
    GAPs in Tea Agroforests 23
    Management Effects on Tea Quality 25
    Evaluation of Botanical Quality through Sensory Discernment 25
    Standardized Tea Preparation Protocol for Sensory Evaluation 26
        Materials for Tasting 26
        Preparation Using the *Gong fu Cha Dao* Method of Multiple Infusions 26
GAPs Framework for Assessing Botanical Quality through Sensory Discernment 27
Acknowledgments 29
References 29

INTRODUCTION

Botanical quality to support human health is highly dependent on the ecological and management conditions of production systems. Concurrently, the cultivation of botanicals and other crops has notable impacts on the environment. In recognition of these linkages, farmers, governments, nongovernmental organizations (NGOs), the private sector, and other stakeholders have developed various sets of good agricultural practices (GAPs) that provide guidance on management practices at each step of production from site selection, planning, and cultivation to postproduction processing, storage, and waste. Although GAPs have become formalized through international, national, and local-level guidelines in recent years, many farming communities worldwide have an extensive history of managing agricultural eco-systems, or agro-ecosystems, for the joint objectives of environmental and human well-being through informal social codes and customs. The management practices of such smallholder communities can be adapted to inform broader and more formalized GAPs supported by the botanical industry, policies, and incentive programs for the widespread cultivation of enhanced plant-based products.

This chapter opens with a brief review of botanical quality and agro-ecosystems to highlight the role of environmental and management factors in producing high-quality botanicals. Next, GAPs in international guidelines are examined before exploring GAPs identified at the community level through primary research with smallholder tea farmers in southwestern China who assess botanical quality on the basis of sensory discernment. The overall objective of this case study is to synthesize lessons learnt from smallholder tea farmers and linkages between agroecological conditions and sensory discernment as an evaluation of botanical quality. This chapter concludes with a generalizable framework that demonstrates links between GAPs and sensory evaluation of botanical quality to support human health outcomes. Drawing from socioecological and chemical ecology theories and the concept of *terroir*, the framework presented here serves as a tool to assess botanicals that have desirable sensory properties and health benefits for humans deriving from the production of secondary metabolites at the agro-ecosystem level. The presented framework highlights that botanical quality starts at the level of the agro-ecosystems for cultivated products before other processes that influence quality including processing, distribution, storage, preparation, and administration. The adoption of this framework by the botanical production industry as well as by formalized GAPs guidelines would likely lead to improved botanical quality to support human health.

BOTANICAL QUALITY

Theory from chemical ecology [1,2] and socioecological systems [3–5] along with the concept of *terroir* is at the nexus of exploring environmental and human well-being and provides an important theoretical foundation for planning for high-quality botanical cultivation. In chemical ecology, plant–environment interactions impact secondary metabolite chemistry of plant products with varying abiotic, biotic, and cultural factors resulting in varying secondary metabolite profiles [6,7]. Secondary metabolites are produced by plants for the ecological role of chemical defense.

Quality of botanicals is largely determined by the types and concentrations of secondary metabolites, also known as phytochemicals, via their effects on sensory profiles and their physiological properties on human consumers [8,9]. The production of secondary metabolites represents a high metabolic cost to plants and their synthesis and concentrations vary on the basis of numerous environmental and management factors that present ecological cues or stressors to spur production [7,10,11]. With regard to human health, the consumption of low concentrations of secondary metabolites may have little or no effect, and at high concentrations, they may be therapeutic or toxic. Understanding the environmental and management factors that impact the presence and concentrations of secondary metabolites is beneficial for their optimum production for high-quality botanicals.

According to socioecological systems theory, human–environment interactions create dynamic feedback loops in which humans both impact and are impacted by ecosystem processes [2]. The concept of *terroir*, deriving from *terre* or land in French, is useful in understanding the production of high-quality botanicals through an emphasis on the complex interactions of environmental and cultural factors that shape agricultural products. *Terroir* highlights the crucial role of human–environment interactions in imparting distinct sensory characteristics and health properties to botanicals. A central tenet is that factors that are beneficial for the environment are also beneficial for human well-being.

Agro-Ecosystem Management

There are many ways in which humans manage agro-ecosystems for the cultivation of botanicals. Agro-ecosystems are a type of coupled natural and human system that results from the manipulation of a natural resource base of biotic and abiotic components for the procurement of crops, botanicals, livestock, and other agricultural products with desirable characteristics. The transformation from an unmanaged system to an agro-ecosystem generally involves a shift from the uncontrolled procurement of resources to controlled procurement through social transactions and decision making along with purposeful breeding of crop species with desired traits such as higher yields, increased nutritional value, improved appearance, stress tolerance, and other factors [12]. Breeding practices focus on the selection of specific genotypes and phenotypes as well as hybridization of traits. Humans have an extensive history of influencing the landscape in order to increase the benefits derived for subsistence and commercial use even before the origins of agriculture approximately 10–11,000 years ago. Paleobotanical evidence in New Guinea indicates that people were manipulating the forest in the late Pleistocene, some 30,000–40,000 years ago, by trimming, thinning, and ring-barking in order to increase the natural stands of taro, bananas, and yams [13]. These examples emphasize that humans have a long history of environmental manipulation and that dichotomous assumptions between natural systems and agricultural landscapes are often not appropriate. Rather, a nature–culture continuum [14,15] better characterizes agro-ecosystems with varying levels of management and intensification.

Botanical quality derived from an agro-ecosystem is highly dependent on its management, inputs, energy flow, nutrient cycling, structural and functional composition, and dynamic interactions between abiotic and biotic components. The biotic factors in agro-ecosystems consist of all living organisms including cultivated plants, livestock, soil microorganisms, pollinators, and herbivores, whereas the abiotic factors of agro-ecosystems consist of physical and chemical components of the environment such as land, water, temperature, moisture, light, and nonliving components of soil. Management factors in agro-ecosystems comprise of cultural practices, perceptions, values, resource rights, and regulations, whereas socioeconomic factors in agro-ecosystems consist of labor and livelihoods among other variables. In addition to interactions among these biotic, abiotic, management, and socioeconomic variables, agro-ecosystems are impacted by exogenous variables including environmental change, market forces and fluctuations, interactions with surrounding ecosystems and cultural systems, and various policies such as farmer subsidies.

In naturally functioning ecosystems, energy flows from plants and other producers capable of capturing solar energy and converting it to chemical energy to consumers at various trophic levels and is lost as heat at each trophic level in various metabolic processes. In addition, in naturally functioning ecosystems, nutrients are stored in the soil reservoir (primarily calcium, phosphorus, sulfur, and potassium) and the atmosphere (primarily carbon, oxygen, and nitrogen) and cycle through the system to maintain life functions of the biotic components and determine productivity. In self-sustaining ecosystems, energy output is balanced by energy input while nutrients stay within the system as they cycle between soil and plants through uptake and decomposition. Agro-ecosystems with low intensification resemble naturally functioning ecosystems with limited management and input. However, in most modern day agro-ecosystems, unlike naturally functioning ecosystems, energy and nutrients are directed out of the system at harvest, and thus, external inputs are most often required to keep these systems functioning. Humans must input energy and nutrients back into the system through management practices such as the application of fertilizer in order to achieve productivity. The way that agro-ecosystems are managed for energy and nutrient input serves to impact the quality of botanicals derived from these systems. Traditionally, agricultural inputs included human and animal labor. With the advent of modern agriculture in the 1930s, inputs shifted to an inclusion of fertilizer, pesticides, growth hormones, and, indirectly, fossil fuel inputs to increase productivity.

In addition, management of structural and functional diversity of agro-ecosystems further influences their stability and subsequent botanical quality. Managing biodiversity in agro-ecosystems is crucial for maintaining ecosystem services, or the benefits obtained from ecosystems, for production. Broadly defined, biodiversity is the variation among living organisms and their environment. It is often operationalized as species richness (the total number of species in a defined space at a given time) and species abundance (the proportion of individuals of each species in relation to the total population size). Biodiversity is fundamental to sustaining life on earth by providing ecosystem services including nutrient cycling, soil protection, pollination, flood control, and genetic resources [16]. Numerous studies have noted the role of biodiversity in maintaining ecosystems and strengthening their resilience in responding to natural and anthropogenic change [17]. Biodiversity in agro-ecosystems is important for the resilience and productivity of farms and farmer livelihoods [18]. Agricultural biodiversity is the product of natural and human selection through interactions among the environment, genetic resources, and farmer management practices. The Convention of Biodiversity (CBD) defines agricultural biodiversity as "all components of biological diversity of relevance to food and agriculture, and all components of biological diversity that constitute the agricultural ecosystems, also named agro-ecosystems: the variety and variability of animals, plants, and microorganisms, at the genetic, species, and ecosystem levels, which are necessary to sustain key functions of the agro-ecosystem, its structure and processes." The CBD acknowledges the distinct cultural dimensions of agricultural biodiversity that set it apart from other components of biodiversity including human activities and management practices, participatory processes, and local knowledge.

Many management practices to enhance yields result in simplified structural and functional diversity of agro-ecosystems, disturb biotic interactions, and ultimately reduce the system's stability. For example, nutrients are lost through monocultural cropping that seeks to enhance consistent product and yields while eroding soil nutrients. Consequently, management of conventional modern agro-ecosystems most often requires external inputs of fertilizers, pesticides, and herbicides to enhance productivity. These external inputs may bioaccumulate or present residues in botanicals with detrimental implications for human health. Understanding the processes that support agro-ecosystems without detrimental effects on environmental and human health helps elucidate GAPs that result in high-quality botanicals.

GOOD AGRICULTURAL PRACTICES

The impact of agro-ecosystem management on environmental and human health has given rise to the development of various sets of GAPs by governments, farmers, NGOs, the private sector, and other stakeholders toward meeting the joint objectives of sustained environmental and human well-being. GAPs provide guidance on management practices at each step of production from farming to postproduction and have become formalized through international and country-level guidelines in recent years. International stakeholders that have developed GAPs include the World Health Organization (WHO) and Food and Agriculture Organization (FAO) of the United Nations. They have identified objectives and goals including (1) production of safe and nutritious food in a manner that is economically efficient and feasible; (2) maintenance and enhancement of the natural resource base that manages on-farm soil, water, crops, fodder, and animals as well as surrounding landscape and wildlife; (3) supporting viable farming enterprises that contribute to livelihoods and human health; and (4) meeting societal demands in culturally appropriate ways. What follows is a summary of GAPs compiled from various sources including the FAO [19] and WHO [20].

Site Selection

Botanical quality is impacted by numerous variables of the site including other plants and living organisms, climate, and human management. Site selection involves identification of a location without risks of contamination that has desirable environmental conditions to influence botanical quality without disturbing the ecological and cultural balance of the surroundings. It is crucial to take into account possible implications of botanical production on the biodiversity, ecosystem health, social aspects, and well-being of the surrounding land use and communities. For example, the introduction of nonnative plants may have a detrimental impact on the local ecology. In addition, the introduction of particular agro-ecosystems may be culturally inappropriate in a certain context.

Managers should evaluate the past history of the site and surrounding land use including the cultivation of previous crops as well as continue to monitor present levels of environmental toxicity of soil, water, and air by any hazardous chemicals. Environmental contamination can result in the uptake of undesirable toxins in plants that are detrimental to human health. The site should also be selected on the basis of environmental conditions that positively influence the secondary metabolite profiles of botanicals including the presence and concentrations of specific secondary metabolites that have desired sensory and health properties for human consumers. Individuals of the same plant species can demonstrate significant differences in botanical quality when cultivated at different locations due to environmental and management variation [7].

CROP PRODUCTION PLANS

GAPs for crop production involve cultivating species and varieties that have characteristics that are optimal for the ecological and cultural aspects of selected sites while providing resilience to environmental variability such as severe weather events and global environmental change. Selected species should reduce the need for agrochemical input while meeting local consumer, cultural, and market needs. On an ecological level, crop production should take a long-term perspective to mitigate risks in the agro-ecosystem by selecting crops that enhance soil fertility, fit within crop rotation, and help with pests and disease control. On a cultural basis, the management plan should promote local cultural identity, livelihoods, and human health. Traditional management practices should be followed where feasible and/or should be better understood through ethnographic research.

Intercropping schemes, crop rotations, planting of perennial polycultures with diverse functions, and integrated pest management are among the most commonly promoted GAPs for long-term crop production that avoid depletion of soil resources. GAPs for annual crops promote careful planning of cultivating crops in a sequential order that includes legumes and livestock for grazing and nutrient cycling to enhance species interactions and enhance soil fertility [21].

SEEDS

Seeds and propagation materials for cultivation should be sourced with a knowledge of their botanical identity, breeding history, performance, and quality. International guidelines for GAPs usually call for genetically modified germplasm to comply with national and local guidelines and be appropriately labeled. For long-term sustainability of botanical supply, it is best for farmers to have the capacity to propagate viable seeds and appropriately store, label, and cultivate this genetic material without the need to rely on purchasing seed.

Soil

Botanical quality is dependent on soil parameters including soil type, drainage, moisture retention, fertility, pH, and appropriate amounts of nutrients and organic matter. GAPs for soils should aim to enhance soil parameters such as fertility and productivity that are optimal for botanical quality through practices that minimize loss of soils, nutrients, and agrochemicals. The health of soils can be maintained through practices focused on carbon buildup including crop rotations with the inclusion of legumes to provide a biological source of nitrogen, application of nutrient-rich manure, integration of livestock and pasture management, integrated pest management, conservation tillage, cover cropping, management of windbreak and water break.

Water Irrigation and Drainage

High-intensity industrial agriculture depletes freshwater resources through extraction for irrigation that results in surface and groundwater runoff. GAPs for water present guidelines to minimize water wastage and ensure safe water for agriculture and community use. Specific GAPs for water management include the creation of structures for water and nutrient catchment, prevention of contamination, monitoring of crop and soil moisture status, and implementing context-specific schedules and strategies for irrigation such as planting of nurse plant guilds that help conserve water resources.

Climate

Climate factors notably impact the quality of botanicals including growth, morphological characteristics, and phytochemical variables [22]. GAPs should take into account resilience of the agro-ecosystem and botanicals to variations of temperature, precipitation, humidity, and duration of sunlight depending on the specific crop variety and its development stage. In addition, GAPs should take into account the possibility of extremes of climate factors.

Harvest and On-Farm Processing and Storage

GAPs for harvesting involve appropriate timing to avoid recent or preferably any application of agrochemicals. Botanicals should be processed in facilities that are safe and hygienic. Lastly, products should be stored in clean containers and packages in determined spaces with appropriate temperature and humidity.

Energy

Agro-ecosystems often require energy input for enhanced production in the form of fertilizers to balance nutrients lost at harvest and fuel to support machinery for production in an efficient manner that reduces human labor. GAPs regarding energy management establish input–output plans for efficient utilization of energy sources, nutrients, and agrochemicals. Energy conservation practices and alternative energy sources (such as wind, solar, and biofuels) should be adopted rather than utilization of nonrenewable fossil fuels that contribute to greenhouse gas emissions. Monitoring of energy use can help track practices and identify inefficiencies.

WASTE MANAGEMENT

GAPs for waste management encourage responsible disposal and minimum output of nonusable and toxic waste. Waste materials should be stored and deposited in a manner to reduce the risk of environmental contamination. Agro-ecosystems should also be managed in a manner that generates minimal waste and alternative uses should be identified for waste that is produced from botanical production.

SOCIAL FACTORS

GAPs involving social factors provide guidelines on income generation opportunities, markets, labor standards, cultural practices, knowledge systems, worldviews, and human well-being. On a cultural basis, production of botanicals should adhere to safe and fair standards for resource managers and other workers and provide attractive income-generating opportunities along with high-quality botanicals to support human well-being.

CASE STUDY OF SMALLHOLDER TEA FARMERS

A growing body of evidence suggests that alternative smallholder agro-ecosystems such as agroforests, home gardens, and various permaculture schemes are valuable models of GAPs that contribute to ecosystem services, plant species conservation, economic risk reduction, food security, and human health [23–26]. However, political, economic, and social dynamics are altering human–environment relationships and are placing pressures on smallholder agrarian systems [27] and on associated botanical quality. There is a need to examine smallholder management practices toward the development of best practices that can be adopted and modified by international guidelines for widespread, context-specific application. The management practices of smallholder farmers of tea agroforests in Yunnan, China, provide a compelling case study of identifying GAPs for the production of high-quality botanicals.

Smallholder farmers of tea agroforests in Yunnan, China, cultivate tea in biodiverse agro-ecosystems for the objective of producing high-quality botanicals to support livelihoods, sensory enjoyment, and human health. Farmers directly evaluate botanical quality of tea products from these systems on the basis of sensory discernment and adjust their agro-ecosystem management practices for desired sensory properties and associated therapeutic values. Some smallholder tea farmers have knowledge of how particular plant characteristics vary with agroecological conditions including shade, altitude, slope, plant–insect interactions, soil type, and hybridization and use this ecological knowledge to manage for their preferences. These practices consequently impact the genetic and secondary metabolite profiles of tea in agro-ecosystems and drive subsequent botanical quality [7].

TEA MANAGEMENT SYSTEMS

Tea in Yunnan, China, is managed in variable systems that can be classified along a continuum of agricultural intensification from low intensity to high intensity including (1) forest tea populations, (2) agroforests, (3) mixed crop systems, (4) organic terrace teagardens from seed, and (5) monocultural terrace teagardens from clonal propagules. Forest tea populations are ecosystems that consist of uncultivated tea plants that are sparsely distributed with other plant species or tea plants that were once cultivated and have become feral. Tea trees in forest populations grow to heights of 15 m or higher and may live for over 100 years [28]. Tea agroforests are agro-ecosystems that have been transformed by thinning forests for the production of tea or where old swidden plots have undergone succession either unintentionally or by intentional human management. Agroforests resemble forest systems in their multistoried vegetative structure with a high canopy layer, mid-level tree layer of tea plants, and herbaceous ground layer [29]. Smallholders prune tea plants in agroforests to spread out their branch formation and reduce their height for increased production and ease of harvest. Mixed crop tea fields consist of tea plants intercropped with crops such as corn plants and wheat. Smallholders manage tea plants in mixed crop fields as shrubs for heightened production. Terrace teagardens, the most intensively managed of tea agro-ecosystems, are open fields that are managed for uniformity, high yield, and efficiency. Tea plants within terrace teagardens are cultivated in compact rows and pruned to waist-high shrubs. These systems may have organic practices and be propagated from seed; in most cases, monoculture tea gardens are cultivated from clonal propagules and use high application of agrochemical inputs of fertilizer, pesticides, and herbicides for productivity.

GAPs in Tea Agroforests

Best practices for tea cultivation by smallholder farmers of agroforests involve management of tea trees in relatively small and biodiverse agro-ecosystems with minimal human input and maintenance of forested edges (fredges) that serve as buffers between fields. Smallholder households manage agroforests that are approximately 0.8–1.5 ha in size and distributed in a complex land-use mosaic of forest, fields, and village settlements. Farmers cultivate approximately 350 tea plants of variable age per hectare of agroforest. These are irregularly distributed among other woody plant species and have several meters of distance between each plant. Tea plants are maintained in tree form between 3 and 9 m tall and are pruned for an expanded crown. The multistoried vegetative structure and biodiverse species composition of tea agroforests provide some of the ecosystem services of a forest system including pest and disease control, modifying microclimate, windbreak, soil fertility, prevention of landslides, and provision of other useful plants such as for food, medicine, ritual, and construction. Tea agroforests do not require agrochemical input for production and smallholders take pride in the resulting tea quality.

Smallholders manage tea agroforests on the basis of cultural practices and traditional ecological knowledge that seeks to mimic a forest environment for high-quality tea production. Following are key management activities within tea agroforests: (1) site selection based on soil type, slope, microclimate, elevation, and other socio-ecological conditions; (2) protection, tolerance, and enrichment planting of tea and other perennial woody plants in a polyculture of different species, functional classes, ages, and sizes to contribute to diversified vegetative structure, composition, seasonal availability of plant products, and cultural use; (3) maintenance of forested edges as buffers between fields to build soils and as a buffer; (4) preservation of herbaceous understory as weed suppressors, soil enrichers, diseases suppressors, and for pollination; (5) selective thinning of woody plant species for intensified tea production; (6) introduction of genetically diverse tea seeds and seedlings with varied morphological and phytochemical profiles from forest populations, neighboring communities, and other complex seed exchange networks; (7) deliberate spacing between individual plants to meet nutrient needs; (8) experimenting with introgression of genetic material; (9) naming of landraces and sharing through community cooperation; (10) manual removal of moss, ferns, and particular epiphytes growing on tea plants; (11) tolerance of specific pests and epiphytes; (12) maintenance of woody plants that serve as natural pesticides; (13) manual fertilizing through a composting technique (known as *wa cha* in Mandarin) of turning over the weedy understory at the base of tea trees and burying of broken branches to promote decomposition and soil fertility; (14) shade management of canopy and sunlight; (15) pruning and crown enhancement of tea plants; (16) social regulation against the application of agrochemicals; (17) conservation of mountaintop forest as a water reservoir; (18) removal of unwanted tea material from genetic contamination; and (19) harvesting of tea buds and leaves through hand-plucking during three contiguous seasons from spring to autumn.

Smallholders of agroforests rely on these management practices coupled with their associated ecosystem services for tea production and do not find the need to apply pesticides, herbicides, artificial fertilizers, hormones, or other agrochemicals. Smallholders report that these management practices impact the growth, yield, taste, therapeutic value, and overall quality of tea [30]. The environmental and management characteristics that are considered to most influence tea quality are soil, shade, slope, aspect, climate (particularly precipitation levels), elevation, nearness to a river, the particular plant species in the agro-ecosystem, age and height of the tea plants, particular landrace or cultivar of the tea plants, propagation through seeds rather than through cuttings, and lack of application of agrochemicals. Smallholders perceive that the maintenance of forest cover and forested edges in their agroforests serves multiple purposes including (1) windbreak for a high-quality tea product; (2) protects tea systems from pests, disease, landslides, floods, and drought; (3) protects soil and captures nutrients to enhance fertility; (4) modulates microclimate; (5) provides diverse plant products including food and construction material; and (6) serves as a habitat and corridor for wildlife and pollinators as well as a space for hunting. Thus, forested areas are encouraged around tea agroforests through fallow and enrichment planting. The presence of specific tea caterpillar species on tea plants is regarded to enrich the therapeutic properties of tea. Similarly, smallholders maintain specific epiphytic orchids on tea plants to impart a desired fragrance and taste to the resulting tea product. In addition, smallholders of tea agroforests believe that the application of agrochemicals destroys soil fertility and the integrity of tea plants as well as adversely influences the taste and therapeutic properties of plants, resulting in deleterious effects on environmental and human health. Despite the encouragement of agrochemicals by extension agents, social norms and regulations have developed in these communities that prohibit the use of agrochemicals in tea agroforests. These local regulations have been strengthened by market demand for tea grown without the use of agrochemicals and price premiums received by farmers for a high-quality botanical product [7].

MANAGEMENT EFFECTS ON TEA QUALITY

Previous research has found that tea agroforests in Yunnan contain greater plant species richness, more complex vegetative structure, higher genetic diversity of tea plants, and improved botanical quality of tea products as determined by sensory characteristics and secondary metabolite concentrations compared to tea products from conventional monoculture tea terraces, mixed crop fields, and forests [7]. Although forest tea populations were found to consist of higher plant species richness than tea agroforests, the latter are characterized by higher genetic diversity of tea plants and by tea products with higher botanical quality [7]. Specifically, the secondary metabolites measured include the key antioxidant catechin compounds and methylxanthines that determine tea quality and give its characteristic health and stimulant properties, respectively. Consequently, consumers are willing to pay high-price premiums for tea procured from tea agroforests and thereby providing attractive economic incentives to farmers to manage tea in such relatively low intensity agro-ecosystems without agrochemical input.

EVALUATION OF BOTANICAL QUALITY THROUGH SENSORY DISCERNMENT

Smallholder farmers, traders, and consumers in Yunnan evaluate botanical quality of tea products primarily through aroma and taste characteristics and perceive this to be linked to production conditions and to impact the therapeutic and other physiological properties for human consumers [31]. For example, master tea tasters perceive that tea with a bitter taste accompanied by a long sweet aftertaste at the back of the throat indicates medicinal properties and a good cultivation environment. Evaluation of botanical quality through secondary metabolite analysis found a positive correlation between tea samples perceived to be healthy with a desired flavor profile during sensory analysis and those with relatively high concentrations of polyphenolic and methylxanthine compounds [31].

Tea tasting for sensory evaluation of botanical quality in Yunnan usually proceeds with the assessment of the upfront aroma experienced by the nasal passage, the flavor experienced in the mouth, the overall mouthfeel, as well as the sensation at the back of the throat. A high-quality infusion of tea from agroforests of Yunnan is regarded as a balance between rich aromas, bitterness, mild astringency, thickness of the infusion, and a powerful sweet lingering aftertaste. Tea astringency is either characterized as tangy with a sharp and puckering action and little aftertaste or as nontangy with a mouth-drying feel and lingering aftertaste. The lingering sweet aftertaste sensed at the back of the throat is referred to as *gaan*. In Yunnan, teas from agroforests are critically judged by the amount of time it takes for *gaan* to emerge, how strong the *gaan* is, and how long it endures. To a much lesser extent, *umami* (*brothy taste*) as well sourness and saltiness may characterize the taste of tea from Yunnan's agroforests. The most prevalent sensory characteristics used by farmers to evaluate tea quality are as follows: (1) total taste, (2) total aroma, (3) total mouth feel, (4) bitterness, (5) sweet aftertaste (back of the throat sweetness), (6) speed of occurrence of sweet aftertaste and duration, (7) astringency, (8) dryness, (9) sourness, and (10) total flavor (a compilation of the previous characteristics) [30]. Botanical quality of tea by smallholder farmers is further described on the basis of physiological sensations such as mentally uplifting and calming, morphological characteristics of tealeaves such as thickness and leaf area, and the color and clarity of infusions.

The *gongfu cha dao* method of brewing tea is widely used in Yunnan to evaluate tea quality. This method is based on multiple infusions of briefly steeping leaves in a *gaiwan* (lidded bowl or in an unglazed clay teapot). This brewing technique functions to release tea's aroma, taste, color, and physiological properties to bring out its nuances. A key component of pu-erh tea tasting is for drinkers to compare and communicate their sensory perceptions of each infusion. The presence and concentrations of secondary metabolites in each infusion may vary significantly from one to another within the same tea [31]. The number of infusions that can be extracted while maintaining desired sensory properties is used to judge the quality of tea from Yunnan's agroforests.

STANDARDIZED TEA PREPARATION PROTOCOL FOR SENSORY EVALUATION

Given the variation of botanical quality on the basis of preparation, it is important to develop standardized preparation protocols for product comparison. The following is an example of a standardized protocol to evaluate the quality of loose green pu-erh tea for sensory analyses. The parameters recorded to develop this protocol include (1) water source, amount, and brewing temperature; (2) type of tea ware; (3) amount of leaf; (4) length of brewing time of each infusion; and (5) number of infusions. One hundred interviews were conducted with expert tea buyers in Yunnan Province of China between 2006 and 2012 to document preparation protocols of green pu-erh tea. Data from the interviews were analyzed to develop the following standardized tea preparation protocol used for sensory analyses to evaluate green pu-erh tea quality.

Materials for Tasting

Tea: 5 g loose green pu-erh tea

Tea ware: Hot water kettle, timer, small white porcelain *gaiwan* (*lidded tea pot*) sized to accommodate at least 80 mL of water, small clear teapot, porcelain strainer, six tiny white porcelain tasting tea cups (one for each sensory evaluator), small plate

Water: Spring water rich in minerals

Palate cleanser: Acidic substances such as lemon juice, preserved plum, or tomato juice

Preparation Using the *Gong fu Cha Dao* Method of Multiple Infusions

1. Warm the tea ware.
2. Heat water to 90°C and pour into the *gaiwan*, teapot, and cups in order to warm the containers before preparing the tea. Discard the water

Infusion #1 (Wash, Wetting, or Opening)

1. Place 5 g of green pu-erh inside the *gaiwan*.
2. Heat water to 90°C.
3. Add 80 mL of heated water to the *gaiwan* and place the lid back on.
4. Immediately start timer for a 45-s infusion.
5. Gently swirl the *gaiwan* (with pot in one hand and lid in the other) twice and use lid to control leaves if need be.
6. Pour the infusion into the teapot.
7. Smell and discard this first infusion, referred to as a *wash*, *wetting*, or *opening* of the leaves; this step is carried out for various reasons including as a ritual, to clean the leaves, to discard part of the caffeine content, as an offering, and to open out the leaves to release the sensory properties.
8. Smell the empty pots.

Infusions #2–#5 (Tasting)

1. Using the *wetted* set of leaves in the *gaiwan*, repeat steps 2–6 above with the exception of infusing the leaves 30 s infusion instead of 45 s as specified above.
2. Pour the second infusion into tasting cups using the porcelain strainer.
3. Smell and sip when the temperature of the teacup feels comfortable for tasting. Reflect and record tasting notes.
4. Smell the empty teacup.
5. Wait for 3.5 min before drinking a mouthful of water.
6. Cleanse the palate.
7. Prepare and taste a third, fourth, and fifth infusion using the same protocol as for the second infusion.
8. Compare the taste profile and intensity between each infusion.

Infusion #6 (Capping)

1. The sixth infusion will be the final infusion for the tasting where the leaves will be *capped* off to extract a bulk of the remaining constituents by infusing in water for 5 min.
2. Pour out the spent leaves from the *gaiwan* onto a plate and observe.
3. You may choose to chew a tea leaf at this time.

GAPs FRAMEWORK FOR ASSESSING BOTANICAL QUALITY THROUGH SENSORY DISCERNMENT

GAPs for the botanical industry as well as those for international guidelines have much to learn from the practices of smallholder farmers that link botanical quality of cultivated products to the agro-ecosystem through sensory discernment. Here, a coupled human and natural systems (CHANS) framework is presented that draws from GAPs of smallholder farmers of tea agroforests to be adopted by the botanical

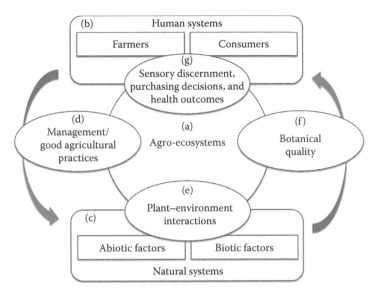

FIGURE 3.1 CHANS framework to examine links between agricultural management and botanical quality. This CHANS framework draws from good agricultural practices (GAPs) of smallholder farmers with agro-ecosystems at the center of this framework (a). The agro-ecosystem consists of human systems (b) and natural systems (c). There are four critical feedbacks in this framework including GAPs and other farmer management (d); plant–environment interactions (e); (f) botanical quality; and (g) sensory discernment of botanical quality and associated purchasing decisions and human health outcomes.

industry for the purpose of planning, cultivating, and assessing for high-quality botanicals (Figure 3.1). This framework highlights that botanical quality starts at the level of the agro-ecosystems for cultivated products before other processes that influences quality including processing, distribution, storage, preparation, and administration. Agro-ecosystems are at the center of this CHANS framework (Figure 3.1a) and consist of human systems (Figure 3.1b) and natural systems (Figure 3.1c), or collectively, CHANS. Accordingly, these human-managed ecosystems impact and are impacted by components of human and natural systems. The human systems of this framework include farmers and consumers and associated interactions. The natural systems of this framework include biotic and abiotic factors and associated interactions and processes. There are four critical feedbacks in this framework including (1) GAPs and other farmer management (Figure 3.1d); (2) plant–environment interactions (Figure 3.1e); (3) botanical quality (Figure 3.1f); and (4) sensory discernment of botanical quality and associated purchasing decisions and human health outcomes. In this framework, the human system impacts the natural system through GAPs and other farmer management practices. Collectively, management practices and plant–environment interactions with both biotic and abiotic factors influence botanical quality. Ultimately, farmers, traders, and consumers are able to directly evaluate botanical quality through sensory evaluation that then impacts consumer purchasing decisions and subsequent income received by farmers. Botanical quality further influences human systems through influencing health outcomes of human consumers.

The incorporation of such a CHANs framework for developing policies and plans for GAPs in the botanical industry will serve to recognize that botanical quality starts at the level of the ecosystem, can be managed for through production practices, and can be directly evaluated by producers, industry agents, and consumers through sensory discernment toward improved botanical quality for human health.

ACKNOWLEDGMENTS

We thank our funding during which we developed the CHANS framework presented here including Program 111 in Ethnobiology of the Chinese Ministry of Education, NIH NIGMS P20GM103474, NSF DDEP OISE-0749961, NSF EAPSI OISE-0714431, and NSF CNH BCS-1313775. In addition, we are grateful to the small-holder farmers of tea agroforests that informed the development of the presented framework.

REFERENCES

1. K. Schepp. Strategy to adapt to climate change for Michimikuru tea farmers in Kenya. AdapCC Report, Deutsche Gesellschaft fur Technische Zusammenarbeit (GTZ) GmbH, Eschborn, Germany, 2009. http://adapcc.org/en/kenya.htm.
2. S.A. Levin. *Fragile Dominion: Complexity and the Commons.* Perseus Books Group, Jackson, TN, 1999.
3. B.H. Walker, L.H. Gunderson, A.P. Kinzig, C. Folke, S.R. Carpenter, and L. Schultz. A handful of heuristics and some propositions for understanding resilience in social-ecological systems. *Ecology and Society* 11(1): 13, 2006.
4. R. Abel and J.R. Stepp. A new ecosystems ecology for anthropology. *Conservation Ecology* 7(3): 12, 2003.
5. O. Mertz, C. Padoch, J. Fox, R.A. Cramb, S.J. Leisz, N.T. Lam, and T.D. Vien. Swidden change in Southeast Asia: Understanding causes and consequences. *Human Ecology* 37: 259–264, 2009.
6. C. Glynn, D.A. Herms, C.M. Orians, R.C. Hansen, and S. Larsson. Testing the growth-differentiation balance hypothesis: Dynamic responses of willows to nutrient availability. *New Phytologist* 176: 623–634, 2007.
7. S. Ahmed, C.M. Peters, L. Chunlin, R. Myer, U. Unachukwu, A. Litt, E. Kennelly, and J.R. Stepp. Biodiversity and phytochemical quality in indigenous and state-supported tea management systems of Yunnan, China. *Conservation Letters* 5(6): 28–36, 2013.
8. F. Poiroux-Gonord, L.P. Bidel, A.L. Fanciullino, H. Gautier, F. Lauri-Lopez, and L. Urban. Health benefits of vitamins and secondary metabolites of fruits and vegetables and prospects to increase their concentrations by agronomic approaches. *Journal of Agriculture and Food Chemistry* 58(23): 12065–12082, 2010.
9. S. Ahmed and J.R. Stepp. Green tea: The plants, processing, manufacturing and production. In: V. Preedy (Ed.), *Tea in Health and Disease Prevention.* Academic Press, Missouri, 2012, pp. 19–31.
10. N.M. van Dam and K. Vrieling. Genetic variation in constitutive and inducible pyrrolizidine alkaloid levels in *Cynoglossum officinale* L. *Oecologia* 99:374–378, 1994.
11. M. Björkman, I. Klingen, A.N. Birch, A.M. Bones, T.J. Bruce, T.J. Johansen, R. Meadow et al. Phytochemicals of brassicaceae in plant protection and human health—Influences of climate, environment and agronomic practice. *Phytochemistry* 72(7): 538–556, 2011.
12. B.B. Simpson and M. C. Ogorzaly. *Economic Botany: Plants in Our World.* 3rd Edition. McGraw Hill, Boston, MA, 2001.

13. C.M. Hladik, O.F. Linares, A. Hladik, H. Pagezy, and A. Semple. Tropical forests, people and food: An overview. In: C.M. Hladik, A. Hladik, O.F. Linares, H. Pagezy, A. Semple, and M. Hadley (Eds.), *Tropical Forests, People and Food: Biocultural Interactions and Applications to Development*, Vol. 13, *Man and the Biosphere* Series, UNESCO-Parthenon, Paris, France, 1993, pp. 3–14.

14. D. McKey, T.R. Cavagnaro, J. Cliff, and R. Gleadow. Chemical ecology in coupled human and natural systems: People, manioc, multitrophic interactions and global change. *Chemoecology* 20(2): 109–133, 2010.

15. M.R. Dove. Transition from native forest rubbers to *Hevea brasiliensis* (Euphorbiaceae) among tribal smallholders in Borneo. *Economic Botany* 48(4): 382–396, 1994.

16. A.E. Magurran. *Measuring Biological Diversity*. Blackwell Publishing, Malden, MA, 2003.

17. L.H. Gunderson and C.S. Holling. *Panarchy: Understanding Transformations in Human and Natural Systems*. Island Press, Washington, DC, 2001.

18. Convention on Biological Diversity. Programme of work on agricultural biodiversity. Decision V/5 of the Conference of the Parties to the Convention on Biological Diversity, Nairobi, Kenya, 2000. http://www.cbd.int/agro/pow.shtml.

19. FAO. Development of a Framework for Good Agricultural Practices. Committee on Agriculture, 2012. http://www.fao.org/docrep/meeting/006/y8704e.htm.

20. WHO. WHO guidelines on good agricultural and collection practices (GACP) for medicinal plants. Geneva, Switzerland, 2003. http://whqlibdoc.who.int/publications/2003/9241546271.pdf.

21. D.G. Bullock. Crop Rotation. *Critical Reviews in Plant Sciences* 11: 309–326, 1992.

22. S. Ahmed, C. Orians, T. Griffin, S. Buckley, U. Unachukwu, A.E. Stratton, J.R. Stepp, A. Robbat, S. Cash, and E. Kennelly. Effects of water availability and pest pressures on tea (*Camellia sinensis*) growth and functional quality. *AoB Plants* 6, 2014. doi:10.1093/aobpla/plt054.

23. C. Potvin, C. Owen, S. Melzi, and P. Beaucage. Biodiversity and modernization in four coffee-producing villages of Mexico. *Ecology and Society* 10(1): 18, 2005.

24. B. Belcher, G. Michon, A. Angelsen, P. Ruiz, M. Ruiz, and H. Asbjornsen. The socio-economic conditions determining the development, persistence, and decline of forest garden systems. *Economic Botany* 59: 245–253, 2005.

25. J. Harlan. Our vanishing genetic resources. *Science* 188: 618–621, 1975.

26. S. Jose. Agroforestry for ecosystem services and environmental benefits: An overview. *Agroforest Systems* 76: 1–10, 2009.

27. P. Koohafkan. Conservation and adaptive management of globally important heritage systems. In: FAO. Globally important, ingenious agricultural systems (GIAHS). *First Stakeholder Workshop and Steering Committee Session*, Rome, Italy, 2002. http://www.bioone.org/doi/abs/10.3969/j.issn.1674-764x.2011.01.004.

28. P.G. Xiao and Z.Y. Li. Botanical classification of tea plants. In: Y.S. Zhen (Ed.), *Tea: Bioactivity and Therapeutic Potential*. Medicinal and Aromatic Plants—Industrial Profiles, Taylor & Francis, London, 2002, pp. 17–34.

29. C. Long and J. Wang. Studies of traditional tea-garden of jinuo nationality, China. In: S. K. Jain (Ed.), *Ethnobiology in Human Welfare*, Deep Publications, New Delhi, India, 1996, pp. 339–344.

30. S. Ahmed. Biodiversity and ethnography of tea management systems in Yunnan, China. Dissertation, Graduate Center, City University of New York, New York, 2011.

31. S. Ahmed, U. Unachukwu, J.R. Stepp, C.M. Peters, L. Chunlin, and E. Kennelly. Pu-erh tea tasting in Yunnan, China: Correlation of drinkers' perceptions to phytochemistry. *Journal of Ethnopharmacology* 132: 176–185, 2010.

4 Using Traditional Taxonomy and Vouchers in Authentication and Quality Control

Wendy L. Applequist

CONTENTS

Introduction..31
Botanical Taxon Circumscription 32
Process of Morphological Authentication..34
  Materials and Methods..34
  Sampling 38
  Limitations 38
Vouchering 39
References...40

INTRODUCTION

Authentication of botanicals by morphology, or the study of a plant's external form, requires no costly technology, but because plant species are defined by their morphological characteristics, it is in fact the most rigorous possible means of identifying a plant. All other methods, as reliable as they can be, are no more than proxies or substitutes for morphological data; indeed, the reliability of other methods can be fully verified only by observing their performance in samples that have been botanically identified by morphology. Moreover, botanical identification of plants is not an exotic skill that only those with advanced degrees in botany can learn. Identifying plants by their appearance, combined with other sensory inputs such as odor, texture, and flavor, is an activity that human beings are naturally adapted to, having relied upon it for survival for hundreds of thousands of years. Most people with a little knowledge and experience in careful observation can rapidly learn to identify the easier botanicals. Therefore, morphological examination ought to be considered the preferred method of authentication whenever a botanical's form and stage of processing are such as to permit it.

BOTANICAL TAXON CIRCUMSCRIPTION

The basic unit of taxonomy, though one that is often not readily defined, is the species. Species are grouped into genera, which in modern times are generally expected to be circumscribed so as to be monophyletic (including all the descendents of some common ancestor and no other species). Under the form of nomenclature that has been used for plants since 1753 [1], the name of a species is a binomial consisting of a genus name and a specific epithet, both Latinized and italicized, which in formal literature is followed, at its first mention, by the abbreviated name of the authority or authorities who were responsible for publishing it, for example, *Nepeta cataria* L. or *Centella asiatica* (L.) Urb. Genera are in turn grouped into families and families into higher-ranked taxa such as orders. Species, in turn, may be further subdivided into subspecies and/or varieties or forms (a usually undesirable rank).

The circumscription of species boundaries is often a matter of taxonomic opinion. Zoologists have historically favored Mayr's biological species concept [2,3], which considers populations to be of the same species if they can interbreed successfully and of different species if they cannot. This is usually reasonable in animals, aside from the fact that gene flow among *good* species and complex situations such as ring species do sometimes occur, but in the absence of breeding studies or detailed genetic data for most species, taxonomists in practice usually use morphological differentiation to define species boundaries. Such a concept is not generally applicable in plants because of their much more varied modes of reproduction. At one extreme, in plants such as oaks (*Quercus*) and hawthorns (*Crataegus*) even distantly related species may hybridize freely [4–6]; at the other extreme, plants such as dandelions (*Taraxacum*) and blackberries (*Rubus*) frequently reproduce asexually [7,8], creating numerous minor but persistently recognizable variants that some have treated as species. Likewise, these biological complexities would complicate the application of any proposed species concept based on genetic descent or similarity.

Hence, the most used concept in plant taxonomy is the taxonomic species concept, which defines species pragmatically as groups of populations that are distinguished from one another by a combination of multiple morphological characters [9,10]; distinguishing species by a difference in only a single character would obviously lead to much overdescription of species that were not biologically meaningful. As the number of meaningful characters understood to separate two groups of specimens can vary depending upon both the quantity of available material and the skill and care of the taxonomist, this definition of species is to some extent inherently subjective. In some problematic genera, even to this day, one taxonomist may recognize several species within a group that another considers to be a single variable species. It will be noted below that those using botanical literature for purposes of plant identification may need to be conscious of such disagreements.

The characters used to define genera and species are primarily macromorphological characters of the aboveground portion of the plant, and especially those of the leaves, flowers, and/or fruits. It is worth noting that if a species is defined by the presence of a suite of vegetative and reproductive characters that together distinguish it from all other species, the most rigorous possible means of identification of material of that species is the confirmation that those characters are present. Certain vegetative characters, such as leaf size, shape, and degree of compounding, display a substantial range of variation within some species. An observed difference that in one group helps to distinguish species may in another group reflect unimportant, environmentally mediated variation; hence, part of the taxonomic revisionary process is becoming familiar enough with a group to recognize which characters are meaningful within that particular group and which are not. Floral morphology is usually of particular importance, but in some groups, for example, umbels (Apiaceae), many species have small flowers that appear similar, and fruit morphology is more informative.

Characteristics of the bark, root, rhizomes, or tubers can be used to aid in the identification of some species. However, these organs are rarely critical to species circumscriptions: they seldom offer enough easily observable characters that vary among species of a genus to be of great use, and they are usually not present on herbarium specimens. For every genus in which closely related species have different anatomy (e.g., *Actaea* rhizomes and roots [11]), there are probably several other genera in which some closely related species have identical anatomy or in which insufficient data exist to determine the situation with certainty.

When noticeable variation exists within a group of populations recognized as a species, especially if that variation is geographically correlated, it may be formally recognized by naming regional variants as subspecies and/or varieties. If both ranks are used, which is sometimes done in treatments of highly variable and widespread species (e.g., *Ptelea trifoliata* [12]), the rank of subspecies is higher than that of variety. If only one rank is used, the choice is mostly a matter of preference and custom; in the United States, there is a historical tendency for Eastern botanists to recognize subspecies and Western botanists to recognize varieties. Unlike the case for species, infraspecific taxa can be supported by a single character that is strongly correlated with geography, and since gene flow is expected within species, intermediate specimens may be relatively common. The strength of evidence favoring recognition of infraspecific taxa, especially in complex classifications, is thus correspondingly lower, and the groups recognized are less distinct, so there is often little need to pay attention to infraspecific classification. In a few botanicals, infraspecific groups correspond to major chemotypes (e.g., sweet fennel, *Foeniculum vulgare* var. *dulce*, vs. bitter fennel, *Foeniculum vulgare* var. *vulgare*), or only one subspecies or variety is preferred for commercial use.

In some genera of useful plants, such as *Crataegus* [5,6], *Lycopus* [13], *Betula* [14], and *Thymus* [15,16], the reality of frequent hybridization among certain species where their ranges overlap must be acknowledged. If two species are largely genetically distinct, individuals with intermediate characteristics should be relatively rare and can easily be identified as hybrids. However, some pairs or groups of related species from genera such as those mentioned form species complexes, hybrid swarms, or broad hybrid zones in which a range of overlapping morphologies may be present and there may be no solid genetic boundaries. These situations are difficult to handle: it would seem inappropriate to identify a large fraction of all specimens of a given species pair as hybrids, but putting a single name on some specimens may be virtually impossible. It should be emphasized that this difficulty is due not to the inadequacies of morphological taxonomy, but to the complexities of biological reality. Even the most sophisticated methods, such as nuclear DNA sequencing, would be unable to put a single name on a plant that actually represents an admixture of two species, and methods that artificially returned a single name, such as sequencing of uniparentally inherited chloroplast DNA, would provide a false sense of precision. For commercial purposes, if all populations from a given region are traditionally used, there is usually no justification for artificially limiting current usage to a narrow portion of the range of morphological variation that is imagined to represent a *pure species*, which may scarcely exist as such in a hybrid zone.

PROCESS OF MORPHOLOGICAL AUTHENTICATION

MATERIALS AND METHODS

Morphological identification requires little equipment; useful basic supplies include fine forceps for manipulating small parts, single-edged razor blades for dissecting some structures, and some means of viewing small characters under magnification. For frequent use, a dissecting microscope is far superior to a hand lens and causes far less physical discomfort; if large amounts of material are to be passed under the microscope, the type in which the microscope body with eyepieces is held above the lab bench on a boom stand is preferable to the type in which the body is mounted above a raised platform. The availability of heated water may facilitate any necessary dissection of structures such as dried fruits, which can be softened by soaking in a little hot water with a tiny droplet of dish soap. Though persons performing botanical identification should be familiar with basic botanical terminology, the acquisition of a thorough reference work on the subject, such as Harris and Harris' *Plant Identification Terminology* [17], is recommended.

Morphological authentication of botanicals consists of several steps, most of which are preparatory and need be performed only once for each species. Preparatory tasks to be undertaken include the following steps: (1) identify relevant literature; (2) ensure that any disagreements regarding species circumscription are understood; (3) determine what other species are likely adulterants; (4) identify characters that distinguish both the species of interest and important adulterants; and (5) if possible, obtain reference specimens or samples of known identity. For each botanical to be identified, a list of criteria for acceptance or rejection of a batch of material should be drawn up or obtained from literature. Prior consideration should be given to a sampling protocol, whose details will vary depending upon batch size and packaging. Routine identification is then simply a matter of comparing samples of a material to the listed criteria and to reference samples. Expert identification of a plant often relies on the rapid and partly subconscious evaluation of its overall appearance or *search image* rather than rote examination of a list of individual characters [18]; however, this is a skill that will develop only with experience of individual species.

The first preparatory step is to identify literature sources that provide information on the appearance of the species of interest, the existence of any similar related species in the region where it is sourced, and any known adulterants. Useful literature includes botanical revisions and monographs, floras, pharmacognostic references and monographs, and publications in the field of plant anatomy. Revisions and monographs are detailed studies of all the species of a single plant group, while floras deal only with the species known to occur in a specified geographic area. All usually provide keys to facilitate specimen identification, which will provide an indication of some characters that are particularly useful in distinguishing groups of species. High-quality taxonomic literature is the ideal source of information when relatively intact aboveground portions of the plant are used. Pharmacognostic literature may supply extremely detailed descriptions of the parts of a species that are most commonly used in commerce. Indeed, the level of detail specified in monographs is sometimes greater than is useful, for example, there is no reason to examine cell shapes in material that is adequately identified by macromorphology. Pharmacognostic literature is also the best source of information on the identity of previously reported adulterants. Publications on plant anatomy may provide useful information about leaf, root, rhizome, or bark anatomy.

When using botanical literature, it should be ascertained whether the group in question is controversial and, if so, whether the species circumscriptions in each publication are consistent with the most widespread opinion or the opinion you favor. There is no obligation to accept all of the taxon circumscriptions used in the most recent treatment, which may not always be the best. For example, recent taxonomic treatments have, by reducing species to the status of subspecies, varieties, or mere synonyms of other species, reduced the number of species of *Sambucus* from over thirty to nine [19], reduced the number of species of *Echinacea* from nine to four [20], and treated the Asian plant commonly called *Astragalus membranaceus* as a mere variety of the rather distinctive European *A. penduliflorus* [21]. Many people, unsure that these radical changes are adequately supported, have continued to use the familiar species circumscriptions and nomenclature. If an author's species circumscriptions are different than those used or assumed by most recent authors, this must be recognized so that morphological descriptions are not misinterpreted. For instance, a description by an author who circumscribes a species broadly would include some morphological variation that those who define the species narrowly would exclude.

In commerce, the use of a customary circumscription of a botanical will generally be required no matter what name or names are put on the taxa involved. For example, if you accept Binns et al.'s four-species classification of Echinacea [20], you must allow only *Echinacea pallida* var. *pallida* to be sold under the Standardized Common Name of *Echinacea pallida* [22]; you may not include material of the former *E. simulata*, not traditionally included in the botanical product *E. pallida*, on the grounds that it is now included within the species *E. pallida* as *E. pallida* var. *simulata*. In the opposite case, if you favored the application of a narrow species concept to a plant that is usually considered a single variable species all of which is usable, logically all of the segregate species should be allowable in the botanical. However, given the fact that regulators evaluating label data might not be familiar with botanical controversies, it would probably be simpler and wiser for those with commercial interests not to utilize such species concepts.

Botanical and pharmacognostic literature should firstly be used to determine what other species are potential substitutes for the desired species. A revision of the

group or a floristic treatment from the material's region of origin should be consulted
to identify related species that occur in the same region and might be commingled,
especially with wild-harvested material. Species that are widespread in cultivation
should also be noted, as they may be accidentally cultivated under the wrong name;
for example, even though star anise (*Illicium verum*) is usually obtained from cultivated plants, adulteration with other species, including the toxic *I. anisatum*, has
occurred repeatedly [23]. If material is purchased from a region in which it has
not historically grown wild or been cultivated, it is particularly important to identify native species that might be more easily and cheaply produced in that region,
inspiring deliberate adulteration. For example, some Chinese material sold as *black
cohosh* (the North American *Actaea racemosa*) has been identified as commonly
cultivated, and possibly occasionally toxic, Chinese species of *Actaea* [24].

Information on the morphology of the relevant organs of the species of interest and
known or likely substitutes should be summarized. A list of characters should be prepared that are expected to be observable in the botanical as purchased and that, taken
together, should distinguish the desired species from all potential adulterants. It is
often useful to begin by carefully studying keys in botanical literature, as these are
(if well written) designed to emphasize the characters that most efficiently distinguish
the included species from one another. Because there is always morphological variation within species, care must be taken to ensure that character lists or descriptions
reflect the full range of variation in acceptable material, rather than only the most
common states. Characters that never occur in the species of interest but that are seen
in adulterants of particular concern should also be enumerated. Pharmacognostic
references and monographs (e.g., the American Herbal Pharmacopoeia monographic
series) may describe means of recognizing common adulterants.

Sometimes, available literature will not provide means of distinguishing partial
material of a species from likely substitutes, but careful independent examination of
herbarium specimens or samples may show that such characters actually exist. For
example, taxonomic literature often does not go into great detail on leaf anatomy,
because intact fertile specimens are more easily identified by other characters. If it is
ultimately not possible to compile a list of observable characters that would rule out
all likely substitutes, then the material of interest is too incomplete or fragmentary to
allow definitive botanical identification and another method of authentication must
be added or substituted.

It is extremely helpful to compare botanicals to reference specimens or samples
of known identity. Written descriptions cannot convey the overall appearance of a
botanical or the distinction between slightly different character states as well as seeing the plants themselves. Reference specimens or samples of known adulterants
can also be very useful. Collection or purchase of herbarium specimens at suitable
reproductive stages can provide an ideal reference material for comparison with
relatively intact leaves, flowers, or fruits. Numerous references on the preparation of
herbarium specimens are available [25–28]. Other parts, such as roots or bark, can
be collected simultaneously with a herbarium specimen to serve as reference materials for authentication of those organs. Authenticated Botanical Reference Materials
(BRMs) of some species can be purchased from a few private companies and from
the National Institute of Standards and Technology. BRMs consist of the parts used

in commerce, sometimes partially processed to reflect forms found in commerce; the identity of processed BRMs should be supported by the existence of a voucher specimen or equally reliable evidence of identity.

Use of reference specimens should always be combined with use of descriptions from literature. Each reference collection can represent only one population or individual of a species, not the full range of variation. For example, the leaves and stems of some species can range from virtually glabrous (hairless) to densely pubescent; if two samples from the extremes of that range were compared, they would look different enough that they could be assumed to be different species if a written description reflecting the full range of variation were not available. Conversely, two species may be distinguished only by small characters that may not immediately be noticed unless a review of literature has called attention to the need to observe them.

Morphological authentication then consists simply of visually examining the sampled material, comparing it to the description and if possible to reference material, and confirming that it possesses the expected characters and that no other species, especially known adulterants, are present in substantial quantity; especially for those in training, the use of a checklist of characters, which can be ticked off in turn, will encourage systematic observation. After the presence of expected characters and the absence of characters of common adulterants have been confirmed, the entire quantity of the sample selected for analysis should be quickly examined to ensure that it is not noticeably heterogeneous. Heterogeneous samples might be a mixture of two different morphologies belonging to the same species, but do require more careful investigation to separately determine the identity of each type of material present. Samples should also be examined for excessive debris or incorrect plant parts, such as portions of stems that may be harvested along with roots; pharmaceutical monographs may set specifications for the maximum allowable quantity of such material.

In commercial settings, a batch-specific record should be created to document, in a manner that will be satisfactory under the applicable regulatory regime, the fact that botanical authentication was performed. The record for each sample authenticated should minimally include the species or taxon, included part(s), source or supplier, batch number, total number of packages and number sampled (if applicable), and the name of the individual who authenticated it and on what date. For legal purposes, it would also be wise to include a means by which the criteria used to define identity can be specified in a shorthand form, such as a blank in which the published or locally generated descriptive standard being used can be named. If a batch of material has an associated herbarium voucher, its identity and location should also be recorded.

Organoleptic analysis can be a very useful adjunct to visual examination, as it is frequently possible for skilled people to recognize certain species, or detect the presence of adulterants, solely by their odor and flavor. Because human languages tend to lack a vocabulary capable of precisely describing those characteristics, organoleptic analyses are inherently dependent upon individual experience, whose development is to some extent limited by the keenness of the individual's sensory perceptions. Personal experience must therefore be gained through the handling of samples that have been identified by other means. Certainly, even for an inexperienced worker, an odor that is very different than that described in pharmacognostic, botanical, or herbal literature should be considered an indicator of possible adulteration or quality problems.

Sampling

No matter what method of authentication of a botanical is used, only a small portion of most batches of material will actually be subjected to scrutiny. When a large quantity of a botanical is to be authenticated, adequate sampling should always be considered. Wholesale batches of some botanicals are typically prepared by combining small amounts of material received from multiple sources. If several packages of such a batch are purchased and the examined sample were to be taken from only one of those, material in the other packages could have come from a different source and be of different quality. Several sampling regimens intended to provide statistically adequate coverage have been devised. The World Health Organization, for example, recommends for bulk material that every package in a batch of up to five packages be sampled, five packages from a batch of 6 to 50 packages, and 10%, rounded up, of the packages in a batch of 51 or more packages [29]; the sample of each package should consist of pooled samples of material from three different portions of the package. When very small batches are purchased, as is often the case when wholesalers buy directly from wild crafters, it will normally be possible and advisable to spread out and casually examine all of the material.

Limitations

Some groups of species are very difficult to identify, either because, as in hybrid swarms, the pattern of variation is not consistent or because the key characters are hard to observe or interpret or are not present on many specimens. Identification of most species in a well-understood flora, however, is quite straightforward—providing that an intact specimen harvested at the most appropriate reproductive stage is available. Commercial samples of botanicals can pose a challenge, because it is common for only single harvested plant parts to be available. If only parts offering limited characters, such as bark or roots, are seen, it can be impossible to distinguish a desired species from its close relatives. Further, commercial botanicals may be purchased already heavily processed. Powdered botanicals are obviously beyond the reach of traditional methods of macromorphological identification; cut and sifted botanicals, depending upon the material involved and the cut size, might or might not be unambiguously identifiable.

Even when botanicals are purchased in a form that cannot be unambiguously authenticated by morphology, so that other assays such as chemical fingerprinting should be performed, morphological examination is still highly recommended as a low-cost first-pass method of quality control. It may be possible to rule out the presence of some adulterants, or alternatively to confirm that they are present, in which case the batch can be rejected without wasting money on further testing. For example, even in finely cut material of skullcap (*Scutellaria*) species, the presence of distinctive nutlets belonging to the toxic adulterant germander (*Teucrium*) can be observed. Heterogeneity in root or bark material may raise suspicions of adulteration. Furthermore, nonbotanical debris or the presence of an excessive quantity of dirt or inappropriate plant parts may be observable.

VOUCHERING

A voucher is a preserved specimen or sample of plant material that represents a single gathering or batch and serves as a permanent record of the identity of that material. In field collections for botanical research, a voucher documents the identity of samples gathered from a single population at a single time; samples collected the next day from the same area, or on the same day from a population five miles away, would call for a different voucher. The purpose of this limitation is to minimize the risk that two similar-looking species will be collected under the same name. For practical purposes, a voucher prepared before or after associated samples are collected is sometimes allowable if there is no possibility whatsoever that the two might fail to represent the same population, such as when a voucher is collected from a specific field of cultivated plants before it is harvested or when samples are repeatedly taken from an individual tree. By necessity a single voucher sample is preserved from a batch of commercial material regardless of whether that batch was derived from multiple sources. However, the voucher sample should include material from multiple packages as needed in order to adequately represent the lot.

The ideal botanical voucher is an intact pressed herbarium specimen that includes leaf, stem, and flowers or fruits. In some cases, growers of cultivated botanicals may be willing to supply a pressed voucher or allow one to be collected before a field is harvested. Simultaneous collection of parts used in commerce, if they are not included on the specimen, supplies the collector with a useful BRM. For partly processed commercial botanicals, a small sample of each batch can be preserved in a sealable plastic bag as a voucher. All plant vouchers should be frozen before storage to kill any insect pests and then kept in a closed cabinet or container to protect them from insects and UV light; extreme humidity should be avoided as it will lead to molding. Vouchers should always be labeled clearly enough that their identity is obvious to other staff who may handle them years later.

Commercial enterprises dealing with botanicals ought to consider preserving vouchers of each batch of raw material for a suitable period of time. It is a common and appropriate practice for commercial facilities to retain samples of batches that are leaving the facility after they have been processed, for example, powdered, extracted, or encapsulated; these are often kept for up to five years, allowing adequate time for any consumer or regulatory complaints about a batch to be received. It seems to be much less common to retain samples of batches of material before processing, yet, because of the loss of identifying features at each stage of processing, the availability of less-processed samples could be useful if the identity of a batch of material were ever to be questioned. If samples taken were small and were not kept indefinitely, it would seem that this practice should not excessively increase costs. Such samples could also ultimately serve as supplementary reference materials, helping to represent the range of expected morphological variation within a species and type of material, or be used as training aids.

Botanicals used in scientific research should normally be documented with vouchers that are permanently preserved and made publicly available for examination by bona fide researchers. Whenever there might plausibly be a question about the correctness of the identification of a botanical test substance, other researchers should be free to examine the supporting evidence, not just for a few years, but for decades to come. Researchers should work with institutional or nearby herbaria to preserve pressed specimens if whole material is collected. Herbarium specimens should be labeled as vouchers for material used in published research to ensure that specimens potentially seen as low value, for example, sterile or cultivated specimens, are not later discarded to save space. For processed botanicals, a collaborating herbarium might be persuaded to curate samples placed in resealable sandwich bags and glued or stapled to an appropriately labeled herbarium sheet. If not, the voucher sample should be stored in the best available place, well labeled, and its location published. If vouchers are kept in an individual lab, consideration should be given to not only how possible requests for loan of material for study will be handled but also how to ensure that the vouchers are preserved and remain accessible after the researcher has departed from the institution.

REFERENCES

1. C. Linnaeus. *Species Plantarum*. Impensis Laurentii Salvii: Stockholm, Sweden, 1753.
2. E. Mayr. *Systematics and the Origin of Species*. Columbia University Press: New York, 1942.
3. E. Mayr, Ed. *The Species Problem*. American Association for the Advancement of Science: Washington, DC, 1957; pp. 1–22.
4. S. Dumolin-Lapègue, B. Demesure, S. Fineschi, V. Le Corre, and R.J. Petit. Phylogeographic structure of white oaks throughout the European continent. *Genetics* 146, 1997: 1475–1487.
5. J.I. Byatt. Hybridization between *Crataegus monogyna* Jacq. and *C. laevigata* (Poiret) DC. in south-eastern England. *Watsonia* 10, 1975: 253–264.
6. T.C. Wells and J.B. Phipps. Studies in *Crataegus* (Rosaceae: Maloideae). XX. Interserial hybridization between *Crataegus monogyna* (Series *Oxyacanthae*) and *Crataegus punctata* (Series *Punctatae*) in southern Ontario. *Canadian Journal of Botany* 67, 1989: 2465–2472.
7. M. Mogie and H. Ford. Sexual and asexual *Taraxacum* species. *Biological Journal of the Linnean Society* 35, 1988: 155–168.
8. J. Einset. Apomixis in American polyploid blackberries. *American Journal of Botany* 38, 1951: 768–772.
9. C.G.G.J. van Steenis, Eds. *Flora Malesiana*, Ser. 1. Noordhoff-Kolff N.V.: Djakarta, Indonesia, 1957; Vol. 5, pp. clxvii–ccxxxiv.
10. V. Grant. *Plant Speciation*, 2nd edn. Columbia University Press: New York, 1981.
11. W.L. Applequist. Rhizome and root anatomy of potential contaminants of *Actaea racemosa* L. (black cohosh). *Flora* 198, 2003: 358–365.
12. V.L. Bailey. Revision of the genus *Ptelea* (Rutaceae). *Brittonia* 14, 1962: 1–45.
13. N.C. Henderson. A taxonomic revision of the genus *Lycopus* (Labiatae). *American Midland Naturalist* 68, 1962: 95–138.
14. A.S. Gardiner and N.J. Pearce. Leaf-shape as an indicator of introgression between *Betula pendula* and *B. pubescens*. *Transactions of the Botanical Society of Edinburgh* 43, 1979: 91–103.
15. R. Morales. Hibridos de *Thymus* L. (Labiatae) en la Peninsula Iberica. *Anales del Jardín Botánico de Madrid* 53, 1995: 199–211.
16. R. Morales. "The history, botany and taxonomy of the genus *Thymus*." In *Thyme: The Genus Thymus*. Medicinal and Aromatic Plants—Industrial Profiles series, E. Stahl-Biskup and F. Sáez, Eds. Taylor & Francis: London, 2002; Vol. 24, pp. 1–43.

17. J.G. Harris and M.W. Harris. *Plant Identification Terminology: An Illustrated Glossary*, 2nd edn. Spring Lake Publishing: Spring Lake, UT, 2001.
18. S. Thayer. *The Forager's Harvest*. Forager's Harvest: Ogema, WI, 2006.
19. R. Bolli. *Revision of the genus* Sambucus. *Dissertationes Botanicae* 223, 1994: 1–227.
20. S.E. Binns, B.R. Baum, and J.T. Arneson. A taxonomic revision of *Echinacea* (Asteraceae: Heliantheae). *Systematic Botany* 27, 2002: 610–632.
21. X.-Y. Zhu. Revision of the *Astragalus penduliflorus* complex (Leguminosae—Papilionoideae). *Nordic Journal of Botany* 23, 2005: 283–294.
22. M. McGuffin, J.T. Kartesz, A.Y. Leung, and A.O. Tucker. *Herbs of Commerce*, 2nd edn. American Herbal Products Association: Silver Spring, MD, 2000; p. 59.
23. F. Saltron, A. Carine, and M. Guerere. Mise en évidence de contamination de badiane de Chine par d'autres espèces d'*Illicium*. *Annales des Falsifications, de l'Expertise Chimique et Toxicologique* 94, 2001: 397–402.
24. S. Foster. Exploring the peripatetic maze of black cohosh adulteration: A review of the nomenclature, distribution, chemistry, market status, analytical methods, and safety concerns of this popular herb. *HerbalGram* 98, 2013: 32–51.
25. D. Bridson and L. Forman, Eds. *The Herbarium Handbook*, 3rd edn. Royal Botanic Gardens, Kew: London, 1998.
26. C.E. Smith, Jr. *Preparing Herbarium Specimens of Vascular Plants*. Agriculture Information Bulletin no. 348; U.S. Department of Agriculture: Washington, DC, 1971.
27. R. Liesner. *Field Techniques Used by Missouri Botanical Garden*. http://www.mobot.org/MOBOT/molib/fieldtechbook/welcome.shtml. Accessed August 27, 2013.
28. J. Hildreth, E. Hrabeta-Robinson, W. Applequist, J. Betz, and J. Miller. Standard operating procedure for the collection and preparation of voucher plant specimens for use in the nutraceutical industry. *Analytical and Bioanalytical Chemistry* 389, 2007: 13–17.
29. World Health Organization. *Quality Control Methods for Medicinal Plant Materials*. World Health Organization: Geneva, Switzerland, 1998; pp. 6–7.

5 The DNA Toolkit
A Practical User's Guide to Genetic Methods of Botanical Authentication

Danica T. Harbaugh Reynaud

CONTENTS

Introduction..43
    DNA Is the Basis of Species Identity...50
DNA Toolkit...51
DNA Sequencing Sample Preparation and Method Validation 60
    DNA Extraction..60
    Gene Selection and Amplification...60
    DNA Sequencing and Reference Material Comparison 62
    Method Validation 63
Conclusions..63
References...64

INTRODUCTION

Accurate botanical species authentication and detection of adulterants are arguably the greatest challenges for raw materials and natural product manufacturers today, especially given the requirement for 100% identity testing in the Food and Drug Administration (FDA)'s dietary supplement cGMPs (21 CFR part 111). After all, the safety and quality of foods, herbal supplements, and other natural products depends on using ingredients that are (1) the correct identity and (2) free of contaminants. Although there are a wide range of technologies developed for authentication and adulterant detection, from microscopy and organolepsis, high-performance thin layer chromatography (HPTLC) and thin layer chromatography (TLC), and Fourier transform infrared (FTIR) and near infrared (NIR) to name a few, still major challenges remain especially regarding differentiation between closely related species and identification of multiple species in a mixture. As a result, there is widespread adulteration in the natural products industry. The combination of an increase in economically motivated adulteration and number of ingredients being sold as extracts and blends rather than in their whole unprocessed form makes the task of identification and detection of contaminants more challenging than ever. As illustrated in

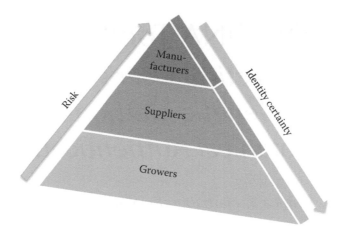

FIGURE 5.1 Risk-Identity Pyramid. The levels on the pyramid represent organizations along the supply chain for natural products, from the growers who farm the plants, to the suppliers who may process, blend, and/or distribute the plant materials, to the manufacturers who produce products using the plant ingredients. On the sides of the pyramid, the arrows indicate the inverse relationship between the certainty that these organizations have in identifying their materials.

the Risk-Identity Pyramid in Figure 5.1, there is an inverse relationship between our ability to confidently identify materials and the risks associated with manufacturing using adulterated ones, as materials are further processed and distributed through suppliers to manufacturers, they become more difficult to identify and are more prone to adulteration. Therefore, having a reliable toolkit that includes the most cutting-edge methods is critical to keep up with the changing times and ensure the safety and efficacy of natural products, especially for product manufacturers.

Authentication using deoxyribonucleic acid (DNA) sequence data is beginning to reveal just how widespread the adulteration issue is. For example, out of 250 randomly selected samples tested at AuthenTechnologies LLC. (Richmond, CA) during 2013 using DNA, a total of 22.4% of the tests were confirmed to be adulterated. Although the figure of 22.4% seems high, we estimate that the real percentage of adulterated materials is actually much greater, especially for blended or more processed materials. As Table 5.1 illustrates, most of the adulterated samples were routine quality assurance/quality control (QA/QC) samples that were identified as being substituted by closely related species (12.8%); this figure was more than double the amount that were adulterated by distantly related substitutes (5.2%). Table 5.2 lists the genera in which we identified adulteration by closely related species. Although some of these taxa are well known for having issues with adulteration, for example, *Cinnamomum* and *Cordyceps* where inferior species are often substituted for more rare and expensive ones, many others were surprising. It is believed that many instances of adulteration by close relatives are due to accidental substitution by species that may look similar or are taxonomically challenging and do not have distinct species boundaries (i.e., *Epimedium* and *Taraxacum*), sometimes as a result of hybridization (i.e., *Glycyrrhiza* and *Passiflora*).

TABLE 5.1
Results from 250 Randomly Selected DNA Species Authentication Tests

Sample Type	Results	Percentage of Test Articles (%)
Routine QA/QC	Correct species identity confirmed	77.6
Routine QA/QC	Closely related substitute identified	12.8
Routine QA/QC	Distantly related substitute identified	5.2
Routine QA/QC	Fungal contamination detected	2.0
Routine QA/QC	Unknown substitute detected	0.8
Out of specification	Distantly related substitute identified	1.6
	Total	100.0

The table includes the type of sample such as a routine QA/QC test (with no indication that the sample was of the incorrect identity) and those that were submitted with prior knowledge of the sample being out of specification using another method. Additionally, the second column describes different categories of results from the DNA authentication tests using DNA sequencing methods. The last column indicates the percentage of test articles that had each of the types of results.

TABLE 5.2
Genera Adulterated by Closely Related Species

Genus	Common Name(s)	Family
Baptisia	Wild Indigo	Fabaceae
Cinnamomum	Cinnamon	Lauraceae
Cistus	Rockrose	Cistaceae
Cordyceps	Cordyceps	Clavicipitaceae
Eleutherococcus	Siberian Ginseng	Araliaceae
Epimedium	Horny Goat Weed	Berberidaceae
Eupatorium/Eutrochium	Gravel root	Asteraceae
Fucus	Bladderwrack	Fucaceae
Glycyrrhiza	Licorice	Fabaceae
Myrica	Bayberry	Myricaceae
Origanum	Oregano, Marjoram	Lamiaceae
Passiflora	Passionflower	Passifloraceae
Plantago	Plantain	Plantaginaceae
Quercus	Oak	Fagaceae
Rhamnus	Buckthorn	Rhamnaceae
Schrophularia	Figwort	Schrophulariaceae
Taraxacum	Dandelion	Asteraceae
Thymus	Thyme, Savory	Lamiaceae
Tribulus	Puncture Vine	Zygophyllaceae
Verbascum	Mullein	Schrophulariaceae
Ziziphus	Jujube	Rhamnaceae
Verbena	Vervain	Verbenaceae

TABLE 5.3
Species Adulterated by Distantly Related Species

Labeled Species	Labeled Family	Labeled Common Name	Identified Species	Identified Family	Identified Common Name
Coriandrum sativum	Apiaceae	Coriander	*Apium graveolens*	Apiaceae	Celery
Eupatorium purpureum	Asteraceae	Gravel root	*Collinsonia canadensis*	Lamiaceae	Stone root
Euphrasia officinalis	Orobanchaceae	Eyebright	*Odontites* sp.	Orobanchaceae	Red Bartisia
Panax quinquefolius	Araliaceae	American Ginseng	*Astragalus membranaceus*	Fabaceae	Astragalus
Petroselinum crispum	Apiaceae	Parsley root	*Pastinaca sativa*	Apiaceae	Parsnip
Picrorhiza kurroa	Plantaginaceae	Kutki	*Musa acuminata*	Musaceae	Banana
Fallopia multiflora	Polygonaceae	Chinese Knotweed	*Silybum marianum*	Asteraceae	Milk Thistle
Ptychopetalum olacoides	Olacaceae	Muira Puama	*Croton echioides*	Euphorbiaceae	Muira Puama
Viburnum opulus	Adoxaceae	Cramp bark	*Acer tataricum*	Sapindaceae	Tatar Maple

In Table 5.3, we have listed examples from the tests that identified the presence of distantly related substitutes. Interestingly, none of the 17 samples were adulterated with their commonly known adulterant. This illustrates the need to use testing that not only detects and identifies close relatives and common adulterants, but also has the ability to detect *unexpected* ones as well. Examples from some of the tests that revealed distantly related adulterants that were unexpected includes substitution of coriander (*Coriandrum sativum*) with celery (*Apium graveolens*) both in the Apiaceae. A particularly unusual one was adulteration of *Picrohiza* with banana root (*Musa acuminata*). We also determined that a sample labeled as *Panax ginseng* was really *Astragalus membranaceus*. Another particularly interesting one was the substitution of Muira Puama (*Ptychopetalum olacoides*) with a plant from a distantly related family, *Croton echioides*, both commonly known as Muira Puama, and with similar aphrodisiac effects. To date, all Muira Puama samples tested at AuthenTechnologies using DNA have been identified as the adulterant, *C. echioides*, including commercially available reference materials. This illustrates the importance of using validated methods of identification, certified reference materials that have been properly authenticated using a valid method, as well as methods that are able to detect and identify unexpected adulterants.

There are myriad DNA-based technologies that have been used widely for decades by academics and law enforcement agencies, from defining species taxonomies to identifying criminals. Yet the natural products community has been slow to adopt DNA-based methods of authentication of ingredients. However, the realization that alternative methods may be unsatisfactory for identification of some species and detection of adulterants, and increasing pressures to comply with federal regulations has pushed more companies to utilize DNA-based species authentication. The first contract-testing laboratory in the United States to specialize in DNA-based technologies for species identification and adulterant detection, AuthenTechnologies LLC, was founded in 2010. Since that time DNA-based methods have quickly become a preferred test for routine quality control across a wide array of natural product ingredient suppliers and manufacturers. As a result, the National Institute of Standards and Technology (NIST) launched the first line of DNA-authenticated Standard Reference Materials for botanical species in 2013 in partnership with AuthenTechnologies. Further development and validation of DNA-based methods through partnership between AuthenTechnologies, the National Institutes of Health (Office of Dietary Supplements and National Center for Complementary and Alternative Medicine), and the FDA's Center for Food Safety and Nutrition is helping to accelerate the use and acceptance of DNA as a routine method in the natural products industry.

DNA is the building block of life and provides a suite of powerful and sensitive methods that can be used for natural product authentication and detection of adulterants. DNA is in every cell of every living or once-living organism on Earth, and it is the only characteristic that remains intact and unchanged throughout its life and beyond. In addition to DNA's extreme specificity, discriminating between the most closely related species and even individuals depending on the specific technique utilized, there are numerous other advantages of using DNA methods for identification. For instance, DNA is not affected by the developmental stage of the organism, nor by the environmental conditions that it lived in, nor by the season it was harvested. DNA-based methods are reliable and powerful especially when applied to single organ specimens where diagnostic taxonomic characters are not present, when applied to powdered materials where the distinguishing characteristics are no longer visible, and when it is difficult to distinguish among closely related and/or morphologically or chemically similar species. Although the morphology and chemistry of an organism can change drastically throughout its life, the DNA remains stable and consistent. Therefore, variation in DNA markers can be interpreted only by the genetic patterns underlying them, which is the foundation of their identity, and can be used to better understand other characteristics, such as their morphology and chemistry.

DNA-based methods have been used widely for plant identification for at least the past two decades, and advancement in technologies in recent years has made the use of DNA faster and more reliable than ever. Some of the early forms of DNA fingerprinting include methods based on hybridization of random pieces of DNA such as restriction fragment length polymorphism (RFLP), amplification of arbitrary DNA such as amplified fragment length polymorphism (AFLP), or site-targeted polymerase chain reaction (PCR). However, recently a more highly reproducible and informative analysis is the comparison of gene sequences from a specific stretch of DNA, or gene, referred to here as DNA sequencing, or barcoding; DNA barcoding is the use of short DNA sequences for the identification of an organism [1,2]. The DNA barcoding method is also widely used in the authentication of animals, and the method has already been validated by the FDA for identifying fish [3,4]. This method has dominated the field of plant taxonomy for the past decade and is beginning to take root as the preferred method of medicinal herb authentication [5,6]. Very recent advancements in sequencing technologies, such as next-generation sequencing (NGS), offer exciting new potential methods for genetic authentication, especially for degraded or mixed materials.

In this chapter, we will explore the use of DNA as the most generally useful basis for determining species identity, and in turn for authenticating unknown raw materials. Because genetic methods reveal the genetic history of the organism, DNA methods can be used to identify material of hybrid origin. This is especially important for medicinal and other commercially grown herbs, where humans have interfered with their reproduction through intentional or inadvertent crosses. Another use of genetic methods is to elucidate the speciation patterns, or phylogenetic patterns, of organisms; a *phylogeny* is a branching diagram or tree that shows the relationships between organisms, not unlike a family tree. Figure 5.2 illustrates an example of a phylogeny of black cohosh (*Actaea racemosa*) and its related species in the genus *Actaea*. By understanding the genetic relationships of species and their dispersal patterns throughout the globe, we are often able to understand the unique chemical profiles within them. In addition to providing great specificity, DNA is also extremely sensitive to contamination and can be used to identify multiple species in a mixture, whether intentional additives to a blend, or intentional or accidental adulteration. Depending on the specific method utilized, DNA is able to detect and identify up to *hundreds* of species in a single sample, even unexpected ones. Because they are highly sensitive, DNA sequences can identify many unexpected adulterants, contaminants, and additives that are present in low levels or chemically inert, including common fillers such as soy flour, as well as nonbotanical contaminants such as bacteria, fungi, and insects.

As with any class of methods, such as morphological or chemical, the specific technique used is differently suited for different types of materials and for detection and identification of adulterants. There are dozens of DNA-based methods, which vary in their utility for natural product ingredient authentication. In this chapter, we will briefly review the traditional use of DNA sequencing data for establishing the basis of species identity, summarize the different types of available genetic methods, and discuss their applicability to botanical authentication. Finally, we will provide a *DNA Toolkit* as a practical user's guide to aid in method selection and development. In this guide, we will discuss many of the aspects that are important to consider when applying these in a commercial setting such as the skill level, development time and cost, analysis time and cost, specificity, repeatability, and ability to detect mixtures, including unexpected and expected contaminants. Finally, we will indicate if there are any known commercial testing laboratories currently offering testing and discuss potential applications of using each technology.

Regardless of the specific DNA technology utilized for authentication or adulterant detection, the question remains whether DNA sequence methods should replace or complement chemical identification methods. Although DNA is well suited for taxonomic identification and for identifying species mixtures such as adulterants, it is unable to identify the plant part, as well as the presence or abundance of chemical components, both of which are critical for the quality and efficacy of the final product. Therefore, depending on the application, most DNA-based methods should be complemented by another analytical method that can, for instance, identify the plant part or the amount of specific marker compounds if necessary. However, DNA

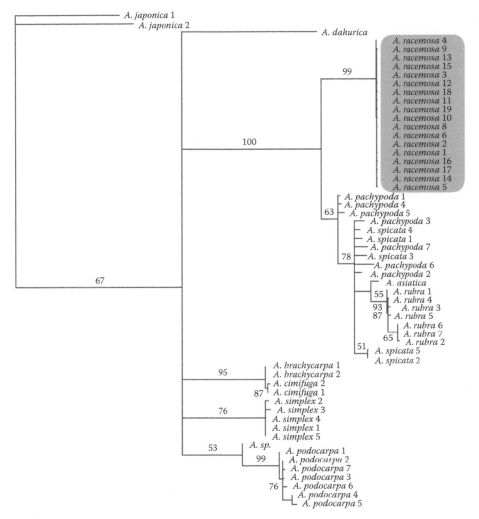

FIGURE 5.2 Phylogeny of black cohosh (*Actaea racemosa*) and related species. A branching diagram using DNA sequence data to elucidate the relationships between multiple specimens of black cohosh (*Actaea racemosa*) and its most closely related species. The numbers on the branches represent the bootstrap (BS) support values out of 100 and the branch lengths are proportional to the amount of genetic change. The numbers after the taxon names are sample numbers, representing independent vouchered collections. The gray box highlights the *Actaea racemosa* samples included in the inclusivity sampling frame, while those not highlighted are the close relatives included in the exclusivity sampling frame.

may be used to replace morphological and microscopic methods of identification, especially in processed materials, which are difficult to perform on a large scale and the quality of the results relies heavily upon the training of the person performing the test. Likewise, it may be used as an alternative to methods such as TLC, HPLC, FTIR, or NIR that are used simply as a species identification tool, and not for characterizing quality or potency. In most cases, the use of DNA may be best used as the initial identity screen before other more costly chemical tests are performed to characterize the quality of the material. In this chapter, we will discuss where alternative identity estimation methods have fallen short, especially with distinguishing between closely related species, and how DNA-based technologies can alleviate these problems.

DNA Is the Basis of Species Identity

Taxonomic botanists have struggled over the centuries to classify plants into real and meaningful groups. Traditionally, species have been identified based on morphological characters. As in the Linnaean system of taxonomy, features of reproductive organs of the flower and the fruit have played central to determining the difference between species. However, due to the inherent variation within organisms and differences in observers' interpretation of this variation (i.e., *lumpers* vs. *splitters*), many species taxonomies have changed over the years. The complex taxonomic history of herbs and changing circumscriptions present a great challenge to the natural products industries that are required to reliably authenticate incoming lots of herbs and produce consistent products. Fortunately, in the past two decades, there has been an important technological advancement that has increased our knowledge of the identity of organisms and has lead to development of more reliable and robust species taxonomies: DNA sequencing. Today, DNA sequencing is the most commonly used method by taxonomic botanists to classify species or to reclassify species that were traditionally circumscribed based on alternative methods such as morphology or chemistry. Thousands of papers have been published on the use of DNA sequence data for species identification and authentication (e.g., see [2,7]). One major benefit of using DNA as the basis for species identity and authentication of natural products is that it is unbiased and can provide a more solid understanding of the identity and relationships of individuals or species that doesn't change—even if the names do.

As botanists unravel the genetic history of species, they are discovering that many of the traditionally used methods using morphology and chemistry do not correlate with the true identity of the species. The problem with basing a system on morphological or chemical characteristics alone is convergent evolution, whereby unrelated groups have similar features due to adaptation to the environment. For example, American cacti and African euphorbias look similar and are both succulent due to adaptations to arid environments, but they are very distantly related [8]. Likewise, identical or similar chemical compounds may arise separately in unrelated lineages [9]. Although families or genera of plants may contain specific classes of compounds [10], these may not be continuously expressed or may be triggered by herbivory, damage, or be dependent on other environmental factors. As a result, certain metabolites that may be present in a given taxon are not shared with their most closely related ones. Inconsistent secondary metabolite profiles mean that the systematic value of chemical characters becomes a matter of interpretation in the same way as traditional morphological characters. The distribution of secondary metabolites has some value for taxonomy, but their occurrence reflects adaptations and particular life strategies embedded in a given phylogenetic framework. However, these data combined with DNA may be the most appropriate taxonomic system with DNA as the *taxonomic scaffold* by which other chemical and morphological characteristics are applied and used in the field for identification [11]. This integrative approach whereby the DNA blueprint is correlated with the morphology and chemistry allows us to make more meaningful and reliable categories of plants that are useful to people, as well as help us to identify strains or lineages that have superior qualities, as in marker-aided selection, commonly used in animal and plant breeding, as well as for botanical drug discovery [12–17]. That said, phylogenetic analyses based on DNA sequences are extremely valuable tools for natural products manufacturers, developers, and researchers not only to authenticate ingredients, but also to understand the variation and patterns in secondary metabolites and activity, as well as other functional or qualitative characteristics of organisms [18], and should play a central role in research and validation studies utilizing *any* botanical identification method.

DNA TOOLKIT

DNA-based methods offer a number of excellent solutions for species identification and detection of adulterants [19]. As with any category of methods (i.e., morphological or chemical), DNA is not appropriate for all types of starting materials. Table 5.4

TABLE 5.4
Materials Appropriate for DNA Testing

Species: Any-botanical, animal, fungal, or bacterial
Form: Any-whole, cut, chopped, sliced, powdered
Part:
   Botanical: Whole, leaves, stems, roots, bark, flowers, stigmas, pollen, aerial parts, herbs, fruits, and seeds
   Animal: Living or once living tissues, organs (e.g., muscle, skin)
   Fungal: Mycelia and fruiting body
   Bacterial: Whole (living or dried)
Type: Fresh, dried, frozen, juiced, puréed

Examples of appropriate ingredients

Dried herbs and spices	Meat products
Fresh seeds and nuts	Liquid extracts
Fruit purées	Pressed oils (e.g., olive oil)
Pea and soy protein	
Juices and concentrates (liquid and dried)	
Grain flours	
Probiotic bacterial cultures (liquid and dried)	
Supplement capsules and tablets	
Food products (except highly refined)	

Examples of inappropriate ingredients

Dried extracts
Distilled oils (e.g., essential oil)
Plant resins
Milk
Oyster shells

lists examples of materials that can be analyzed using DNA. In general, DNA testing can be used on any species and most forms, as long as there is some cellular material intact. Examples include dried and fresh botanical, fungal, animal, and bacterial samples as well as purées, juices, and select supplement capsules and tablets and finished products such as foods that are not highly processed other materials that often contain DNA are pressed oils (i.e., olive oil) and liquid extracts+tinctures. Examples of materials that do not contain DNA are powdered herbal extracts, essential oils and resins, as well as animal milk and shells such as from oysters.

There are a wide range of DNA-based technologies each with their own advantages and shortcoming for a particular application such as the specificity; precision; repeatability; upfront development cost and time; ability to work on processed materials; difficulty of data analysis; and ability to detect unexpected adulterant, contaminant, or substitute species. Therefore, it is imperative to weigh each of these aspects when selecting the most appropriate method for a particular purpose. Therefore, in this section I will highlight several of the most widely used genetic methods for species identification studies and examine the features of each method that are most relevant to natural products researchers and manufacturers. Features of the genetic tools are rated using my own ordinal scale based on experiences as outlined in Table 5.5, to indicate the level of effort, time, and/or cost.

1. DNA hybridization
   a. *Description:* Comparison of random pieces of DNA such as in RFLP.
   b. *Skill:* 2.
   c. *Development time:* 3.
   d. *Development cost:* 3.
   e. *Analysis time:* 3.
   f. *Analysis cost:* 3.
   g. *Specificity:* 3.
   h. *Repeatability:* 3.
   i. *Degraded DNA analysis:* 1.
   j. *Mixture detection-expected:* 1.
   k. *Mixture detection-unexpected:* 1.
   l. *Commercial availability:* Unknown.
   m. *Discussion:* DNA hybridization techniques were once popular for plant genetic studies [20–23], but have been replaced by more reliable

---

TABLE 5.5

Scale Used for Rating DNA Tool Features

Rating	Description
0	Very low
1	Low
2	Moderate
3	High
4	Very high

---

technologies such as DNA sequencing analysis. In RFLP, a restriction enzyme is used to produce DNA fragments of different lengths. In theory, the patterns of DNA fragments vary between different strains or species. The fragments are separated through electrophoresis on an agarose gel. The DNA fragment profile is then transferred to a matrix and hybridized with a fluorescently labeled DNA probe and polymorphisms in fragments are detected by the presence or absence of fragments that fluoresce. Depending on their design, RFLP assays may be highly specific, accurate, and reproducible. It may be able to detect adulterants; however, their identity may not be determined. However, the RFLP profile obtained may be contingent on developmental stage, organ sampled, or environmental conditions depending on the enzyme(s) selected for the restriction reaction. The upfront development time and cost and ongoing sample analysis cost and time are very high. Additionally, it requires high-quality DNA, so it is not a good candidate for processed or degraded materials.

2. Arbitrary PCR
    a. *Description:* PCR amplification of DNA loci using arbitrary primers as in random amplified polymorphic DNA (RAPD), AFLP, and inter-simple sequence repeat (ISSR) analysis.
    b. *Skill:* 2.
    c. *Development time:* 2.
    d. *Development cost:* 2.
    e. *Analysis time:* 2.
    f. *Analysis cost:* 1.
    g. *Specificity:* 2.
    h. *Repeatability:* 1.
    i. *Degraded DNA analysis:* 2.
    j. *Mixture detection-expected:* 0.
    k. *Mixture detection-unexpected:* 0.
    l. *Commercial availability:* Unknown.
    m. *Discussion:* PCR-based methods involve the use of primers that select a specific region of the genome and amplify them using a thermal cycler or PCR machine. In general, PCR-based methods are advantageous because only very small quantities of DNA are necessary for analysis. In arbitrary PCR methods, short arbitrary oligonucleotide primers are used to generate a large number of random PCR products. Polymorphism in the products is due to random mutations, which are visualized on a gel with a fluorescent DNA stain. In some methods, such as AFLP, the use of restriction enzymes is combined with PCR of random fragments. These methods require no knowledge of DNA sequences or relationships between individuals or species; therefore, there is little upfront development time and cost. Although these methods are efficient for screening multiple sites across the genome, and can differentiate between individuals within a single species, this category of methods can be prone to issues with repeatability. For instance,

degraded DNA can affect the amplification patterns. Additionally, they are not designed to identify multiple species in a mixture because it is difficult to distinguish between variation in the target and closely related species. Although this category of methods is relatively quick and affordable, it can lead to generation of data that is very complex and difficult if not impossible to analyze and understand patterns, including making species identifications. However, AFLP and RAPD studies have been performed on a number of medical plant groups, especially Traditional Chinese Medicinal herbs [24–29]. This category of methods is best utilized for analysis within a single species or within a set of closely related species, such as to understand relationships between populations, varieties, or cultivars.

3. Site-targeted PCR
   a. *Description:* Amplification and detection of particular loci using specific primers either through traditional PCR, as in cleaved amplified polymorphic sequence, direct amplification of length polymorphisms, amplification refractory mutation system, simple sequence repeats, sequence characterized amplified region, and through the use of quantitative PCR (qPCR) techniques.
   b. *Skill:* 3.
   c. *Development time:* 4.
   d. *Development cost:* 3.
   e. *Analysis time:* 2.
   f. *Analysis cost:* 1.
   g. *Specificity:* 4.
   h. *Repeatability:* 4.
   i. *Degraded DNA analysis:* 4.
   j. *Mixture detection-expected:* 4.
   k. *Mixture detection-unexpected:* 2.
   l. *Commercial availability:* Yes.
   m. *Discussion:* There are a wide variety of PCR-based techniques that utilize primers that amplify specific gene regions or loci. Each one of them varies slightly in their overall skill level, development time and cost, and analysis time and cost, specificity, and repeatability. However, in general, PCR-based methods are extremely reliable in that they are repeatable and can be used on a wide variety of starting materials including lightly processed and cooked finished products and some botanical extracts because they can be designed to amplify very small fragments of DNA. One major advantage of this technique is that it is generally robust against DNA degradation, as primers are designed to amplify small fragments of DNA (~100–200 bp). Additionally, it is highly sensitive and useful for screening multiple potential contaminants or adulterants quickly and affordably.

   Site-targeted PCR has been used to detect adulterants in several important Traditional Chinese Medicines [30,31]. Similar site-targeted

PCR methods have been used to detect species in botanical raw materials [32–35] and, as well as identify the origin of plant species in vegetable oils from olive to soybean oil [36–40] and fruit species in juices [41]. Additionally, one study identified plant species in liquid botanical extracts using this method [42]. Unless careful and extensive validation studies are performed however, to examine the specificity and sensitivity of the PCR primers, and necessary precautions are taken to eliminate contamination, these highly sensitive tests may lead to erroneous results.

The particular assay's ability to discriminate between different species of interest, as well as detect low levels of DNA in a mixture, depends heavily on the design of the assay. For instance, more *universal* primers can be used to detect a large set of organisms (i.e., all plants) or species-specific primers can be used for detection of one species or variety. However, the sensitivity of the primers to detect low levels of contaminants is affected by the primer design and specific PCR cycling parameters used. Additionally, if a particular plant species is excluded from the primer design, they may not be detected. Therefore, expected adulterants can reliably be detected with this method. However, if more universal primers are not also used, unexpected adulterants may be missed; these universal primers, however, tend to be less sensitive to the very specific ones. Although a lot of upfront cost and development time is required to design and validate the PCR primers, the cost and time to test samples and analyze the results once developed is minimal. One of the key components to designing a valid PCR-based method is to first validate the specificity of the gene regions, using DNA sequencing analysis, as described in DNA Sequencing: Sanger section.

In addition to species authentication, site-targeted PCR methods are used more widely in commercial applications for the detection of specific allergens, such as wheat or peanut. Additionally, similar site-targeted methods are used for the detection of genetically modified organisms (GMOs).

4. Chips and arrays
   a. *Description:* Simultaneous detection of a predetermined set of genes or species using one of a number of traditional microarray and emerging chip technologies.
   b. *Development time:* 4.
   c. *Development cost:* 4.
   d. *Analysis time:* 2.
   e. *Analysis cost:* 3.
   f. *Specificity:* 3.
   g. *Repeatability:* 3.
   h. *Degraded DNA analysis:* 3.
   i. *Mixture detection-expected:* 4.
   j. *Mixture detection-unexpected:* 0.

    k. *Commercial availability:* Unknown.
    l. *Discussion:* DNA chip and microarray technologies are extremely
        powerful in that they can be used to analyze a high number of genes
        in several samples simultaneously. There are a number of chip and
        array technologies, the specifics of which are outside the scope of this
        chapter. However, depending on their design, chips and arrays can be
        highly specific, accurate, and reproducible. Although they are relatively
        time-consuming and expensive to develop, the relative ease of per-
        forming routine testing once they are developed makes them suitable
        for automation. However, they are difficult to modify once developed.
        These technologies are typically resistant to problems with degraded
        DNA. Like site-targeted PCR, a predetermined set of species must be
        designed into the assay in order to detect them. Therefore, the specific-
        ity and the accuracy of the assays for detecting specific species can
        vary from assay to assay and must be validated extensively before being
        put into use. Applications of this technology include screening for the
        presence of target and adulterant species in incoming raw materials or
        finished products and have been used for authentication of medicinal
        herbs and in herbal drug research [43–47].
 5. DNA sequencing: next generation
    a. *Description:* Amplification of a select gene followed by elucidation of
        the identity and arrangement of bases using one of a number of cutting-
        edge sequencing technologies called NGS, including pyrosequencing
        (454), MiSeq (Illumina), sequencing by ligation (SOLiD), and ion semicon-
        ductor (Ion Torrent), among others.
    b. *Development time:* 2.
    c. *Development cost:* 2.
    d. *Analysis time:* 4.
    e. *Analysis cost:* 4.
    f. *Specificity:* 4.
    g. *Repeatability:* 4.
    h. *Degraded DNA analysis:* 3.
    i. *Mixture detection-expected:* 4.
    j. *Mixture detection-unexpected:* 4.
    k. *Commercial availability:* Yes.
    l. *Discussion:* There are a number of NGS technologies that vary widely
        in their specific abilities and cost. However, in general, NGS methods
        produce huge quantities of data relatively rapidly. Although for the
        amount of DNA sequencing data they generate in a relatively short
        period of time the cost per base sequenced is quite low, many thou-
        sands if not millions of DNA base pairs can be sequenced with these
        methods, making the cost per sample much higher than other methods.
        However, depending on the application, NGS can be a particularly pow-
        erful tool and well worth the expense. For instance, NGS can produce
        upward of 10,000 short DNA sequences from a single sample, which
        can allow for identification of hundreds of species, even unexpected

ones. One of the drawbacks to NGS methods is that they require highly skilled analysts to perform data analysis and comparison to reference sequences. In general, NGS technologies are well suited for degraded DNA because they amplify and sequence relatively short fragments of DNA. If universal primers are used, there is little upfront development cost and time; however, the extensive time and money spent to perform each analysis and analyze the data makes NGS technologies not suitable for routine authentication testing, until bioinformatics pipelines are optimized. Additionally, before NGS methods are used, the specific gene regions used for identification must be validated for specificity as described below such as through using traditional Sanger sequencing, and the proper reference materials must be obtained. There are numerous potential applications to NGS using universal gene regions, including authentication of ingredients, complex blends, and finished products, as well as product de-formulation and detection of adulterants, allergens, and filth. This technology has also been useful for identification of complex microbial communities in environmental and plant samples [48] as well as for detection of plant and animal adulterants in Traditional Chinese Medicine [49]. This technology has been useful for label claims and detection of unlabeled ingredients, as illustrated in the Certificate of Analysis from NGS testing (Figure 5.3) of a finished dietary supplement capsule in which 16% of the DNA sequences originated from soy, which was not listed on the label.

6. DNA sequencing: Sanger
   a. *Description:* Amplification of a select gene region followed by elucidation of the identity and arrangement of bases using traditional Sanger sequencing, also known as the chain-termination method.
   b. *Development time:* 2.
   c. *Development cost:* 2.
   d. *Analysis time:* 4.
   e. *Analysis cost:* 4.
   f. *Specificity:* 4.
   g. *Repeatability:* 4.
   h. *Degraded DNA analysis:* 3.
   i. *Mixture detection-expected:* 3.
   j. *Mixture detection-unexpected:* 3.
   k. *Commercial availability:* Yes.
   l. *Discussion:* Sanger DNA sequencing based on the chain-termination method has been used widely throughout academic and medical research institutions since the 1980s. Despite the fact that NGS technologies are now emerging, Sanger sequencing still remains the most popular sequencing method, especially for species identification and taxonomic studies and remains the most well-suited choice for routine authentication of ingredients, from all organisms including plants, animals, fungi, and bacteria. DNA sequencing, though Sanger sequencing technology, is highly specific, accurate, and reproducible. The protocols

Latin Name	Family	Common Name	On Label	Percentage (%)
Allium cepa	Alliaceae	Green onion	Yes	0.10
Allium sativum	Alliaceae	Garlic	Yes	0.10
Alternaria alternata	Pleosporaceae	Leaf spot fungus	No	0.17
Amaranthus sp.	Amaranthaceae	Amaranth	Yes	0.00
Apium graveolens	Apiaceae	Celery	Yes	0.17
Asparagus officinalis	Asparagaceae	Asparagus	Yes	0.12
Beta vulgaris	Chenopodiaceae	Beet	Yes	0.30
Brassica oleracea	Brassicaceae	Broccoli, kale, red cabbage, Brussels sprouts, cauliflower	Yes	9.79
Capsicum annuum	Solanaceae	Bell pepper	Yes	0.10
Cicer arietinum	Fabaceae	Garbanzo bean	Yes	0.22
Cladosporium sp.	Davidiellaceae	Mold	No	0.05
Cucumis sativus	Cucurbitaceae	Cucumber	Yes	1.95
Daucus carota	Apiaceae	Carrot	Yes	1.41
Fagopyrum esculentum	Polygonaceae	Buckwheat	Yes	0.00
Fragaria sp.	Rosaceae	Strawberry	Yes	1.04
Fusarium sp.	Nectriaceae	Soil fungus	No	0.17
Gibellulopsis nigrescens	Plectosphaerellaceae	Black fungus	No	0.05
Glycine max	Fabaceae	Soy	No	16.15
Helianthus annuus	Asteraceae	Sunflower seed	Yes	0.07
Lactobacillus bulgaricus	Lactobacillaceae	Lactobacillus bacteria	Yes	0.00
Linum usitassimum	Linaceae	Flax seed	Yes	0.05
Lycopersicon esculentum	Solanaceae	Tomato	Yes	0.00

Medicago sativa	Fabaceae	Alfalfa	Yes	11.08
Petroselenium crispum	Apiaceae	Parsley	Yes	5.59
Phaseolus vulgaris	Fabaceae	Kidney bean	Yes	0.05
Phoma sp.	Incertae sedis	Soil fungus	No	0.20
Phomopsis longicolla	Valsaceae	Mold	No	0.05
Pleospora sp.	Pleosporaceae	Plant fungus	No	0.05
Prunus avium	Rosaceae	Cherry	Yes	0.30
Rubus fruticosus	Rosaceae	Blackberry	Yes	0.35
Rubus idaeus	Rosaceae	Raspberry	Yes	0.32
Saccharomyces cerevisiae	Saccharomycetaceae	Brewer's yeast	Yes	21.37
Salvia hispanica	Lamiaceae	Chia seed	Yes	0.05
Sesamum indicum	Pedaliaceae	Sesame seed	Yes	0.00
Spinacea oleracea	Amaranthaceae	Spinach	Yes	28.47
Vaccinium corymbosum	Ericaceae	Blueberry	Yes	0.10
Vigna angularis	Fabaceae	Adzuki bean	Yes	0.00

FIGURE 5.3 Next-generation sequencing test results using 454 pyrosequencing from a finished product. The table in this figure indicates the Latin name, Family, and Common name of all species identified in a finished dietary supplement capsule. Additionally, the fourth column indicates whether or not the species was listed on the label. In the final column, the percentage of sequences originating from each species is indicated.

are straightforward and amenable to automation. The upfront development time and cost, as well as sample analysis time and cost, is relatively low, especially when using universal primers. Degraded DNA can
be a problem if large fragments are being amplified; however, methods
can be modified to amplify multiple shorter overlapping fragments to
counteract degradation. Data analysis is usually simple and DNA standards are not needed with each test sample. Taxonomic discrimination
is dependent on the markers targeted, but a well-designed sequencing
assay can be highly discriminatory. When universal primers are used,
unexpected adulterants can easily be detected. Additionally, mixtures
and hybrids can also be detected by observing overlapping bases in the
DNA chromatogram. Therefore, based on these criteria authentication
of species based on DNA sequencing (using Sanger sequencing technology) is arguably the most appropriate routine method for commercial applications, primarily raw material ingredient authentication and
detection of adulterants. There are numerous published papers on using
DNA sequencing to detect adulteration in raw materials (see [19]) and
finished products, from dietary supplements to teas [50–53]. See next
sections for more information on sample preparation and validation of
DNA sequencing methods.

DNA SEQUENCING SAMPLE PREPARATION AND METHOD VALIDATION

DNA EXTRACTION

As with any DNA-based method, the first step in the process is DNA extraction. There
are numerous published methods, for instance, using cetyl trimethylammonium bromide or using silica columns, as well as a variety of commercially available kits that
use silica columns or glass-coated magnetic beads. However, the suitability of the
method or kit depends on the starting material, and the nature of the DNA extraction procedure can dramatically affect the success of the amplification reactions.
For instance, many medicinal plants contain an abundance of secondary metabolites
(polysaccharides, tannins, essential oils, phenolics, alkaloids, and waxes), which
if not properly removed can interfere with the PCR and produce a negative result.
Additionally, many of the materials used commercially have not been specifically
dried or stored in a way that preserves the DNA (such as rapid desiccation using
silica gel). Therefore, the DNA of commercial materials is often degraded and minor
adjustments to standard protocols may be necessary.

GENE SELECTION AND AMPLIFICATION

Choosing the desired gene region to sequence is the most critical step in using DNA
sequences for plant identification; the gene must be variable enough to distinguish
between the target and nontarget species. Although for animals, a single gene or

DNA barcoding region, cytochrome oxidase I is useful for authentication; because of the heterogeneous rates of evolution between plant groups, different gene regions are necessary in different groups to resolve relationships. For example, in a recent validation study funded by the National Center for Complementary and Alternative Medicine, testing eight different potential regions across five commonly used medicinal herbs found that a unique combination of two loci was best suited for each plant group (Reynaud, unpublished). For instance, in some groups, particular loci would not amplify at all or had little or no variation, while in other groups, they were easily amplified and contained significant variation. Because the base mutations that are used for species identification occur randomly in the genome, it is difficult to predict which regions are going to contain the features necessary for species resolution. Therefore, testing and validating multiple candidate regions is critical for DNA sequence authentication. In some cases, a single gene region may be sufficient for identification. However, multiple genes from different parts of the genome (i.e., chloroplast or nuclear) are necessary for independent verification of the identity and to ensure that different genes have not undergone different evolutionary histories, such as hybridization [54]. Fortunately, intraspecific hybrids can easily be identified visually using DNA sequencing methods as long as nuclear genes are analyzed as they provide evidence from both parents, as opposed to chloroplast genes, which are typically maternally inherited. Figure 5.4a illustrates what a DNA chromatogram looks like when a hybrid, or mixture of very closely related species with equal concentrations, is present. Figure 5.4b shows an example of a DNA chromatogram from a sample that is contaminated by a more distantly related species; the low level of underlying *noise* is due to the presence of a secondary DNA sequence that is of a different length from the dominant signal.

Once the appropriate gene regions are selected, amplification of the genes is performed using a thermal-cycler, or PCR machine. In addition to the design of the primers used to select the specific regions, a number of other factors can affect the success of amplification including the enzymes used in the reaction, as well as

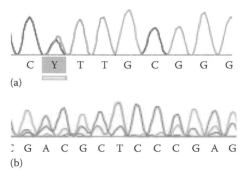

(a)

(b)

FIGURE 5.4 DNA chromatograms showing evidence of hybridization and contamination. (a) Hybridization or contamination by a close relative is evident by two overlapping bases (as indicated by *Y*, where both *C* and *T* are present) at specific bases in a sequence. (b) Contamination by a more distant relative is evident by multiple overlapping bases along all or most of the sequence.

the temperature and cycling times used in the PCR machine. All of these factors must be tested in a proper validation study to examine their affect on the sensitivity and specificity of the methods.

DNA Sequencing and Reference Material Comparison

Once PCR products are amplified, the DNA is sequenced using Sanger sequencing. Nowadays, capillary electrophoresis machines are the most widely used for sequencing the DNA. The resultant DNA chromatograms can be obtained electronically from the sequencing machine and used for further analysis and identification. Comparison of the DNA sequences to authentic reference materials is a critical step in performing authentication of a test sample. As with any method, the reliability and accuracy of a method depends on the quality of the reference materials, both of the target and nontarget species. One major benefit to using DNA sequences for reference materials is that huge databases of reference sequences can be built to which test samples can be compared, as opposed to alternative methods whereby reference materials must be run alongside the test sample, which limits the scope of references that can be used. Currently, AuthenTechnologies has initiated development of the Herbal Reference Barcode (HERB™) Database of validated DNA reference sequences obtained from herbarium vouchers for plant authentication and detection of adulterants, in partnership with major herbaria around the United States including its partner at the Center for Herbal Identity at the University of California, Berkeley. What separates DNA from chemical or other analytical methods is that it is the *only* method that can analyze herbarium vouchers; because DNA is consistent throughout all organs of a plant, we are able to obtain DNA reference sequences from tiny leaf fragments from the vouchers. This enables us to obtain reference materials not only from commonly traded herbs, but also from rare ones and other close relatives not typically available for purchase through other reference material suppliers. Most importantly, by having the ability to base DNA identification methods on herbarium material that can unambiguously be authenticated by a botanist, using the important taxonomic features, DNA-based methods are extremely scientifically robust.

Once sequences from authentic reference materials are obtained for the appropriate gene regions, the DNA sequence of test samples is compared in a number of ways to make an identity confirmation. Typically, the first step is to produce an aligned DNA matrix, which allows us to compare homologous bases in the DNA sequence. Although there are a number of computer algorithms that can help with producing alignments, it is imperative to check each alignment visually, as this is critical for making an accurate identification. For materials in which there is little sequence divergence within the target species, and the nontarget species are easily distinguished, taxonomic identification can be made by visually inspecting the DNA sequences and identifying specific bases that uniquely discriminate the target species. For those species with more variation that are difficult to assess visually, test materials can be authenticated by building a phylogenetic tree, or branching diagram, using one of a number of different algorithms. Because identification cannot be made if a test sample falls outside the range of variation of the reference materials, it is imperative to validate methods using a wide range of samples from within the target and closely related species, as described subsequently.

METHOD VALIDATION

It is critical when using DNA sequencing for species authentication that the gene regions are validated for specificity, such as by following AOAC's Guidelines for Botanical Identification Method Validation [55]. These validation procedures are based on testing a wide sampling of the target and most problematic nontarget species (which in the case of genetics are the most closely related ones) to ensure that the genes differentiate between them a desired percentage of the time (i.e., probability of identification [POI] exceeds 95%). Additionally, the range of potential variation that can be expected in the target species must be characterized, in order to not falsely reject authentic materials. Gene specificity must be validated before other methods of genetic analysis are used, such as site-specific PCR or even NGS methods. Fortunately, because DNA sequences are not affected by environmental conditions or plant part, the range of variation within species is often negligible. Therefore, once gene regions are carefully selected, the reliability of these methods to correctly identify species is exceptional. In fact, for all of the validation studies completed at AuthenTechnologies to date, POI is 100% or nearly so for all genes selected after careful pre-validation studies have been performed to examine the rates of variation within genera (unpublished data). Figure 5.2 illustrates a validation study of black cohosh (*Actaea racemosa*) and its close relatives based on nuclear ribosomal DNA (nrDNA) data, in which approximately 20 herbarium vouchers of *Actaea racemosa* were sampled from across its geographic range, as well as multiple exemplars from its most closely related species. In this figure, all samples of black cohosh are nearly indistinguishable genetically, indicated by the cluster of *A. racemosa* specimens with no resolution between them. However, the group of *A. racemosa* is well supported (bootstrap support value of 99) and distinct from all other closely related species of *Actaea* analyzed. Therefore, this figure demonstrates the utility of DNA sequence data to unequivocally identify black cohosh and exclude all other close relatives.

CONCLUSIONS

Accurate authentication of natural products and the ingredients used to formulate them is the foundation for any quality control process. However, because of the rampant rates of adulteration and the fact that many materials used by manufacturers are powdered, processed, or blended, the already difficult task of identification can become impossible unless alternative and/or cutting-edge technologies are considered and integrated into QC processes, or used to replace methods that are not suitable. In this chapter, we shed light on the *real* rates of adulteration in the marketplace using DNA and on the issue of substitution by close relatives, which often go undetected by alternative methods. We briefly reviewed the importance of using DNA as the foundation for developing and validating *any* methods of identification and understanding patterns in chemical and morphological variation, because DNA is the *basis* for species taxonomy. Finally, in the *DNA Toolkit*, we reviewed several of the most applicable genetic methods for species authentication and detection of adulterants. As with any category of methods, genetic methods vary in their overall reliability and careful selection of the most appropriate method for the intended purpose is necessary. Whether it is authenticating incoming lots of raw materials, verifying label claims on finished products, developing authenticated reference materials, or identifying the most closely related species to include in validation studies of other methods, DNA-based methods offer a wide array of extremely valuable methods that should be considered as part of any authentication or quality control process.

REFERENCES

1. W.J. Kress, K.J. Wurdack, E.A. Zimmer, L.A. Weigt, and D.H. Janzen. Use of DNA barcodes to identify flowering plants. *Proceedings of the National Academy of Sciences of the United States of America* 102, 2005: 8369–8374.
2. N.J. Sucher and M.C. Carles. Genome-based approaches to the authentication of medicinal plants. *Planta Medica* 74, 2008: 603–623.
3. H.F. Yancy et al. A protocol for validation of DNA-barcoding for the species identification of fish for FDA regulatory compliance. *Laboratory Information Bulletin* 24, 2008: 1–25.
4. S.M. Handy et al. A single-laboratory validated method for the generation of DNA barcodes for the identification of fish for regulatory compliance. *Journal of AOAC International* 94, 2011: 1–10.
5. M.T. Cimino. Successful isolation and PCR amplification of DNA from National Institute of Standards and Technology Herbal Dietary Supplement Standard Reference Material powders and extracts. *Planta Medica* 76, 2010: 495–497.
6. J. Ma, S.L. Chen, M.E. Thibault, and J. Ma. Enhancing quality control of botanical medicine in the 21st century from the perspective of industry: The use of chemical profiling and DNA barcoding to ensure accurate identity. *HerbalGram* 97, 2012: 58–68.
7. D.T. Harbaugh and B.G. Baldwin. Phylogeny and biogeography of the sandalwoods (*Santalum*, Santalaceae): Repeated dispersals throughout the Pacific. *American Journal of Botany* 94, 2007: 1028–1040.
8. A. Bennici. The convergent evolution in plants. *Rivista di Biologia* 96, 2003: 485–489.
9. E.D. Brodie. Convergent evolution: pick your poison carefully. *Current Biology* 20, 2010: R152–R154.
10. M. Wink. Evolution of secondary metabolites from an ecological and molecular phylogenetic perspective. *Phytochemistry* 64, 2003: 3–19.
11. D. Tautz, P. Arctander, A. Minelli, R.H. Thomas, and A.P. Vogler. A plea for DNA taxonomy. *Trends in Ecology & Evolution* 18, 2003: 70–74.
12. B. Sidjimova, S. Berkov, S. Popov, and L. Evstatieva. Galanthamine distribution in Bulgarian *Galanthus* species. *Pharmazie* 58, 2003: 936–937.
13. S. Berkov, B. Sidjimova, L. Evstatieva, and S. Popov. Intraspecific variability in the alkaloid metabolism of *Galanthus elwesii*. *Phytochemistry* 65, 2004: 579–586.
14. N. Rønsted, G.D. Weiblen, W. Clement, N. Zerega, and V. Savolainen. Reconstructing the phylogeny of figs (*Ficus*, Moraceae) to unravel the origin of fig-wasp mutualisms. *Symbiosis* 45, 2008: 45–56.
15. N. Rønsted et. al. Can phylogeny predict chemical diversity and potential medicinal activity of plants? A case study of Amaryllidaceae. *BMC Evolutionary Biology* 12, 2012: 182.
16. M.M. Larsen, A. Adersen, A.P. Davis, M.D. Lledo, A.K. Jäger, and N. Rønsted. Using a phylogenetic approach to selection of target plants in drug discovery of acetylcholinesterase inhibiting alkaloids in Amaryllidaceae tribe Galantheae. *Biochemical Systematics and Ecology* 38, 2010: 1026–1034.

17. C.H. Saslis-Lagoudakis et al. The Use of Phylogeny to Interpret Cross-Cultural Patterns in Plant Use and Guide Medicinal Plant Discovery: An Example from *Pterocarpus* (Leguminosae). *PLoS One* 6(7), 2011: e22275. doi:10.1371/journal .pone.0022275.
18. I. Schmitt and F.K. Barker. Phylogenetic methods in natural product research. *Natural Product Reports* 26, 2009: 1585–1602.
19. G. Heubl. New aspects of DNA-based authentication of Chinese medicinal plants by molecular biological techniques. *Planta Medica* 76, 2010: 1963–1974.
20. M. Yamazaki, A. Sato, K. Saito, and I Murakoshi. Molecular phylogeny based on RFLP and its relation with alkaloid patterns in *Lupinus* plants. *Biological & Pharmaceutical Bulletin* 16, 1993: 1182–1184.
21. N. Trifi-Farah and M. Marrakchi. *Hedysarum* phylogeny mediated by RFLP analysis of nuclear ribosomal DNA. *Genetic Resources and Crop Evolution* 48, 2001: 339–345.
22. N. Mori, T. Moriguchi, and C. Nakamura. RFLP analysis of nuclear DNA for study of phylogeny and domestication of tetraploid wheat. *Genes & Genetic Systems* 72, 1997: 153–161.
23. N.J. Gawel, R.L. Jarret, and A.P. Whittemore. Restriction fragment length polymorphism (RFLP)-based phylogenetic analysis of *Musa*. *Theoretical and Applied Genetics* 84, 1992: 286–290.
24. M. Yamasaki, A. Sato, K. Shimomura, K. Saito, and I. Murakoshi. Genetic relationships among *Glycyrrhiza* plants determined by RAPD and RFLP analyses. *Biological & Pharmaceutical Bulletin* 17, 1994: 1529–1531.
25. J.P. Loh, R. Kiew, A. Kee, L.H. Gan, and Y.Y. Gan. Amplified fragment length polymorphism (AFLP) provides molecular markers for the identification of *Caladium bicolor* cultivars. *Annals of Botany (London)* 84, 1999: 155–161.
26. K.T. Chen et al. Identification of *Atractylodes* plants in Chinese herbs and formulations by random amplified polymorphic DNA. *Acta Pharmacologica Sinica* 22, 2001: 493–497.
27. M. Zhang, H.R. Huang, S.M. Liao, and J.Y. Gao. Cluster analysis of *Dendrobium* by RAPD and design of specific primer for *Dendrobium candidum*. *Zhongguo Zhong Yao Za Zhi* 26, 2001: 442–447.
28. W.Y. Ha, P.C. Shaw, J. Liu, F.C. Yau, and J. Wang. Authentication of *Panax ginseng* and *Panax quinquefolius* using amplified fragment length polymorphism (AFLP) and directed amplification of minisatellite region DNA (DAMD). *Journal of Agricultural and Food Chemistry* 50, 2002: 1871–1875.
29. J.J. Qi, X.E. Li, J. Song, A.E. Eneji, and X. Ma. Genetic Relationships among *Rehmannia glutinosa* cultivars and varieties. *Planta Medica* 74, 2008: 1846–1852.
30. T. Feng, S. Liu, and X.-J. He. Molecular authentication of the traditional Chinese medicinal plant *Angelica sinensis* based on internal transcribed spacer of nrDNA. *Electronic Journal of Biotechnology* 13, 2010: 1–10.
31. C.Z. Wang, P. Li, G.-Q. Jin, and C.-S. Yuan. Identification of *Fritillaria pallidiflora* using diagnostic PCR and PCR-RFLP based on nuclear ribosomal DNA internal transcribed spacer sequences. *Planta Medica* 71, 2005: 384–386.
32. C. Howard, P.D. Bremner, M.R. Fowler, B. Isodo, N.W. Scott, and A. Slater. Molecular identification of *Hypericum perforatum* by PCR amplification of the ITS and 5.8S rDNA region. *Planta Medica* 75, 2009: 864–869.
33. N. Techen, I.A. Khan, Z. Pan, and B.E. Scheffler. The use of polymerase chain reaction (PCR) for the identification of *Ephedra* DNA in dietary supplements. *Planta Medica* 72, 2006: 241–247.
34. N. Techen, Z. Pan, B.E. Scheffler, and I.A. Khan. Detection of *Illicium anisatum* as adulterant of *Illicium verum*. *Planta Medica* 75, 2009: 392–395.

35. V. Joshi, N. Techen, B.E. Scheffler, and I.A. Khan. Identification and differentiation between *Hoodia gordonii* (Masson) Sweet ex Decne., *Opuntia ficus indica* (L.) P. Miller, and related *Hoodia* species using microscopy and PCR. *Journal of Herbs, Spices & Medicinal Plants* 15, 2009: 253–264.

36. M. Hellebrand, M. Nagy, and J.T. Mörsel. Determination of DNA traces in rapeseed oil. *European Food Research and Technology* 206, 1998: 237–242.

37. N. Gryson, F. Ronsse, K. Messens, M. De Loose, T. Verleyen, and K. Dewettinck. Detection of DNA during the refining of soybean oil. *Journal of the American Oil Chemists' Society* 79, 2002: 171–174.

38. M. Busconi, C. Foroni, M. Corradi, C. Bongiorni, F. Cattapan, and C. Fogher. DNA extraction from olive oil and its use in the identification of the production cultivar. *Food Chemistry* 83, 2003: 127–134.

39. C. Breton, D. Claux, I. Metton, G. Skorski, and A. Bervilleä. Comparative study of methods for DNA preparation from olive oil samples to identify cultivar SSR alleles in commercial oil samples: Possible forensic applications. *Journal of Agricultural and Food Chemistry* 52, 2004: 531–537.

40. S. Kumar, T. Kahlon, and S. Chaudhary. A rapid screening for adulterants in olive oil using DNA barcodes. *Food Chemistry* 127, 2001: 1335–1341.

41. M.-A.L. Clarke, J.J. Dooley, S.D. Garrett, and H.M. Brown. An investigation into the use of PCR-RFLP profiling for the identification of fruit species in fruit juices. FSA Final Report Q01111. CCFRA Project 98200, 2008.

42. J. Novak, S. Grausgruber-Groger, and B. Lukas. DNA-based authentication of plant extracts. *Food Research International* 40, 2007: 388–392.

43. P.Y. Tsoi, H.S. Wu, M.S. Wong, S.L. Chen, W.F. Fong, P.G. Xiao, and M.S. Yang. Genotyping and species identification of *Fritillaria* by DNA chip technology. *Acta Pharmaceutica Sinica* 4, 2003: 185–190.

44. T. Li, J. Wang, and Z. Lu. Accurate identification of closely related *Dendrobium* species with multiple species-specific gDNA probes. *Journal of Biochemical and Biophysical Methods* 62, 2005: 111–123.

45. Y.B. Zhang, J. Wang, Z.T. Wang, P.P. But, and P.C. Shaw. DNA microarray for identification of the herb of *Dendrobium* species from Chinese medicinal formulations. *Planta Medica* 69, 2003: 1172–1174.

46. W.Y. Lin, L.R. Chen, and T.Y. Lin. Rapid authentication of *Bupleurum* species using an array of immobilized sequence-specific oligonucleotide probes. *Planta Medica* 74, 2008: 464–469.

47. P. Chavan, K. Joshi, and B. Patwardhan. Review DNA microarrays in herbal drug research. *Evidence-Based Complementary and Alternative Medicine* 3(4), 2006: 447–457. doi:10.1093/ecam/nel075.

48. A.R. Ottesen et al. Baseline survey of the anatomical microbial ecology of an important food plant: *Solanum lycopersicum* (tomato). *BMC Microbiology* 13, 2013: 114. http://www.biomedcentral.com/1471-2180/13/114.

49. M.L. Coghlan, J. Haile, J. Houston, D.C. Murray, N.E. White, P. Moolhuijzen, M.I. Bellgard, and M. Bunce. Deep sequencing of plant and animal DNA contained within traditional Chinese medicines reveals legality issues and health safety concerns. *PLoS Genetics* 8(4), 2012: e1002657. doi:10.1371/journal.pgen.1002657.

50. S.G. Newmaster, M. Grguric, D. Shanmughanandhan, S. Ramalingam, and S. Ragupathy. DNA barcoding detects contamination and substitution in North American herbal products. *BMC Medicine* 11, 2013: 222.

51. D.P. Little. The use of DNA barcode techniques to identify the constituents of herbal dietary supplements. *Planta Medica* 78, 2012: IL11.

52. D.A. Baker. DNA barcode identification of black cohosh herbal dietary supplements. *Journal of AOAC International* 95, 2012: 1023–1034.
53. S. Chen et al. Validation of the ITS2 region as a novel DNA barcode for identifying medicinal plant species. *PLoS One* 5, 2010: e8613.
54. D.T. Harbaugh, H. Oppenheimer, K. Wood, and W.L. Wagner. Taxonomic revision of the Hawaiian red-flowered sandalwoods (*Santalum*) and discovery of an ancient hybrid lineage. *Systematic Botany* 35, 2010: 827–838.
55. AOAC International; Guideline Working Group. AOAC guidelines for validation of botanical identification methods. *Journal of AOAC International:* 95, 2012: 268–272.

6 Metabolic Profiling and Proper Identification of Botanicals
A Case Study of Black Cohosh

Tiffany Chan, Shi-Biao Wu, and Edward J. Kennelly

CONTENTS

Introduction 69
Identification of Black Cohosh 70
    Morphology 70
    Taxonomy 70
    Molecular Systematics 73
Authenticity of Black Cohosh 73
    Methods and Technique Applied in the Authenticity of Black Cohosh 73
    Markers That Distinguish *A. racemosa* from Other American *Actaea* Species 74
    Markers That Distinguish *A. racemosa* from Asian *Actaea* Species 76
        Cimifugin and Derivatives 76
        Triterpene Glycosides 77
        Alkaloid 77
        Formononetin 77
Quality of Black Cohosh 84
    Quality of Black Cohosh Commercial Products 84
    Pricing and Quality of Black Cohosh Commercial Products 85
Conclusion, Future Prospects, and Challenges 86
Acknowledgments 86
References 86

INTRODUCTION

"This interesting remedy was a decided favorite with the early Eclectic practitioners, and to this day holds a very prominent place among the remedies originally placed before the medical profession by our school," was the way J.U. Lloyd and H.W. Felter described the importance of black cohosh in Eclectic medicine in the 1800s [1].

Actaea racemosa L. (syn. *Cimicifuga racemosa* L.), also known as black cohosh, black snake root, squaw root, rattle root, bugbane, rattleweed, and macrotys [1,2], is a plant native to North America and carries a rich history of medicinal use. Native Americans used black cohosh and caught the attention of early settlers who then incorporated it into their everyday lives as a treatment for common discomfort such as malaise, malaria, rheumatism, kidney function irregularities, menstrual abnormalities, and sore throat [3–5]. These uses were then well documented by a group called the Eclectics, a circle of physicians in which Lloyd and Felter were participants, who praised the plant: "This is a very active, powerful, and useful remedy and appears to fulfill a great number of indications" [1].

Black cohosh is now used most commonly for women's health, especially for the treatment of menopausal hot flashes [6,7]. Its products rank in the top 10 of botanical dietary supplement sales in the United States [8], and interest in black cohosh has risen since 2003 when reports cited health risks, including cardiovascular problems, for women receiving hormone replacement therapy [9–11]. Black cohosh is often used by menopausal women to treat hot flashes because it is perceived to be a safer alternative to long-term estrogen replacement therapy, and not estrogenic [12].

There has been confusion with the identification of black cohosh and its closely related *Actaea* species. Some black cohosh products were found to be adulterated with other related plants from the same genus [13–15]. Certain issues have been raised concerning black cohosh safety [16]. Identifying the correct *Actaea* species is essential for good manufacturing practices of safe black cohosh products. This chapter delves into metabolic profiling, correct identification of black cohosh from other related species, as well as the quality of black cohosh products.

IDENTIFICATION OF BLACK COHOSH

MORPHOLOGY

Black cohosh is 1 of 28 herbaceous perennial plant species of the *Actaea* genus [17]. It is characterized by large compound leaves closer to the rhizomes and basal leaves near the top. White flowers typically appear during the summer and are comprised of groups of long white stamen that circle a central white stigma (Figure 6.1). These flowers arise from a long raceme, and together, they can measure more than a meter high. The fruit consists of a dry follicle and has one carpel with seeds. The plant gives off a sweet smell that is attractive to insects. The rhizomes (underground stems) and their associated roots are the parts of black cohosh most widely used medicinally [18].

TAXONOMY

Actaea taxonomically belongs to the Ranunculaceae (buttercup) family, consisting of 28 herbaceous perennial species growing in the temperate regions of the northern hemisphere. In the eighteenth century, Linnaeus considered *Actaea* and *Cimicifuga* as separate genera, based upon visual differences in flower-to-carpel

FIGURE 6.1 Drawing of *Actaea racemosa* (black cohosh).

ratio [17]. However, both genera were lumped together eventually by Nutall in 1818, who referenced similarities seen in the fruit [17]. This notion was scientifically supported by the twentieth century research that described the two genera as having the same micromorphology, phytochemical qualities, serological reactions, and similar pollen types [17]. Compton added to this taxonomic question with his own research on *Actaea* plant species' deoxyribonucleic acid (DNA) sequences and morphology [17]. Compton ultimately lumped the genera *Actaea*, *Cimicifuga*, and *Souliea* into one genus, *Actaea* [17]. However, there are still researchers that disagree with the merging of *Actaea* and *Cimicifuga* due to morphological differences in fruit [19,20]. According to Compton et al. [17], 19 species originate from eastern Asia, eight species originate from North America, and one species comes from Europe (Table 6.1). None of the 27 sister species (excluding *A. racemosa*) of black cohosh is commonly used for menopausal symptoms [21].

There are eight North American *Actaea* species, with five being distributed along the east coast from Ontario and Maine to southern Georgia and as west as Missouri [22], while the remaining three typically growing in the western United States. *Actaea racemosa*, *A. podocarpa*, *A. pachypoda*, *A. rubra*, and *A. cordifolia* constitute the five eastern species, and some of these species look alike, growing within overlapping regions, and therefore may be misidentified [23]. Black cohosh is obtained by the dietary supplement industry both by cultivation and wild crafting [24]. To tell them apart, researchers and wild crafters use morphological markers,

TABLE 6.1
List of *Actaea* Species and Their Original and Selected Synonyms

Location	Region	Species	Selected Synonyms
		Actaea racemosa L.	*Cimicifuga racemosa* (L.) Nutt
North America	East coast	*Actaea podocarpa* (DC.) Elliott	*Cimicifuga americana* Michx
		Actaea pachypoda Elliot	
		Actaea rubra (Aiton) Willd.	
		Actaea cordifolia (DC.) Torr. et A. Gray	*Cimicifuga rubifolia* Kearney
	West coast	*Actaea laciniata* S. Watson	
		Actaea elata (Nutt.) Prantl	
		Actaea arizonica S. Watson	
Asia		*Actaea heracleifolia* Kom.	
		Actaea mairei (H. Lév.) J. Compton	
		Actaea dahurica Turcz. ex Fisch. & C.A. Mey.	
		Actaea brachycarpa P.G. Xiao	
		Actaea simplex (DC.) Wormsk	
		Actaea cimicifuga L.	*Cimicifuga foetida* L.
		Actaea yunnanensis P.G. Xiao	
		Actaea taiwanensis (J. Compton, Hedd. & T.Y. Yang) Lufgerov	
		Actaea biternata Prantl.	
		Actaea japonica Thunb.	
		Actaea purpurea (P.G. Xiao) J. Compton	
		Actaea bifida (Nakai) J. Compton	
		Actaea matsumurae (Nakai) J. Compton & Hedd	
		Actaea yesoensis J. Compton & Hedd.	
		Actaea asiatica H. Hara	
		Actaea frigida Wall.	
		Actaea kashmiriana (J.Compton et Hedd)	
		Actaea spicata L.	
		Actaea vaginata (Maxim.) J. Compton	
Europe		*Actaea europaea* (Schipcz.) J. Compton	

but if the plants are not in flower or fruit, they may be difficult to distinguish using these characteristics.

There are 18 Asian *Actaea* species that currently exist (Table 6.1). *A. cimifuga*, *A. dahurica*, and *A. heracleifolia* are among the most popular for their healing qualities in traditional Chinese medicine. *Shengma* (Eng.), (Chinese: 升麻), is the standard Chinese name designated to multiple Asian species of *Actaea* [13,14]. Though treating menopausal symptoms is not considered one of the traditional uses of *Shengma*, many black cohosh products today are nonetheless adulterated with some of these Asian species.

MOLECULAR SYSTEMATICS

DNA fingerprinting has been used to distinguish black cohosh from closely related species [25]. For example, amplified fragment length polymorphisms (AFLP) generated 262 markers that existed within *Actaea* species examined. In juxtaposition with *A. pachypoda*, *A. cordifolia*, and *A. podocarpa*, black cohosh had only one clear fingerprint, whereas the other sympatric species had multiple. That said, *A. racemosa* is unique and does not share the same set of markers with any other *Actaea* species [25]. In addition to DNA fingerprinting, phytochemical fingerprinting has been used on this plant, and this chapter will summarize the research conducted in our laboratory and others on the authenticity of black cohosh.

AUTHENTICITY OF BLACK COHOSH

METHODS AND TECHNIQUE APPLIED IN THE AUTHENTICITY OF BLACK COHOSH

The structural components of triterpene glycosides make liquid chromatography-mass spectrometry (LC-MS) a viable method for analysis. Triterpene glycosides have weak chromophores and are therefore only capable of absorbing low levels of UV-light. For this reason, methods that employ photodiode array (PDA) or ultraviolet detectors are not ideally suited for the identification of black cohosh triterpene glycosides. Despite this, cimiracemoside F is detectable using PDA because of its low UV absorbance [13]. Some researchers use thin-layer chromatography (TLC) for triterpene glycosides identification [26]. One benefit to this method is the cost-effectiveness and use of color reagents for triterpenoid detection, but the results can be difficult to interpret due to low resolution and higher limits of detection when compared to other chromatographic methods. Nuclear magnetic resonance (NMR) is another technique used for the identification of metabolites from black cohosh, but it is rarely used as an in-line detector with high-performance liquid chromatography (HPLC) [27].

Many methods to identify triterpene glycosidic markers in black cohosh and in other *Actaea* species exist. Evaporative light scattering detection (ELSD) has been used as a detector to study black cohosh triterpene glycosides, resulting in the separation and quantification of cimiracemoside A, 26-deoxyactein, and actein [28]. Li et al. reported a validated and reproducible HPLC-PDA-ELSD method to detect and analyze both triterpene glycosides and polyphenolics [29]. Afterward, mass spectrometric techniques were used in a number of black cohosh studies, and He et al. developed a liquid chromatography with positive atmospheric pressure chemical ionization mass spectrometry [LC/(+)APCIMS] and LC-PDA/MS/ELSD methods to detect markers in black cohosh [30,31]. In 2005, Wang et al. reported a LC/turbo ion spray (TIS)-MS method for the *fingerprint profiling* of some *Cimicifuga* herbs and other black cohosh commercial products [32].

Our group has employed reversed-phase HPLC using diode array detection (RP-HPLC-PDA) [33], HPLC-PDA [34], LC-MS/MS [13,34], and most recently LC-MS-TOF [14] in our studies to detect and identify the markers compounds of black cohosh, its related species, and some commercial products.

Recently, our group focused on the use of HPLC coupled with time-of-flight high-resolution mass spectrometry (LC-MS-TOF) to analyze herbal medicines and metabolites. LC-MS-TOF provides accurate mass data to give chemical formulas and has low limits of detection. These data allow for the identification of many minor compounds, and statistical tools like principle component analysis (PCA) can be used to find marker compounds. For example, we used PCA to analyze the HPLC-TOF-MS total ion current (TIC) chromatograms of *Actaea* species [14]. To the best of our knowledge, this is the first report to apply multivariate statistical analysis for the study of black cohosh markers.

MARKERS THAT DISTINGUISH *A. RACEMOSA* FROM OTHER AMERICAN *ACTAEA* SPECIES

Actaea racemosa is a challenging plant to identify because it shares many morphological features with other *Actaea* species, but there are phytochemical methods that can aid in the correct identification. There are seven species, other than *Actaea racemosa*, that are native to North America, three of which are based on the west coast: *A. laciniata*, *A. elata*, and *A. arizonica* (Table 6.1). The other four American species can grow in the same geographical area as *A. racemosa* and therefore may be confounded with black cohosh.

Phytochemical studies have identified marker compounds that can be used to distinguish black cohosh from some of the remaining American species (Table 6.2). According to our study in 2011, cimigenol and hydro-shengmanol (Figure 6.2 and Table 6.2) are two marker compounds present in black cohosh that are not detected in *A. laciniata*, *A. podocarpa*, *A. pachypoda*, and *A. rubra* [14].

Another method is looking at how black cohosh's polyphenolic content ratios differ from those of its sister American *Actaea* species. Our study in 2007 looked at the presence of eight major polyphenols in *A. racemosa* and other American *Actaea* species [33]. These polyphenols include caffeic acid; ferulic acid; isoferulic acid; fukinolic acid; and cimifugic acids A, B, E, and F (Figure 6.3). *Actaea racemosa* and *A. rubra* showed all eight major polyphenols but the proportions of these compounds were significantly greater in *A. rubra* than they were in black cohosh. Cimifugic acid F and isoferulic acid were not detected in *A. pachypoda* and *A. podocarpa*, respectively [33]. However, our follow-up study in 2011 was able to detect caffeic acid, ferulic acid, isoferulic acid, fukinolic acid, and cimifugic acids A and B in *A. pachypoda* and *A. podocarpa* [34] (Figure 6.3 and Table 6.2). Only trace amounts

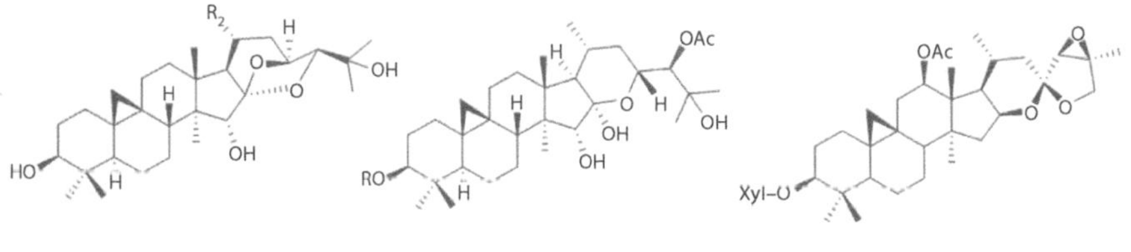

Cimigenol: $R_1 = R_2 = H$
Cimiracemoside C: $R_1 = $ alpha-L-Ara; $R_2 = CH_3$

Hydro-shengmanol: R = xyl or ara

23-Epi-26-deoxyactein (27-deoxyactein)

FIGURE 6.2 Selected triterpene glycoside markers in *Actaea racemosa*.

TABLE 6.2
Selected Marker Compounds for Actaea Species Identification

	Triterpene glycosides										Polyphenols							Chromones			Alkaloid
	Cimigenol	Hydro-shengmanol	Ambalo-ciliigenol	Shengmanol	16,23-dihroxyacteigenol	Anhydro-hydro-shengmanol	Cimiracemoside C	Cimiracemoside F	23-epi-26-deoxyactein (26-deoxyactein)	Cimicifugoside H-1	Caffeic acid	Ferulic acid	Isoferulic acid	Fukinolic acid	Cimicifugic acid A	Cimicifugic acid B	Cimicifugic acid E	Cimifugin	Fran-O-β-D-glycosylfukiigenin	Diervoic acid	Cimicifuga V
East coast American Actaea species																					
Actaea racemosa	✓[14]	nd[14]	nd[14]	nd[14]	nd[14]	nd[14,34]	✓[14,34]	Overlap [34]	✓[34]	nd[14,34]	✓[33,34]	Trace[33] nd[34]	✓[33,34]	✓[33,34]	✓[33,34]	✓[33,34]	✓[33,34]	nd[14]	nd[34]	nd[14]	nd[14]
Actaea podocarpa	nd[14]	✓[14]	nd[14]	nd[14]	nd[14]	✓[14] Overlap [34]	✓[14]	nd[14]	✓[34]	Overlap [34]	✓[34]	✓[34]	✓[34]	✓[34]	✓[34]	✓[33]	✓[33]	✓[34]	✓[34]	nd[34]	nd[14]
Actaea pachypoda	nd[14]	nd[14]	nd[14]	nd[14]	nd[14]	nd[14]	nd[14]	nd[14]	✓[34]	nd[14,34]	✓[34]	nd[34]	✓[34]	✓[34]	✓[34]	✓[33]	Trace [33]	✓[34]	✓[34]	nd[34]	nd[14]
Actaea rubra	nd[14]	nd[14]	nd[14]	nd[14]	nd[14]	nd[14]	✓[14]	✓[34]	✓[34]	✓[14]	✓[33]	✓[33]	✓[34]	✓[33]	✓[34]	✓[33]	✓[33]	✓[14]	✓[34]	nd[34]	nd[14]
Actaea cordifolia	✓[14]	nd[14]	nd[14]	nd[14]	nd[14]	✓[14] nd[34]	✓[34]	✓[34]	Overlap [34]	✓[14]	✓[34]	✓[34]	✓[34]	✓[34]	✓[34]	nt	nt	✓[14,34]	✓[34]	✓[34]	nd[14]
West coast American Actaea species																					
Actaea laciniata	nd[14]	nd[14]	nd[14]	✓[14]	nd[14]	nd[14]	nd[14]	✓[14]	nd[14]	Overlap [34]	✓[34]	nd[34]	✓[34]	✓[34]	✓[34]	nt	nt	nd[14] nd[14]	✓[34]	nd[34]	nd[14]
Actaea elata	nd[14]	nd[14]	✓[14]	nd[14]	nd[14]	nd[14]	nd[14]	nd[14]	✓[14]	✓[34]	✓[34]	✓[34]	✓[34]	✓[34]	✓[34]	nt	nt	✓[14,34]	✓[34]	✓[34]	nd[14]
Actaea arizonica	nd[14]	nd[14]	nd[14]	nd[14]	nd[14]	nd[14]	nd[14]	✓[14]	✓[34]	✓[14]	✓[34]	nd[34]	✓[34]	✓[34]	✓[34]	nt	nt	nd[14] Trace [34]	✓[34]	nd[34]	✓[14]
Asian Actaea species																					
Actaea heracleifolia	nd[14]	nd[14]	nd[14]	nd[14]	nd[14]	✓[14] Overlap [34]	✓[14]	Overlap [34]	✓[34]	Overlap [34]	✓[34]	✓[34]	✓[34]	✓[34]	✓[34]	✓[27]	✓[27]	✓[14,34]	✓[34]	✓[34]	✓[14]
Actaea marei	nd[14]	nd[14]	nd[14]	nd[14]	nd[14]	✓[14]	✓[14]	Overlap [34]	✓[34]	✓[34]	✓[34]	✓[34]	✓[34]	✓[34]	✓[34]	nt	nt	✓[14,34]	✓[34]	✓[34]	✓[14]
Actaea dahurica	nd[14]	nd[14]	nd[14]	nd[14]	nd[14]	✓[14] nd[34]	Overlap [34]	Overlap [34]	✓[34]	Overlap [34]	✓[34]	✓[34]	✓[34]	✓[34]	✓[34]	nt	nt	✓[14,34]	✓[34]	✓[34]	✓[14]
Actaea brachycarpa	nd[14]	nd[14]	nd[14]	✓[14] nd[34]	nd[14]	✓[14] nd[34]	✓[14]	Overlap [34]	✓[34]	Overlap [34]	✓[34]	✓[34]	✓[34]	✓[34]	✓[34]	nt	nt	✓[14,34]	✓[34]	✓[34]	✓[14]
Actaea simplex	nd[14]	nd[14]	nd[14]	✓[14] nd[34]	nd[14]	✓[14]	Overlap [34]	✓[34]	✓[34]	Overlap [34]	✓[34]	✓[34]	✓[34]	✓[34]	✓[34]	nt	nt	✓[14,34]	✓[34]	✓[34]	✓[14]
Actaea cimicifuga	nd[14]	nd[14]	nd[14]	nd[14]	nd[14]	✓[14]	✓[14]	Overlap [34]	✓[34]	✓[34]	✓[34]	✓[34]	✓[34]	✓[34]	✓[34]	nt	nt	✓[14,34]	✓[34]	✓[34]	✓[14]
Actaea yunnanensis	nd[14]	nd[14]	nd[14]	nd[14]	nd[14]	✓[14]	✓[14]	Overlap [34]	✓[34]	✓[34]	✓[34]	✓[34]	✓[34]	✓[34]	✓[34]	nt	nt	✓[14,34]	✓[34]	✓[34]	✓[14]

Selected marker compounds for Actaea species identification. Detected compounds in Actaea species are labeled "✓." Compounds that are not detected are labeled "nd" and compounds that were not tested are labeled "nt." Overlap refers to compounds that were not baseline separated during HPLC-PDA analysis. Constituent amounts that are less than 0.005% are labeled trace. Suggested marker I compounds [14] used for black cohosh identification from other American Actaea species are highlighted in light gray (▨). Suggested marker II compounds [14] used for black cohosh identification from Asian Actaea species are highlighted in dark gray (▨). These results are derived from our studies conducted in 2007 [33], 2010 [34], 2011 [14], and 2012 [27].

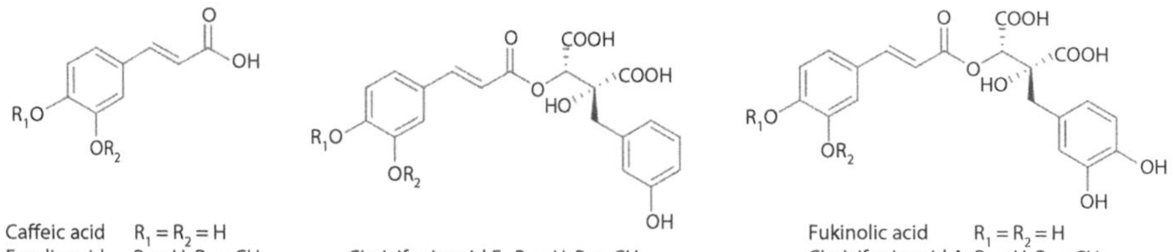

Caffeic acid $R_1 = R_2 = H$
Ferulic acid $R_1 = H, R_2 = CH_3$
Isoferulic acid $R_1 = CH_3, R_2 = H$

Cimicifugic acid E $R_1 = H, R_2 = CH_3$
Cimicifugic acid F $R_1 = CH_3, R_2 = H$

Fukinolic acid $R_1 = R_2 = H$
Cimicifugic acid A $R_1 = H, R_2 = CH_3$
Cimicifugic acid B $R_1 = CH_3, R_2 = H$

FIGURE 6.3   Selected polyphenol markers in *Actaea racemosa*.

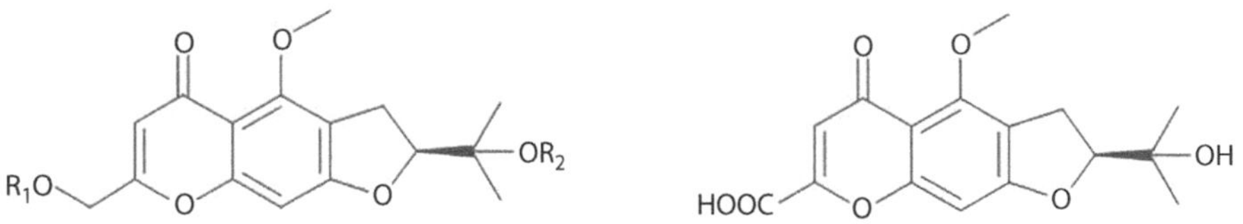

Cimifugin: $R_1 = H, R_2 = H$
Prim-*O*-glucosylcimifugin: $R_1 = glu, R_2 = H$

Divaricatacid

FIGURE 6.4   Selected markers in Asian *Actaea* species.

of isoferulic acid and cimifugic acid B were observed in *A. pachypoda* and *A. podocarpa*, respectively. Black cohosh has significant levels of both isoferulic acid and cimifugic acid B, and thus, the presence of both of these markers can be used as for black cohosh identification [34].

Additionally, some of the west coast American *Actaea* species contain cimifugin, a chromone that is characteristic of Asian *Actaea* species and is absent in black cohosh [34] (Figure 6.4). This, as our laboratory has proposed, could suggest a close evolutionary relationship between *Actaea* species from western America and Asia, as explained through biogeography [34].

MARKERS THAT DISTINGUISH *A. RACEMOSA* FROM ASIAN *ACTAEA* SPECIES

Actaea racemosa grows primarily in the eastern United States, and the practice of wild crafting this botanical has become a threat to its survival in the wild [35]. Therefore, due to the limited supply and high cost of wild crafted black cohosh, dietary supplements in the market may be adulterated with cheaper Chinese *Actaea* species that are not black cohosh [14,34]. Although black cohosh is cultivated in the United States and Europe, closely related *Actaea* species, already commonly used in Traditional Chinese Medicine, are cultivated in China and tend to be considerably less expensive [13]. Correct identification of black cohosh is the critical first step in good manufacturing of black cohosh products. To properly identify this plant of interest, there are unique bioactive marker compounds that can be used to assist in its proper identification.

Cimifugin and Derivatives

Cimifugin and its derivatives are a group of linear dihydrofurochromone-class compounds (Figure 6.4 and Table 6.2). This group, which includes cimifugin, prim-*O*-glycosylcimifugin, and divaricatacid, is only present in Asian species and certain western American rare species [14,30,34]. This group of marker compounds is perhaps the best indicator that suggests a black cohosh product has been adulterated with an Asian *Actaea* species.

Triterpene Glycosides

In addition to cimifugin being the main marker that separates all the Asian species from black cohosh, many investigations tried to find distinct markers within black cohosh that would identify the plant. Many researchers investigated whether certain cimiracemosides could be used for this purpose. Earlier work dating from 2006 from our laboratory investigated black cohosh sourcing and authenticity in 11 products available in the market by way of TLC, HPLC, paired with selected ion monitoring LC-MS. Our findings concurred with He et al. [30], who reported that cimiracemoside C (Figure 6.2 and Table 6.2) is a unique black cohosh marker for identification. It was concluded that black cohosh authenticity is marked by the presence of triterpene glycoside cimiracemoside C and marked by the absence of cimifugin [13]. However, one of our later studies found cimiracemoside C in Asian *Actaea* species and concluded that the marker was not sufficient to differentiate between black cohosh and Asian species [13].

Initially, 23-epi-deoxyactein (Figure 6.2 and Table 6.2) was used as a marker to single out black cohosh from other *Actaea* species. However, our studies found 23-epi-deoxyactein present in black cohosh and in Asian species like *A. dahurica* and *A. cimicifuga* (*C. dahurica* and *C. foetida*) [32]. Each species displayed different amounts of this compound. 23-Epi-deoxyactein quantity levels, and not the compound's identity alone, can serve as a determining factor for differentiating unadulterated black cohosh products. Adulterated products that contain other *Actaea* species had greater or lesser amounts of 23-epi-deoxyactein listed on their product labels than authentic black cohosh products [32].

Alkaloid

In 2011, our study tested commercial black cohosh products and determined cimicifine A, a triterpene alkaloid, is a marker that characterizes Asian *Actaea* species [14] (Figure 6.5 and Table 6.2). This compound was detected in four Asian *Actaea* species, notably *A. heracleifolia*, *A. dahurica*, *A.marei*, and *A. yunnanensis*. Other studies have investigated certain alkaloids, like dopargine, salsolinol, *N*ω-methylserotonin, cimipronidine, and other cimipronidine congeners, that are possible agonists to corresponding receptors, and these compounds may be the bioactive constituents that help to relieve menopausal hot flashes (Figure 6.5) [36,37].

Formononetin

Formononetin is an isoflavone also known as 7-hydroxy-3-(4-methoxyphenyl) chromen-4-one and some have reported its presence in black cohosh [38–40] (Figure 6.6). Its very presence in black cohosh, and therefore whether it would be a good marker for this species, has been disputed (Table 6.3). Currently, there are four peer-reviewed studies that claim to have found this compound in black cohosh roots and rhizome extracts [39–43], using a combination of methods that include TLC,

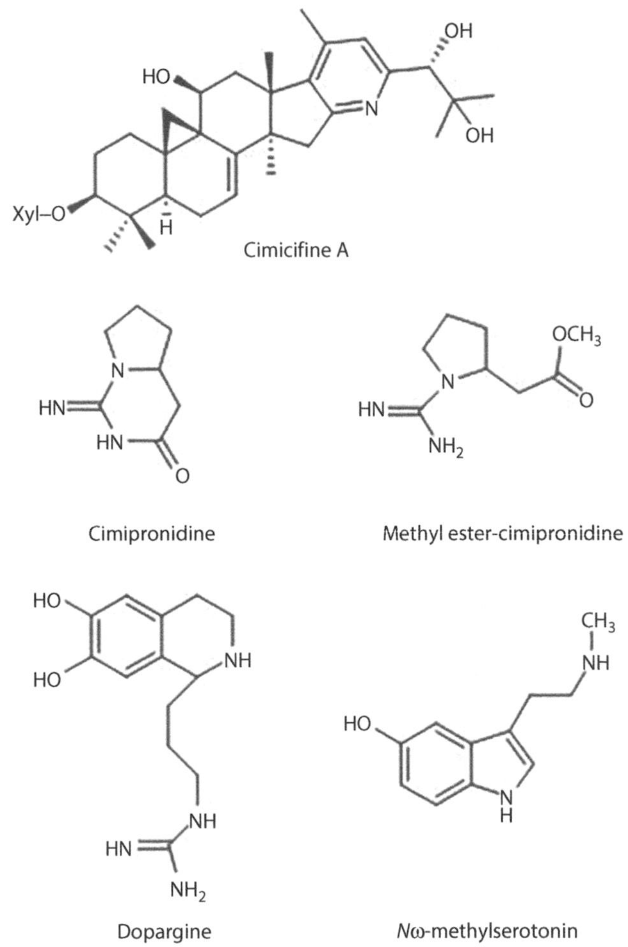

FIGURE 6.5 Alkaloids in *Actaea racemosa*.

HPLC-UV, infrared (IR), MS, NMR, TLC-fluorescent densitometry, HPLC-PDA, gas chromatography (GC)-MS, and LC-MS-MS. On the other hand, eight studies have not detected this compound in black cohosh using HPLC-PDA, LC-MS-MS, ELSD, TLC, ultra-performance liquid chromatography (UPLC)-UV/ELSD, and UPLC-MS [13,29,34,44–48]. Studies reporting formononetin have suggested that varying results may be due to plant selection differences [43], different collection locales [43], HPLC-PDA, and LC-MS's limits of detection [41]. In two separate studies [13,47], our laboratory has been unable to detect formononetin in black cohosh at a 0.6 ng limit of detection or 0.0759 ppm (based on methanol density) by HPLC-PDA [13,47], which is more sensitive than TLC-densitometry's 0.08 ppm level of detection [41]. Furthermore, LC-MS provides additional means of identification including mass (parent ions and fragments) and UV spectra. These conflicting results strongly suggest that formononetin is not a reliable black cohosh marker and may not even be present in this species.

TABLE 6.3

Formononetin and Black Cohosh

Year	Formononetin Detected (Yes/No)	Method Used	Plant Source	BC Extract Type	Level of Detection	Detection Amount	Conclusion	References
1985	Yes	Formononetin detected in methanol extract and binded to estrogen receptor in *in vitro* rat experiment. TLC, HPLC-UV, IR, MS, and NMR.	Black cohosh: roots and rhizomes (methanol extract).	Methanol	UV spectra: two maxima at 298 and 245 nm; after vapor treatment with NH_3, blue fluorescence, indicator of formononetin HPLC: n/a	n/a	Estrogenic activity is observed. There was a decrease in luteinizing hormone (LH) in ovariectomized rats due to the binding of substances found in *Cimicifuga* extract to estrogen receptors.	[39,40]
1996	No	Extracts were fractionated by column chromatography, TLC, and then analyzed by IR and NMR spectroscopy.	Black cohosh: Roots and rhizomes.	n/a	n/a	n/a	Biochanin A was detected.	[44]

(Continued)

TABLE 6.3

(Continued) Formononetin and Black Cohosh

Year	Formononetin Detected (Yes/No)	Method Used	Plant Source	BC Extract Type	Level of Detection	Detection Amount	Conclusion	References
1997	No	HPLC	Black cohosh: Extracts and five medicines consisting of BC rhizomes (ethanol extracts).	Ethanol (60% v/v), isopropanol (40% v/v).	1 µg mL^{-1}	Perhaps the sample is below the 1 µg mL^{-1} detection limit or not present in extract.	Formononetin was either below limit of detection or absent.	[45]
2002	No	TLC, HPLC-PDA detector, and mass spectrometer.	Black cohosh: Roots and rhizomes in 13 populations from the eastern United States, Remiferin, CimiPure (dried extract).	Roots and rhizomes extracted overnight in 80% methanol. 20 tablets of Remiferin grounded and extracted overnight in 80% methanol.	n/a	None	Formononetin was absent in 13 populations of black cohosh.	[46]

Year		Method	Sample					Ref.
2002	No	HPLC, a PDA detector, and an ELSD.	Black cohosh: Liquid extract, powered extract, milled plant material and commercial product.	Methanol	26–55 ng	None	Formononetin was not detected.	[29]
2004	Yes	TLC-fluorescent densitometry.	Rhizomes and roots methanolysis products.	Methanol	0.08 ppm	3.1–3.5 µg/g	Formononetin was present in black cohosh extracts. HPLC is unable to detect compounds such as formononetin because of high level of detection and is not sensitive enough for plant and drug material analysis.	[41]
2006	No	HPLC-PDA detection and ion monitoring LC-MS.	Dietary supplements.	Two tablets were dissolved in 5–10 mL of methanol.	0.6 ng	None	Formononetin was absent.	[13]

(Continued)

TABLE 6.3

(Continued) Formononetin and Black Cohosh

Year	Formononetin Detected (Yes/No)	Method Used	Plant Source	BC Extract Type	Level of Detection	Detection Amount	Conclusion	References
2006	No	TLC, HPLC-PDA, and LC-MS.	Black cohosh: Roots and rhizomes.	Methanol and aqueous methanol.	0.6 ng, approximately 0.0759 ppm	None	Formononetin was absent in black cohosh extracts. HPLC's level of detection is lower than TLC-fluorescent densitometry. HPLC is a more adequate in determining the presence of formononetin.	[47]
2007	Yes	HPLC and GC-MS.	Rhizomes and roots.	n/a	n/a	n/a	There was a reasonable match of formononetin in the black cohosh rhizomes, but formononetin is absent in the roots, lyophilized samples, and young rhizomes.	[42]

Year		Method	Source	Solvent			Notes	Ref
2009	No	UPLC-UV/ELS and UPLC-MS.	Rhizomes from two populations of A. racemosa, dietary supplements.	Methanol	0.01 µg/mL	None	No formononetin was found.	[48]
2011	No	HPLC-PDA and LC-MS.	Black cohosh products.	Each extract dissolved in 70% methanol.	n/a	None	No formononetin was found. Supported experimental data in previous reports in 2006.	[34]
2012	Yes	LCMS-MS with a microbore C18 column, combined with ion pairs of m/z 269-253, 269-225, and 269-197.	Rhizomes	Methanol, ethanol, and isopropanol.	<125 ng/g	<125 ng/g in methanol extracts	Formononetin was found only in the crude methanol BC extract. Believes that other laboratories are unsuccessful in retrieving compound because of plant selection differences and different locations. Believes that HPLC is not an effective method in detecting formononetin because of the compound's low detectability.	[43]

n/a = not available.

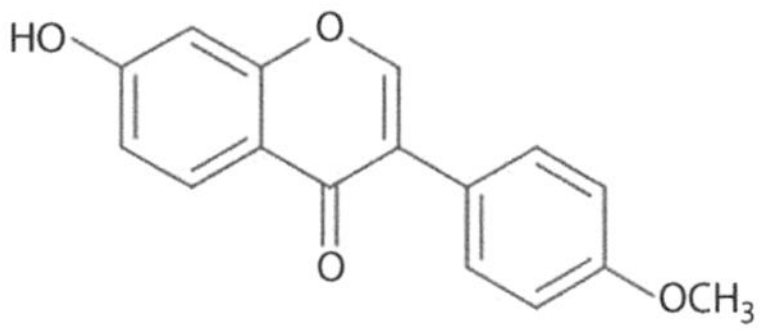

Formononetin

FIGURE 6.6 Formononetin.

QUALITY OF BLACK COHOSH

QUALITY OF BLACK COHOSH COMMERCIAL PRODUCTS

Previously, studies from our laboratory have examined the presence of black cohosh in commercial products. In comparing the plant's rhizome and root extracts with 11 black cohosh products, we found three products adulterated with Asian species and one showed a combination of multiple *Actaea* species and also displayed small traces of black cohosh [13]. Another subsequent examination of 38 black cohosh products was done in 2012 [8], and it was found that approximately 34% of products were adulterated, proportionally similar to the results found in Jiang's black cohosh product collection [13]. The chromatograms of different products were compared to that of black cohosh and their nonconcordance hinted at adulteration [13].

Moreover, unadulterated black cohosh products were found to have significant variability, with different amounts of triterpene glycosides and polyphenols [13]. Triterpene glycoside proportions varied 44-fold among products. Polyphenol content varied 64-fold among eight of the tested products containing black cohosh [13].

These differences in triterpene glycoside and polyphenol amounts in different products could lead to consumers ingesting highly variable amounts of compounds. For example, the labels on some of the tested products direct consumers to take one capsule per day. In comparing capsules of two different brands, we found one of them had 3.8 times the amount of triterpenoid glycosides and 1.8 times the polyphenols as the other. One brand of tablets exceeded another by 4.8 times the amount of triterpenes and 4.7 times the amount of polyphenols present [13].

Our work suggests that black cohosh product manufacturers may routinely use 23-epi-26-deoxyactein as the primary marker compound for black cohosh identification, but it is not a useful marker for species differentiation because a number of *Actaea* species contain this compound (Table 6.2) [13].

It was initially proposed by He et al. [30] and by our laboratory [13] that cimiracemoside C is a marker compound found only in black cohosh but our findings later refuted this claim due to its identification in other *Actaea* species (Table 6.2) [34]. We have hence developed a method that can identify unadulterated black cohosh products: if a type I marker compound is found within the product and type 2 markers are absent, then the black cohosh product is unadulterated (Figure 6.7 and Table 6.2) [14]. Type I markers are cimigenol and hydro-shengmanol, which were markers present only in *Actaea racemosa* and not in any other sympatric American *Actaea* species. Type II markers include cimifugin and its derivatives that identify Asian *Actaea* species [14].

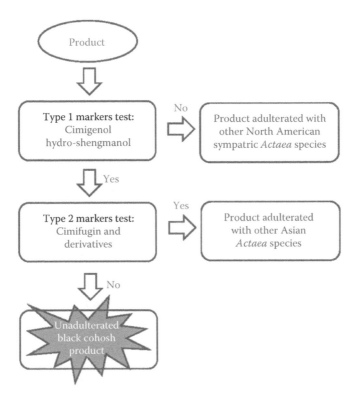

FIGURE 6.7 Scheme to distinguish unadulterated black cohosh products.

PRICING AND QUALITY OF BLACK COHOSH COMMERCIAL PRODUCTS

Since the publication of the Women's Health Initiative on the dangers of hormone replacement therapy [9–11], black cohosh products have ranked among the top 10 herbal products being sold in the United States; for example, in 2012, black cohosh products ranked 7th [49]. The United States Food and Drug Administration (FDA) regulates black cohosh dietary supplements under the Dietary Supplement Health and Education Act [8]. However, FDA funding for overseeing the safety of supplements has decreased over the years, and the number of adulterated products has increased [50].

Consumers are faced with the challenge of choosing suitable unadulterated black cohosh supplements on their own. According to one of our collaborative studies with Nagler and other researchers, customers' choice in supplements is correlated with key words on the label [51]. These key words also correlate with the pricing of products. Words such as *guarantee* have a negative effect on customers and tend to have a lower price value, while *safe* accompanied by a list of possible side effects attract more buyers and is higher in cost [51]. Customers also rely on words such as *certified*, *claims*, and *standardized*, which most often are misleading terms that sell unauthentic tablets [8]. It was found from this study that 13 out of 38 products, or 34.2%, did not contain black cohosh. These statistics were similar to the ones found in our past study, where 27% of tested products did not contain black cohosh [13].

For this reason, it is important for customers to understand the risks associated with purchasing adulterated black cohosh products. Our 2012 collaborative article with Nagler stresses the importance of being smart consumers [8]. Awareness of adulteration in products is important and can ultimately limit the purchase of adulterated products sold in the market.

CONCLUSION, FUTURE PROSPECTS, AND CHALLENGES

We speculate that adulteration is not primarily due to improper identification, but that certain suppliers are substituting cheaper Asian *Actaea* species for more expensive, American grown black cohosh. Those in the dietary supplement industry who produce end products are required to be aware of the constituents within their products for proper labeling. Due to this issue of adulteration, proper methods that can discern black cohosh from its closest relatives are needed in the industry.

The first step for good manufacturing practices in the botanical dietary supplement industry should be properly identifying black cohosh and not confusing it with related *Actaea* species. Botanists are able to differentiate black cohosh among species by its morphological features, but if the plants are not in flower or fruit, they may be difficult to distinguish using these characteristics. However, if black cohosh has been adulterated for economic reasons with related *Actaea* species, once an extract is made, it is next to impossible to differentiate it unless more costly chemical methods are employed.

Our laboratory and others have created validated chemical techniques, using HPLC-PDA and LC-MS, to identify black cohosh. Although these methods are useful, they may be prohibitively expensive for small dietary supplement manufacturers to use on a routine basis. Other less expensive methods to distinguish extracts of black cohosh from other related species should be developed. The adulteration of black cohosh with related *Actaea* species may have series public health impact, because some reported cases of toxicity have been associated with adulterated black cohosh products.

ACKNOWLEDGMENTS

We thank Ms. Ghislaine Kersten for her original pencil drawing of black cohosh. We thank Devhra BennettJones, Archivist, Lloyd Library & Museum (Cincinnati, OH) for her help in obtaining materials for this review. Work has been funded, in part, by the National Institutes of Health-National Center for Complementary and Alternative Medicine (Grants P50-AT00090 and R21AT002930). The content of this book chapter is solely our responsibility and does not necessarily reflect the official views of NIH-NCCAM.

REFERENCES

1. J.U. Lloyd and H.W. Felter. *King's American Dispensatory*, The Ohio Valley Company: Cleveland, OH, 1905; p. 529–530.
2. S. Foster. Black cohosh: *Cimicfuga racemosa*, a literature review. *HerbalGram* 45, 1999: 35–49.

3. B. Barton. *Collections for an Essay towards a Material Medica of the United States*, Way and Groff: Philadelphia, PA, 1798.
4. C.S. Rafinesque. *Medical Flora or Manual of the Medical Botany of the United States of North America*, Atkinson & Alexander: Philadelphia, PA, 1828.
5. T.L. Dog, K.L. Powell, and S.M. Weisman. Critical evaluation of the safety of *Cimicifuga racemosa* in menopause symptom relief. *Menopause* 10, 2003: 299–313.
6. M. Blumenthal. *The ABC Clinical Guide to Herbs*, American Botanical Council: Austin, TX, 2003.
7. M. Blumenthal, C. Cavaliere, and P. Rea. Herbal supplement sales in United States show growth in all channels. *HerbalGram* 78, 2008: 60–63.
8. M. Nagler, F. Kronenberg, E.J. Kennelly, B. Jiang, and C.H. Ma. The use of indicators for unobservable product qualities: Inferences based on consumer sorting. *International Journal of Marketing Studies* 4, 2012: 19–34.
9. J.E. Rossouw, G.L. Anderson, R.L. Prentice, A.Z. LaCroix, C. Kooperberg, M.L. Stefanick, R.D. Jackson et al. Risks and benefits of estrogen plus progestin in healthy postmenopausal women: Principal results from the Women's Health Initiative randomized controlled trial. *Journal of the American Medical Association* 288, 2002: 321–333.
10. G.L. Anderson, M. Limacher, A.R. Assaf, T. Bassford, S.A. Beresford, H. Black, D. Bonds, R. Brunner, R. Brzyski, and B. Caan. Effects of conjugated equine estrogen in postmenopausal women with hysterectomy: The Women's Health Initiative randomized controlled trial. *Journal of the American Medical Association* 291, 2004: 1701–1712.
11. J.E. Manson, J. Hsia, K.C. Johnson, J.E. Rossouw, A.R. Assaf, N.L. Lasser, M. Trevisan et al. Estrogen plus progestin and the risk of coronary heart disease. *New England Journal of Medicine* 349, 2003: 523–534.
12. R. Lupu, I. Mehmi, E. Atlas, M.-S. Tsai, E. Pisha, H.A. Oketch-Rabah, P. Nuntanakorn, E.J. Kennelly, and F. Kronenberg. Black cohosh, a menopausal remedy, does not have estrogenic activity and does not promote breast cancer cell growth. *International Journal of Oncology* 23, 2003: 1407–1412.
13. B. Jiang, F. Kronenberg, P. Nuntanakorn, M.-H. Qiu, and E.J. Kennelly. Evaluation of botanical authenticity and phytochemical profile of black cohosh products by high-performance liquid chromatography with selected ion monitoring liquid chromatography-mass spectrometry. *Journal of Agricultural and Food Chemistry* 54, 2006: 3242–3253.
14. C. Ma, A.R. Kavalier, B. Jiang, and E.J. Kennelly. Metabolic profiling of *Actaea* species extracts using high performance liquid chromatography coupled with electrospray ionization time-of-flight mass spectrometry. *Journal of Chromatography A* 1218, 2011: 1461–1476.
15. D.A. Baker, D.W. Stevenson, and D.P. Little. DNA barcode identification of black cohosh herbal dietary supplements. *Journal of AOAC International* 95, 2012: 1023–1034.
16. J.M. Betz, L. Anderson, M.I. Avigan, J. Barnes, N.R. Farnsworth, B. Gerde'n, L. Henderson et al. Black cohosh, considerations for safety and benefit. *Nutrition Today* 44, 2010: 155–162.
17. J.A. Compton, A. Culham, and S.L. Jury. Reclassification of *Actaea* to include *Cimicifuga* and *Souliea* (Ranunculaceae): Phylogeny inferred from morphology, nrDNA ITS, and cpDNA trnL-F sequence variation. *Taxon* 47, 1998: 593–634.
18. E. Small and P. Catling. *Canadian Medicinal Crops*, Monograph Publishing Program: Ottawa, Ontario, Canada, 1999.
19. M.H. Hoffmann. The phylogeny of *Actaea* (*Ranunculaceae*): A biogeographical approach. *Plant Systematics and Evolution* 219, 1999: 251–263.
20. H.-W. Lee and C.-W. Park. New Taxa of *Cimicifuga* (Ranunculaceae) from Korea and the United States. *Novon* 14, 2004: 180–184.

21. Z. Tian, R.L. Pan, Q. Chang, J. Si, P.G. Xiao, and E. Wu. *Cimicifuga foetida* extract inhibits proliferation of hepatocellular cells via induction of cell cycle arrest and apoptosis. *Journal of Ethnopharmacology* 114, 2007: 227–233.
22. J.L. Egert. Medicinal Plant Fact Sheet: *Cimicifuga racemosa*/Black cohosh. A collaboration of the IUCN Medicinal Plant Specialist Group, PCA-Medicinal Plant Working Group, and North American Pollinator Protection Campaign. PCA-Medicinal Plant Working Group: Arlington, VA, April 2007.
23. H.A. Gleason and A. Cronquist. *Manual of the Vascular Plants of Northeastern United States and Adjacent Canada*, 2nd Edition, New York Botanical Garden: New York, 1991.
24. K. Schlosser. Counting cohosh. *North Carolina Wildflower Preservation Society* 14, 2002: 1–4.
25. N.J.C. Zerega, S. Mori, C. Lindqvist, Q. Zheng, and T.J. Motley. Using amplified fragment length polymorphisms (AFLP) to identify black cohosh (*Actaea racemosa*). *Economic Botany* 56, 2002: 154–164.
26. S.M. Verbitski, G.T. Gourdin, L.M. Ikenouye, J.D. McChesney, and J. Hildreth. Detection of *Actaea racemosa* adulteration by thin-layer chromatography and combined thin-layer chromatography-bioluminescence. *Journal of AOAC International* 91, 2008: 268–275.
27. S.H. Yim, H.J. Kim, S.H. Park, J. Kim, D.R. Williams, D.W. Jung, and I.S. Lee. Cytotoxic caffeic acid derivatives from the rhizomes of *Cimicifuga heracleifolia*. *Archives of Pharmacal Research* 35, 2012: 1559–1565.
28. M. Ganzera, E. Bedir, and I.A. Khan. Separation of *Cimicifuga racemosa* triterpene glycosides by reversed phase high performance liquid chromatography and evaporative light scattering detection. *Chromatographia* 52, 2000: 301–304.
29. W. Li, S. Chen, D. Fabricant, C.K. Angerhofer, H.H.S. Fong, N.R. Garnsworth, and J.F. Fitzloff. High-performance liquid chromatographic analysis of black cohosh (*Cimicifuga racemosa*) constituents with inline evaporative light scattering and photodiode array detection. *Analytica Chimica Acta* 471, 2002: 61–75.
30. K. He, G.F. Pauli, B. Zheng, M. Roller, and Q. Zheng. *Cimicifuga* species identification by high performance liquid chromatography-photodiode array/mass spectrometric/evaporative light scattering detection for quality control of black cohosh products. *Journal of Chromatography A* 1112, 2006: 241–254.
31. K. He, B. Zheng, C.H. Kim, L. Rogers, and Q. Zheng. Direct analysis and identification of triterpene glycosides by LC/MS in black cohosh, *Cimicifuga racemosa*, and in several commercially available black cohosh products. *Planta Medica* 66, 2000: 635–640.
32. H.-K. Wang, N. Sakurai, C.Y. Shih, and K.-H. Lee. LC/TIS-MS fingerprint profiling of *Cimicifuga* species and analysis of 23-epi-26-deoxyactein in *Cimicifuga racemosa* commercial products. *Journal of Agricultural and Food Chemistry* 53, 2005: 1379–1386.
33. P. Nuntanakorn, B. Jiang, H. Yang, M. Cervantes-Cervantes, F. Kronenberg, and E.J. Kennelly. Analysis of polyphenolic compounds and radical scavenging activity of four American *Actaea* species. *Phytochemical Analysis* 18, 2007: 219–228.
34. B. Jiang, C. Ma, T. Motley, F. Kronenberg, and E.J. Kennelly. Phytochemical fingerprinting to thwart black cohosh adulteration: A 15 *Actaea* species analysis. *Phytochemical Analysis* 22, 2010: 339–351.
35. Anonymous. Review of four species for potential listing on the convention on international trade in endangered species appendix II, Sustainable Development and Conservation Biology Program of the University of Maryland. 1999. http://www.nps.gov/plants/MEDICINAL/pubs/cites-a.htm. Accessed April 17, 2014.
36. T. Gödecke, D.C. Lankin, D. Nikolic, S.-N. Chen, R.B. van Breemen, N.R. Farnsworth, and G.F. Pauli. Guanidine alkaloids and Pictet-Spengler adducts from black cohosh (*Cimicifuga racemosa*). *Journal of Natural Products* 72, 2009: 433–437.

37. D.S. Fabricant, D. Nikolic, D.C. Lankin, S.-N. Chen, B.U. Jaki, A. Krunic, R.B. van Breemen, H.H.S. Fong, N.R. Farnsworth, and G.F. Pauli. Cimipronidine, a cyclic guanidine alkaloid from *Cimicifuga racemosa*. *Journal of Natural Products* 58, 2005: 1266–1270.
38. A.L. Ososki and E.J. Kennelly. Phytoestrogens: A review of the present state of research. *Phytotherapy Research* 17, 2003: 845–869.
39. H. Jarry and G. Harnischfeger. Studies on the endocrine efficacy of the constituents of *Cimicifuga racemosa*: 1. Influence on the serum concentration of pituitary hormones in ovariectomized rats. *Planta Medica* 51, 1985: 46–49.
40. H. Jarry and G. Harnischfeger. Studies on the endocrine efficacy of the constituents of *Cimicifuga racemosa*: 2. in vitro binding of constituents to estrogen receptors. *Planta Medica* 51, 1985: 316–319.
41. A. Panossian, A. Danielyan, G. Mamikonyan, and G. Wikman. Methods of phytochemical standardisation of rhizoma *Cimicifuga racemosa*. *Phytochemical Analysis* 15, 2004: 100–108.
42. E.D. Freeburg, L.-N. Olazabal, R. Hannigan, and F. Medina-Bolivar. Extraction and identification of formononetin from black cohosh (*Actaea racemosa*) utilizing gas chromatography coupled mass-spectroscopy and ultraviolet detection of high performance thin layer chromatography. Abstract of Papers. ACS National Meeting: Chicago, IL, 2007.
43. H. Al-Amier, S.J. Eyles, and L. Craker. Evaluation of extraction methods for isolation and detection of formononetin in black cohosh (*Actaea racemosa* L.). *Journal of Medicinally Active Plants* 1, 2012: 6–12.
44. J. McCoy and W. Kelly. Survey of *Cimifuga racemosa* for phytoestrogenic flavonoids. *212th American Chemical Society National Meeting*, Orlando, FL, August 25–29, 1996.
45. D.M. Struck, M. Tegtmeier, and G. Harnischfeger. Flavones in extracts of *Cimicifuga racemosa*. *Planta Medica* 63, 1997: 289–290.
46. E.J. Kennelly, S. Bagett, P. Nuntanakorn, A.L. Ososki, S.A. Mori, J. Duke, M. Coleton, and F. Kronenberg. Analysis of thirteen populations of black cohosh for formononetin. *Phytomedicine* 9, 2002: 461–467.
47. B. Jiang, F. Kronenberg, M.J. Balick, and E.J. Kennelly. Analysis of formononetin from black cohosh (*Actaea racemosa*). *Phytomedicine* 13, 2006: 477–486.
48. B. Avula, Y.H. Wang, T.J. Smillie, and I.A. Khan. Quantitative determination of triterpenoids and formononetin in rhizomes of black cohosh (*Actaea racemosa*) and dietary supplements using UPLC-UV/ELS detection and identification by UPLC-MS. *Planta Medica* 75, 2009: 381–386.
49. A. Lindstrom, C. Ooyen, M.E. Lynch, and M. Blumenthal. Herb supplement sales increase 5.5% in 2012: Herbal supplement sales rise for 9th Consecutive Year; turmeric sales jump 40% in natural channel. *HerbalGram* 99, 2013: 60–65.
50. J. Wechsler. Ensuring quality for dietary supplements. *Pharmaceutical Technology*. 2007. http://www.pharmtech.com/pharmtech/Article/Ensuring-Quality-for-Dietary-Supplements/ArticleStandard/Article/detail/445555. Accessed April 17, 2014.
51. M.G. Nagler, F. Kronenberg, E.J. Kennelly, and B. Jiang. Pricing for a credence good: An exploratory analysis. *Journal of Product & Brand Management* 20, 2011: 238–249.

7 A Model for Nontargeted Detection of Adulterants

James Harnly, Joe Jabolonski, and Jeff Moore

CONTENTS

Introduction.. 91
Sample Complexity and Modeling Tools.. 92
    Sample Complexity 92
    Statistical Tools 93
        Chemometrics... 93
        Classical Statistics 94
Materials and Methods... 95
    Samples 95
    Sample Preparation 95
    Instrumentation 95
    Sample Analysis... 96
    Data Processing.. 96
Results and Discussion 97
    Is the Sample Adulterated?.. 97
    Where to from Here?... 101
Conclusion 105
Acknowledgments.. 105
References.. 105

INTRODUCTION

The basic approach to nontargeted detection of adulterants is simple; build a model for authentic materials and determine whether the test sample fits in the model. As peter Scholl (pers. comm.) at US Food and Drug Administration (FDA) has pointed out, this is like a *Sesame Street* question where young children are shown a triangle mixed with three squares and asked "Which one doesn't fit?" Or, a slightly better analogy, the children are shown three squares and asked if the triangle is the same. The difference between the *Sesame Street* question and the predicament of modern science is that the spectra or chromatograms of the authentic samples are much more complex than the squares and triangles of *Sesame Street*. The human eye is no longer an adequate tool for detecting differences. Modern tools, such as classical statistics and chemometrics, are required.

There is truly no such thing as a nontargeted analysis. All analyses are either specifically or broadly targeted. Specific targeting is possible when the adulterant is known, for example, adulteration of American ginseng with Asian ginseng. In this case, all the sample preparation, instrumental parameters, and data processing can be optimized for detection of the adulterant. Broad targeting occurs when an adulterant is unknown and starts when the analyst decides whether to analyze the solid sample (most likely the powdered solid) or an extract of the solid and which instrumentation to use. Analysis of solids tends to focus on the macrocomponents that swamp the spectral contributions of the microcomponents. Extraction can eliminate the contribution of macrocomponents but leads to further broad targeting with respect to the choice of solvents. A polar or nonpolar solvent, or any polarity in between, can be used to provide focus on specific microcomponents, and the choice of analytical method will provide further focus. True, nontargeted analysis requires multiple broadly targeted methods.

Nontargeted analysis is an inaccurate term for another reason. The name implies the detection of an unspecified substance that may not be present. Thus, the process is intent on proving a negative result for possibly thousands of compounds in a complex chromatogram or spectrum. In reality, a nontargeted analysis is meant to prove that, within a prespecified statistical limit, the test sample matches the chromatograms or spectra of the authentic materials. Thus, a nontargeted analysis can be more correctly described as a test of authentication. However, because nontargeted analysis is a popular term, it will continue to be used in this study.

In this study, we will discuss the complexity of the samples, discuss the tools that are available for building models for authentic materials, and provide an example of a simple and direct nontargeted analysis. Although the sample of interest is skim milk powder (SMP), examined for protein adulteration, the approach is applicable to any botanical material. We will systematically apply principal components analysis (PCA), soft independent modeling of class analogy (SIMCA), and analysis of variance (ANOVA) to the data to search for adulteration. ANOVA will be applied to each variable to further refine the search and improve the signal-to-noise ratio.

SAMPLE COMPLEXITY AND MODELING TOOLS

SAMPLE COMPLEXITY

Any food or botanical will produce a complex chromatogram or spectrum consisting of thousands of variables regardless whether the solid or an extract is tested. Some of these variables will be more useful than others for constructing a model of the authentic materials. This raises the question as to whether selected features (variables) can be used to simplify the model and reduce data storage. However, selected features of the authentic samples may differ from key features of the adulterant and allow it to remain undetected. In other words, a model based on authentic materials cannot anticipate the characteristics of an unknown material. So the answer is *no*, there can be no reduction of features. The most reliable proof of no adulteration will be obtained by considering the full set of data and by selecting multiple methods.

STATISTICAL TOOLS

Chemometrics

Brerton points out that there are two types of chemometric methods: supervised and unsupervised [1]. The only truly unsupervised method is PCA which makes no assumptions about the identity of the samples and bases all calculations on the entire data set. Supervised methods require a priori identification of the samples and are used either for classification and/or regression (calibration). Common examples of supervised methods are SIMCA and partial least squares-discriminant analysis (PLS-DA). SIMCA is a modeling method that fits a PCA model to one or more specified data classes (subsets of the input data) and determines their similarity. The word *soft* in SIMCA means that the unknown material can be identified as belonging to more than one class or to none of the classes. PLS-DA is a classification method that requires identification of all the classes in a data set. It is considered a *hard* method because an unknown material can be assigned to only one class and cannot be classified as belonging to none of them.

Classification methods, in general, are not applicable to nontargeted analyses because specific classes of adulterated samples cannot be identified. However, use of a modeling method to characterize just one class of data is very useful. SIMCA can be used to construct a model for only one class of samples and then determine whether a test sample is statistically similar. The spectra or chromatograms of an identified set of authentic samples are analyzed by PCA and a model is constructed based on the principal component (PC) scores of the samples and the loadings for the variables. The PC scores of the test samples are then computed based on their spectra or chromatograms and the loadings of the authentic samples. The PC scores of the samples are compared to those of the authentic samples and determined to be similar or different at a specified statistical level.

Application of SIMCA to one class (authentic samples) is illustrated in Figure 7.1. An authentic data set (stars) consisting of 12 samples and 2 variables (plotted on the X and Y axes) is modeled by a single PC vector. Adulterated samples are represented by the open circles. Because data are mean centered, the PC vector passes through the origin (0,0) of the plot. The fit of a data point to the model is determined by a projection from the sample perpendicular to the PC vector. The square of the distance of the projected position on the PC vector to the vector center is called the Hotelling T^2 statistic and is the variance accounted for by the model. The square of the distance from the data point to the PC vector is called the Q statistic and is the variance not accounted for by the model. This latter statistic was developed for process control and can now be found as an option on most commercially available chemometric programs [1].

In this example, the authentic samples have varying Hotelling T^2 values and low Q values. The adulterated samples have Hotelling T^2 values that fall within the limits of the model but have large Q values. A mean and standard deviation can be calculated for each statistic and used to establish a 95% confidence limit. Thus, the Hotelling T^2 and Q statistic can be used to quantify the level of similarity/dissimilarity. Based on the Hotelling T^2 values in Figure 7.1, most of the adulterated samples would be judged as fitting the model. However, based on the Q statistic, the adulterated

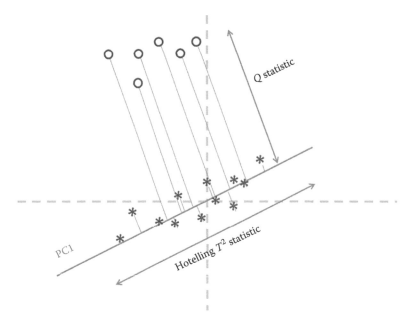

FIGURE 7.1   Application of SIMCA to one class of data. A PCA model based on one PC is constructed for the authentic data (*) and used to evaluate the authenticity of the test data (o).

samples obviously fall outside the model and would be judged as adulterated. The large orthogonal distance indicates that they possess chromatographic or spectral features that were not present in the authentic samples or lack features that were present. Thus, the Q statistic is an extremely useful measurement tool for nontargeted analysis of adulterants.

Classical Statistics

Classical statistical analysis can be applied to an entire chromatogram or spectrum (all variables), to specific regions, or to individual variables. Harrington et al. described a method called ANOVA-PCA [2]. He demonstrated that subarrays could be derived from the original data array using ANOVA calculations that isolated each experimental factor. PCA of the subarrays led to score plots with clusters separated solely on the X-axis. Harnly et al. showed that the subarrays could be used to perform classical ANOVA for the whole spectra by pooling the variance for all the variables [4]. Thus, an F test can be applied to the full spectrum or chromatogram or to individual variables.

Regardless whether classical statistics or PCA is used, the total variance for a data set consisting of a thousand or more variables can be quite large. The size of the total variance determines the size of the variance of an individual component that is needed to establish statistical difference between the authentic and test samples, either by separation on a PCA score plot or by an F test using ANOVA. Isolation of smaller groups of variables reduces the size of the total variance and increases the probability of recognizing a significant deviation on the part of a test sample. The logical extrapolation of this approach is an examination of each variable. This is possible using the data matrices developed using ANOVA-PCA. Instead of summing the variance for the entire spectrum or chromatogram and computing a single F value, F values can be computed for each variable and used to isolate regions or individual variables of high significance. This process is identical to a t-test filter that has been previously described [4].

MATERIALS AND METHODS

SAMPLES

Twelve authentic SMP samples and 8 authentic SMP samples adulterated with 0.5% soy protein isolate (SPI) were analyzed by ultra high performance liquid chromatography (UHPLC) as previously described [5]. All samples of SMP and the SPI adulterant were furnished by the US Pharmacopeia (Rockville, MD). SPI, Lot A-027, 86.95% protein, was originally obtained by FDA from Shandong Sinoglory Health Food Co., Ltd., Liaocheng City, People's Republic of China. SMP samples adulterated with 0.5% SPI constituted the test samples.

SAMPLE PREPARATION

Samples were prepared for UHPLC analysis as previously described [5]. Briefly, 50.0 mg of SMP was weighed into a 15-mL polypropylene tube, 10 mL of denaturing buffer (6M guanidine-HCl, 20 mM dithiothreitol, 5 mM trisodium citrate, pH 7) was added to the sample, and the tube was immediately vortexed and shaken on a Geno-Grinder for 15 min at 1000 RPM. Samples were diluted by a factor of 3 with the dilution solvent to produce samples with nominal SMP concentrations of 1.67 mg/mL in a microcentrifuge tube. The samples were vortexed and placed in a refrigerator at 4°C–8°C for minimum of 30 min. Samples were checked for precipitation and then centrifuged at 12,000 RPM for 10 min at 10°C. SMP adulterated with 0.5% SPI was prepared in the same way by adding 0.25 mg of SPI to the 50.0 mg of SMP in the initial step.

INSTRUMENTATION

As previously reported [5], an Acquity UPLC system with an e-Lambda photodiode array (PDA) wavelength range 190–800 nm with 1.5 µL titanium flow cell was used for UHPLC analysis (Waters Corp., Milford, MA). The column was a BEH C4 (150 × 2.1 mm, 1.7 µm packing) and the guard column was a Vanguard (BEH C4) from Waters Corp. The Acquity Sample Manager module was temperature controlled and the sample loop volume was 10 µL. Mobile phase A1 was water + 0.1% trifluoroacetic acid (TFA). Mobile phase B1 was acetonitrile + 0.1% TFA. The sample manager weak flush solvent was 10/90 acetonitrile/water and the strong flush solvent was methanol + 0.1% formic acid. The sample compartment temperature was 10°C, the inject volume was 7.5 µL, and the column temperature was maintained at 50°C. The PDA collected spectra from 210 to 800 nm with a resolution of 1.2 nm at a sampling rate of 5 Hz and a filter time constant 0.4 s.

Sample Analysis

As previously described, sample sets typically consisted of 14–20 injections [5]. The column was flushed with methanol, followed by 70/30 acetonitrile/water (column storage solvent) after each sample set. A few *conditioning injections* of dilution buffer and SMP samples were made prior to collection of chromatograms for inclusion into the data sets used for statistical analysis. The chromatographic data system was Masslynx© (Waters Corp.). Chromatographic data were collected 300 times a minute for 32 min (from 3 to 25 min) to provide a profile of 9600 points.

Data Processing

Data from individual UHPLC runs were downloaded to Excel (Microsoft, Bellingham, WA) and combined to provide a file consisting of 20 samples (12 authentic and 8 test) by 9600 variables. Retention times for the individual chromatograms were aligned using correlation optimized warping (COW) from Camo Software (Oslo, Norway). For this study, the data file was imported into Solo (Eigenvector Research, Inc., Wenatchee, WA). For PCA and SIMCA, data were preprocessed by multiplicative scatter correction (MSC) and normalization (sum of the squares of all the variables was set equal to 1.0) of each sample chromatogram and mean centering of each variable in the data set. PCA used two PCs to permit two-dimensional display of the data and SIMCA used only a single PC for the model.

ANOVA was performed in Excel using a data file imported from Solo after preprocessing (MSC, normalization, and mean centering). The preprocessed data array, designated as A1 consisted of 20 samples (rows) by 9600 variables (columns). ANOVA was performed as follows:

1. The samples in A1 were sorted; authentic samples were in the first 12 rows, test samples were in the next 8 rows (rows 13–20).
2. Each value in A1 was squared and placed in a second array (A2). All the values in A2 were summed to provide the total sum of squares.
3. A third array (A3) was constructed consisting of the class (authentic and test samples) means. That is, for each variable, the values in rows 1–12 were averaged and the average inserted for each sample. The same was done for each variable for rows 13–20.
4. Each value in A3 was squared and placed in array A4. All the values in A4 were summed to provide the sum of squares between classes.
5. Array A5 was constructed by subtracting array A3 from A1 to provide the class residuals.
6. Each value in A5 was squared and placed in array A6. All the values in A6 were summed to provide the sum of squares within classes.
7. The F value was computed for the entire data set by dividing the sum of squares between classes by the sum of squares within classes:

$$F = \frac{\text{sum A4}/1}{\text{sum A6}/18} \tag{7.1}$$

where 1 and 18 are the degrees of freedom between and within classes, respectively.

8. F values for individual variables were computed in a similar fashion. Thus, Equation 1 becomes

$$F = \frac{\text{sum } A4_{5215}/1}{\text{sum } A6_{5215}/18} \tag{7.2}$$

where:

$A4_{5215}$ corresponds to variable (column) 5215 in array A4

$A6_{5215}$ corresponds to variable 5215 in array A6

In this manner, F values can be computed for each variable, a range of variables (e.g., 5100 to 5500), or the whole array (variables 1–9600).

RESULTS AND DISCUSSION

IS THE SAMPLE ADULTERATED?

In this study, 12 authentic SMP samples were used to construct a model and 8 test samples were examined to determine if they were adulterated. Average chromatograms for 12 authentic SMP samples and 8 test samples are shown in Figure 7.2. Each chromatogram ran for 35 min and 9600 data points were collected between

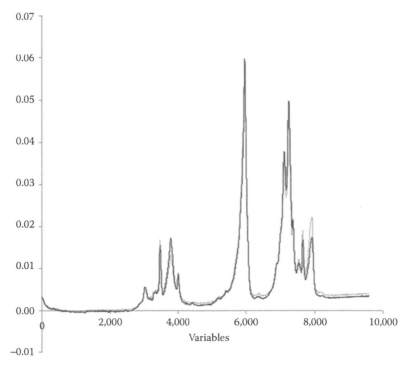

FIGURE 7.2 Typical chromatograms for authentic SMP (black trace) and test (gray trace) samples.

3 and 35 min. The horizontal axis is presented in variables (1–9600) rather than retention times (3–35 min). Prior to averaging, all the chromatograms were subjected to retention time alignment using COW from Camo Software (Oslo, Norway). In addition, each chromatogram was subjected to MSC and normalized (sum of the squares of all the variables was set equal to 1.0). It can be seen that there is almost perfect overlap of the two chromatograms. The only visible differences were at variables greater than 7500.

Figure 7.3 shows the PCA score plot of the 20 chromatograms (12 authentic and 8 test samples) averaged in Figure 7.2. As above, the data set (20 samples by 9600 variables) was aligned with respect to retention time, corrected by MSC, and normalized. In addition, the entire data set was mean centered (the mean of each variable was subtracted from each individual value). The first two PCs accounted for approximately 85% of the total variance. There is no obvious pattern to the data. Examination of a three-dimensional plot that incorporated the third PC (an additional 4% of the variance) also failed to show any pattern. There was no indication of adulteration of the test samples. Observed differences in the two traces were found in regions (variables greater than 7500) with high variability and were not found to be significant.

The loadings for PC1 and PC2 are shown in Figure 7.4. The maximum and minimum values for both PCs correspond to the times of the peak maxima in the chromatograms in Figure 7.2. This is not surprising because PCA is based on variance and not the signal-to-noise ratio. Consequently, the variable with the largest

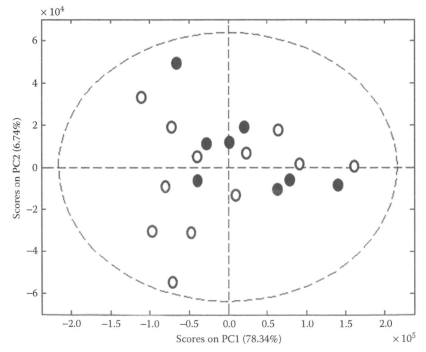

FIGURE 7.3   PCA score plot for full chromatograms of authentic SMP (o) and test (•) samples.

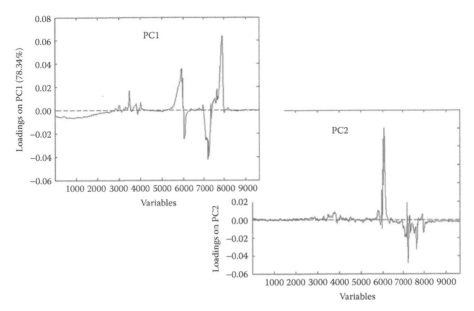

FIGURE 7.4 Loadings for PC1 and PC2 for PCA score plot for full chromatograms of authentic SMP and test samples shown in Figure 7.3.

magnitude, hence the largest variance, will have the greatest influence on the score plots of the samples. There was no indication of adulteration of the test samples.

Figure 7.5 shows a plot of the Q statistic versus the Hotelling T^2 statistic obtained using SIMCA. A model was constructed for the 12 authentic SMP samples using only PC1. The model loadings were used to compute PC scores for the test samples. All the authentic samples (open circles) fall within the 95% confidence limits as do all 8 preparations of the test sample (solid circles). There was no indication of adulteration of the test samples.

The final test was to apply ANOVA to the whole data set and systematically to each individual variable. ANOVA for the whole data set is shown in Table 7.1. It can be seen that the F test is not significant. Only 6% of the total variance was found between the means of the authentic and test samples. The remainder of the variance was found within each class. The F values versus each variable are plotted in Figure 7.6. Two regions show statistically significant F values ($F > 10.0$): between 3500 and 4000 and between 5100 and 5500. The first region corresponds to minor peaks in the chromatogram (Figure 7.2) but the second, with the largest F values, corresponds to an area of relatively little signal. Maximum F values were seen for variables 5215 and 5435.

ANOVA for variable 5215 is shown in Table 7.2. It can be seen that 68% of the total variance is found between the means of the authentic and test samples. The F value of 37.5 is statistically significant with a probability less than 0.001. As shown in Figure 7.6, the surrounding variables are well correlated with the F value systematically increasing from 5100 to 5215 and from 5500 to 5435. The width of the chromatographic region with high F values matched observed peak widths

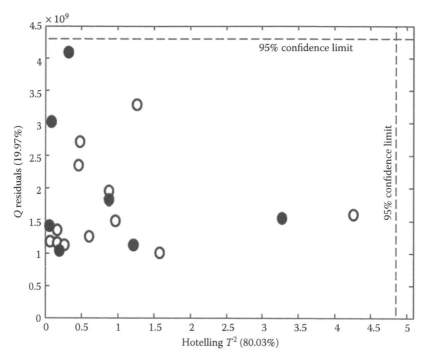

FIGURE 7.5 Q statistic versus Hotelling T^2 statistic for authentic SMP (o) and test (•) samples shown in Figure 7.3. Model was based on authentic samples.

TABLE 7.1

ANOVA for Full Chromatograms

	n	*df*	Variance	Percentage of Grand Mean Residuals (%)	Mean Variance	*F* Value	*P*
Grand mean residuals	20		1517×10^8	100			
Between authentic and test	2	1	88×10^8	6	88×10^8	1.05	0.99
Within authentic and test	20	18	1429×10^8	94	79×10^8		

establishing that these high values were not simply random excursions in the data (i.e., noise).

Figure 7.7 shows an expanded view of the chromatographic region for variables 5100–5500. Figure 7.7a shows all the samples after MSC correction and normalization. Figure 7.7b shows the averages of the same data; the lower trace is the average of the 12 authentic SMP samples and the upper trace is the average of the 8 preparations of the test sample. Only the average plots show a discernable difference between the authentic and test samples.

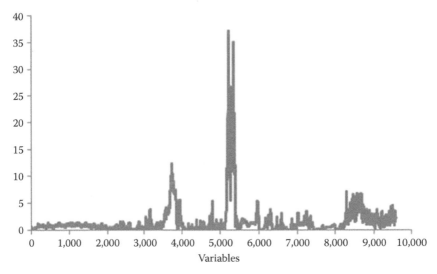

FIGURE 7.6 *F* values for individual points of full chromatograms for authentic SMP and test samples.

TABLE 7.2
ANOVA for Variable 5215

	n	df	Variance	Percentage of Grand Mean Residuals (%)	Mean Variance	F Value	P
Grand mean residuals	20		200×10^3	100			
Between authentic and test	2	1	135×10^3	68	135×10^3	37.05	<.001
Within authentic and test	20	18	65×10^3	32	3.6×10^3		

Figure 7.8 presents the PCA score plot for variables 5100–5500. With one exception, all of the test samples lie above and to the right of the authentic samples. Examination of plots with PC3 and PC4 established that the test sample to the lower left is in fact an outlier. Figure 7.9 shows the SIMCA for variables 5100–5500. All the test samples lie above the 95% confidence limit. Thus, Figures 7.7 through 7.9 and Table 7.2 show beyond doubt that the test samples are adulterated. It does not fit the authentic model. These results also emphasize the importance of examining the variables individually or in small groups. The large total variance of the entire population will mask small but significant variations.

WHERE TO FROM HERE?

The previous exercise showed that the test samples were adulterated. That is the end of the story unless the analyst has additional knowledge. There is no information in

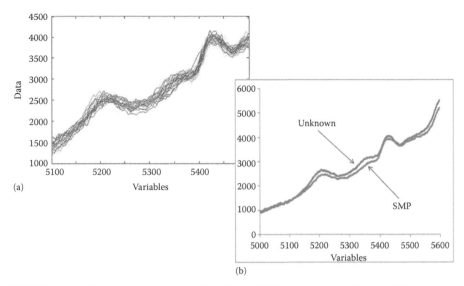

(a)

(b)

FIGURE 7.7 Chromatograms (variables 5100–5500) for (a) all authentic SMP and test samples and (b) average for authentic and test samples.

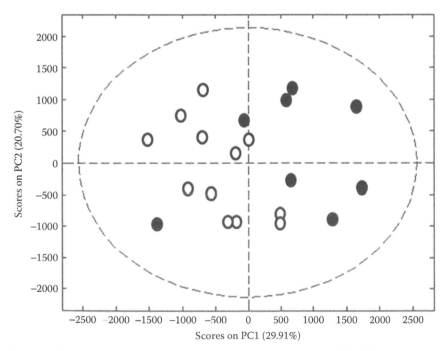

FIGURE 7.8 PCA score plot for partial chromatographic region (5100–5500) of authentic SMP (o) and test (•) samples.

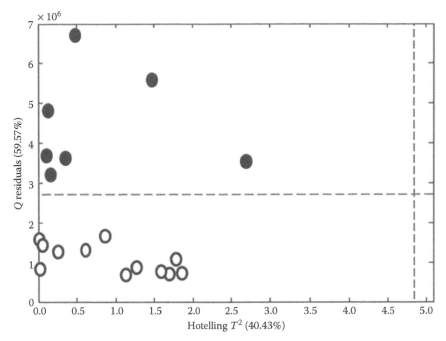

FIGURE 7.9 Q statistic versus Hotelling T^2 statistic for authentic SMP (o) and test (•) samples shown in Figure 7.8. Model was based on authentic samples.

the data set that will allow quantification of the level of adulteration. Depending on the sensitivity of the adulterant for detection at 215 nm, the adulterant could range from a fraction of a percent to a very large percent. Because detection at 215 nm is largely due to refraction and is not specific absorbance, the chances are the adulteration is at a very low percent.

Identification and quantification of the adulterant at this point is dependent on the availability of a library of spectra and/or retention times for common protein adulterants or the special experience or knowledge of the analyst. In either case, the spectral variation for variables 5100–5500 (retention times of 20.0–21.3 min) can be identified as characteristic of SPI (Figure 7.10). An expanded view of variables from 5000 to 6800 is shown in Figure 7.11. The detection limit for SPI at variable 5215 is 0.2% adulteration [5]. The level of SPI in the test sample is approximately 0.5%.

In Figure 7.11, the absorbance of the SPI between variables 5100 and 5500 is clearly shown and the region between 6300 and 6800 would also appear to be distinctive. This latter region did not register for ANOVA in Figure 7.6. Absorbance due to refraction should be linear with respect to concentration at low absorbances. Thus, it is not surprising that the less intense spectral differences between variables 5000 and 6800 did not show up in the ANOVA plot (Figure 7.6). However, it is puzzling that the less significant, but still obvious F values between variables 3500 and 4000 in Figure 7.6 do not appear to correspond to any visible spectral differences in Figure 7.10.

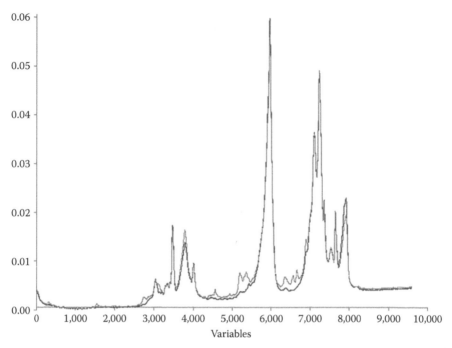

FIGURE 7.10 Average chromatograms for SMP samples (black trace) and SMP samples adulterated with 10% SPI (gray trace).

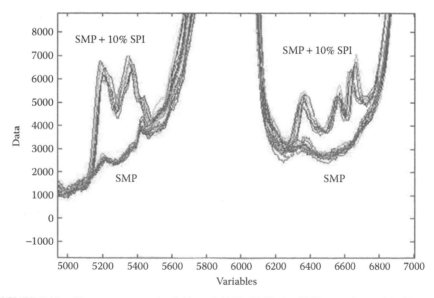

FIGURE 7.11 Chromatograms (variables of 4900–7000) for SMP samples and SMP samples adulterated with 10% SPI.

CONCLUSION

Nontargeted detection of adulterants is simply a matter of building a model based on authentic materials and then determining whether the test material fits that model. A systematic approach using PCA, SIMCA, and ANOVA of the full data set presented in this study failed to indicate any protein adulteration. Systematic ANOVA of each variable revealed chromatographic regions where deviations between the authentic and test samples occurred. PCA, SIMCA, and ANOVA of the selected regions revealed that protein adulteration had occurred. Reducing the total variance of the data set by analyzing individual variables and subregions was key to detecting small variations from the authentic model. This study shows that relatively unsophisticated chemometric and classical statistical methods can be used for nontargeted detection of adulterants.

ACKNOWLEDGMENTS

One of the authors (JH) thanks the Agricultural Research Service of the US Department of Agriculture and the Office of Dietary Supplements at the National Institutes of Health for support of this research.

REFERENCES

1. Richard Brerton. *Chemometrics for Pattern Recognition.* West Sussex: Wiley, 2009.
2. Peter Harrington, N. Viera, J. Espinoza, J. Kien, R. Romero, and Al Yergey. Analysis of variance-principal component analysis: a soft tool for proteomic discovery. *Analytica Chimica Acta.* 544, 2005:118–127.
3. Pei Chen, James Harnly, and Gene Lester. Flow injection mass spectral fingerprints demonstrate chemical differences in rio red grapefruit with respect to year, harvest time, and conventional versus organic farming. *Journal of Agricultural and Food Chemistry* 58, 2010:4545–4553.
4. James Yuk, Kristina L. McIntyre, Christian Fischer, Joshua M. Hicks, Kimberly L. Colson, Ed Lui, Dan Brown, and John T. Arnason. Distinguishing Ontario ginseng landraces and ginseng species using NMR-based metabolomics. *Analytical and Bioanalytical Chemistry* 13, 2013:4499–4509.
5. Joseph Jablonski, James Harnly, and Jeffrey Moore. A non-targeted UPLC-UV method with classical and multi-variate data analysis to detect adulteration of skim milk powder with foreign proteins. *Journal of Agricultural and Food Chemistry.*

8 The Promise of Class Prediction

How Multivariate Statistics Can Help Determine Botanical Quality and Authenticity

Stephan Baumann

CONTENTS

Introduction..107
Materials and Methods...109
   General Workflow 109
   Cranberry Juice Experimental (GC/MS Workflow)......................................110
   EVOO Experimental (GC/QTOF Workflow)..111
   Identification of Compounds...112
   Multivariate Data Processing: Alignment 112
   Multivariate Data Processing: Data Filtering...115
   Multivariate Data Processing: Statistical Analysis......................................116
   Developing Classification Models 119
Conclusions..122
References...123

INTRODUCTION

The goal of this chapter is to determine how multivariate statistics can play a role in determining botanical quality and authenticity. The most common approach for determining sample quality is to combine tests for known degradation products with a tasting panel's sensory evaluation (Figure 8.1). Some examples of degradation products used for sample quality are pyropheophytin, a degradation product of chlorophyll, and 1-octen-3-ol, which is known to contribute to a musty odor. Neither chemical tests nor sensory panel results are ideal on their own. Most of the chemical tests only screen for a small number of well-characterized attributes— fermented, rancid, musty, and so on. Tasting panels can capture nuanced information about sample characteristics that are difficult to determine analytically, but are expensive and slow: expensive because it requires convening trained experts to evaluate sample quality and slow because a tasting panel is only capable of working for a limited time before their acuity begins to suffer. Multivariate statistics can offer a cheaper, more efficient approach for determining botanical quality and authenticity.

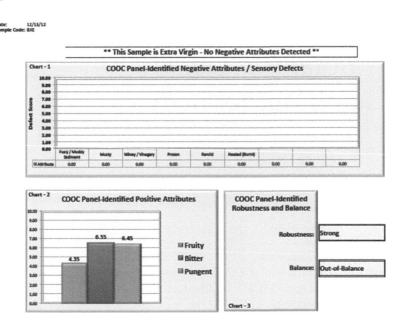

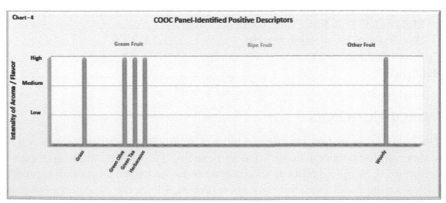

FIGURE 8.1 This is an example of a good sensory test. It tracks not only negative attributes (top third, chart 1) but also classifies olive oil according to positive attributes (middle, chart 2) and flavor descriptors (bottom, chart 4). Caution must be taken with subtle sensory attributes because they can be masked by other components and misclassified descriptors are difficult to identify through statistics. This is where the robustness and the balance classifiers (middle, chart 3) are important.

There has been a steady increase in the quality and quantity of analytical instrumentation available to researchers. Likewise, the improvement in computing power has made it possible to develop sophisticated software tools for mining analytical data. One of the more promising tools is multivariate statistics. This statistical approach allows us to extract useful information from complex data sets in which hidden variables are often present. Until recently, these approaches have been too expensive or complex for researchers to perform by themselves, instead the data have been handed over to dedicated statisticians or have never been fully investigated for *hidden* information [1,2]. But now there are both commercial and shareware options that allow researchers to use statistical approaches for determining attributes such as sample quality and building sample class prediction (SCP) models.

The goal of this chapter is to show how combining high-quality sensory panel data with statistical analysis of untargeted analytical data can help identify hidden variables and determine some of the chemical components responsible for botanical flavor attributes.

MATERIALS AND METHODS

GENERAL WORKFLOW

We show two gas chromatography/mass spectrometry (GC/MS)-based analytical projects that have benefited from multivariate statistics and demonstrate two separate analytical workflows based on the mass accuracy of the instrument. The first example involved extra virgin olive oil (EVOO) and an Agilent 7200 Accurate Mass gas chromatography/quadrupole time-of-flight (GC/QTOF) coupled to an Agilent 7890A GC system (Agilent Technologies; Santa Clara, CA). In this study, we constructed a simple model that predicted whether an olive oil would pass the sensory test. In the second project, we evaluated cranberry juice aliquots for fermented and musty flavors using the Agilent 6850 GC equipped with a CTC PAL autosampler (LEAP Technologies; Carrboro, NC) configured for headspace and coupled to the single-quadrupole Agilent 5975 GC/MS. Both experiments began with a sensory test to provide the original classification. Both gas-phase experiments used SCP to automatically determine volatile flavor characteristics. They differed significantly in the tools that provided the more informative data.

The botanicals were analyzed in random order to ensure that run order effects did not influence the overall outcome. The small number of samples in these experiments can lead to bias. However, this is only one of the factors that can lead to erroneous results. In fact, having a statistically relevant but limited sample size makes it easier to police the data and draw meaningful results. For example, subtle characteristics were missed by the sensory panel in strong flavored EVOO but found in light flavored olive oil. It was only through cluster analysis, discussed later, that these relationships were found.

Retention time locking (RTL) was applied. The concept behind RTL is very simple—GC separations are based on the partial pressure of analytes in the gas phase. The more volatile components have a higher partial pressure in the carrier gas and elute more quickly off the analytical column. As long as the phase type and the ratio between the film thickness of the stationary phase and the column volume remains the same, differences in dead volume, column length, and column diameter can be compensated for by adjusting the chromatographic conditions and the column flow. This means that it is possible to create an analytical method on one instrument and transfer it to instruments with different dead volumes, column lengths, and column diameters. By using RTL principals, it is possible to create libraries that contain chemical information to collaborate or deny tentative identifications by electron ionization (EI) spectra.

The concept of retention indexes (RI) is very similar except that no fixed analytical method is used, instead a homologous series of compounds, typically alkanes or esters, are run with each batch of samples and the retention times are converted into a scale based on the relative retention time of the homologous series of compounds.

CRANBERRY JUICE EXPERIMENTAL (GC/MS WORKFLOW)

Alternatively, the cranberry juice project was done via headspace solid-phase microextraction (HS-SPME). Sample aliquots were taken from 15 tankers of cranberry juice. Two-gram aliquots of cranberry juice were transferred to 20 mL headspace vials. Six grams of sodium chloride was added to increase the HS-SPME efficiency. A 1 cm conditioned 50/30 μm DVB/Carboxen/PDMS StableFlex SPME fiber was used (Supelco; Bellefonte, PA). Samples were subjected to 15 min of dynamic HS-SPME extraction at 80°C, and the fibers were desorbed for 5 min in the GC inlet at 250°C. The split vent trap was then opened to flush out the inlet, and the fiber was allowed to condition for an additional 7 min after each injection. The GC oven temperature began at 60°C with a hold time of 1 min. The oven was then ramped to 240°C at 3°C/min. The cranberry analytical method comes from the RTL food flavor database [3].

In the cranberry study, the total ion current (TIC) traces revealed little difference between the control and the fermented and musty flavors (Figure 8.2). Cranberry juice aliquots were deconvoluted by automated mass spectral deconvolution and

Visual inspection of the total ion chromatogram (TIC)

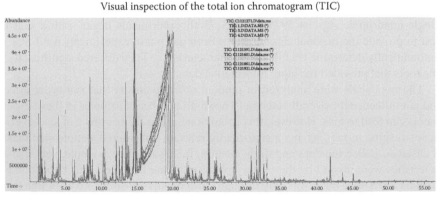

FIGURE 8.2 In this slide, we see an overlay of the TIC for all of the cranberry juice samples. Visually we only see small differences between individual samples but the TIC reveals little about significant differences between the control and the fermented and the musty flavors. Moreover, it is very tedious to go through these samples manually to find the compounds that may be different between samples. Fortunately, there are tools to do this work for us.

identification system (AMDIS; build 2.7). It was developed by the National Institute of Standards and Technology (Gaithersburg, MD) with the support of the Department of Defense to aid in the detection of chemical weapons in complex matrices. It was later made available to the scientific community in general. The process is the same with both unit and exact mass deconvolution. It starts with a noise analysis for each ion chromatogram. It then uses these results to determine a model peak shape for each chromatographic component. A narrow retention time window around the expected retention time from the RTL food flavor database increases the confidence of the identifications by introducing chromatographic information as an orthogonal technique for identification. The retention times and peak shape information allow for the identification of components that have similar peak shapes and apexes in adjacent scans. AMDIS generates spectra for both the identified and unidentified compounds. Chromatographic deconvolution helps identify compounds near their detection limits in complex matrices [4].

This completes the initial stage of data processing. The next step is to export the ELU and FIN files created by AMDIS for multivariate statistical analysis.

EVOO Experimental (GC/QTOF Workflow)

In total, 10 olive oil samples were obtained from the UC Davis Olive Center. All of these samples had been subjected to International Olive Council (IOC) sensory test using a panel sanctioned by the IOC to determine if they passed or failed the criteria for EVOO. The samples were kept in the dark at room temperature. The samples were diluted 1:10 in cyclohexane (Sigma-Aldrich; St. Louis, MO). A cold split injection was used, and the inlet temperature was ramped from 50°C to 300°C to minimize thermal decomposition. The column was a 30 m × 0.25 mm × 0.25 μm HP-5MSUI (J&W Scientific; Folsom, CA). The initial oven temperature was 45°C with a 4.25 min hold time. The oven was then ramped at 5°C/min until 75°C with no temperature hold. This was done in order to better separate the early eluting peaks. The temperature was then ramped to 320°C at 10°C/min with a 10 min isothermal temperature hold [5]. A slow temperature ramp was used in order to separate isobaric compounds.

The accurate mass deconvolution built into MassHunter Unknowns Analysis (build B.06; Agilent Technologies, Santa Clara, CA) follows the same general approach as AMDIS. With accurate mass data it takes longer in that it generates at least 10 times more ion chromatograms than unit mass deconvolution, all of which need to undergo noise and peak shape analysis. Time-of-flight instruments typically have background noise so it is advantageous to set either a deconvolution relative area filter of 0.1% of the largest peak or an absolute signal filter to avoid collecting noise and low abundance co-eluting chemical interferences (Figure 8.3).

The deconvoluted spectra are then searched against the accurate mass RTL library. The process is similar to the one used by AMDIS except that accurate mass offers an additional degree of confidence that identifications are accurate. Both the identified and unidentified components are exported as compound exchange format (CEF) files. The CEF file format contains abundance, spectra, and information related to the library identification. This file type can be directly imported into the multivariate statistic package.

Chromatographic deconvolution

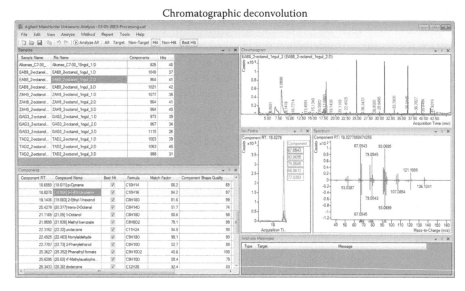

FIGURE 8.3   It is important to inspect the results of chromatographic deconvolution as a quality control check. The parameters for deconvolution need to be verified and the identifications need to be verified for accuracy. There are many parameters involved that can impact the results.

IDENTIFICATION OF COMPOUNDS

Note that it is not necessary to know the identity of the compounds used to build a classification model. In fact, it is preferable to forgo identification over introducing dubious information into the model building process. Chromatography is an oft overlooked tool that can be used to improve the confidence around identifications. This is usually done using an RTL library or RI data [6]. These tools can help identify compounds that give similar EI spectra but have different chromatographic characteristics, often structural isomers. Conversely, information is lost by not obtaining an overview of the compounds present in the sample. Of course, gaining an understanding of the chemical milieu in a sample leads to an understanding of those chemical components that adversely affect the sensory qualities. To deal with these contradictory needs, it often make sense to process the data both as identified and as unidentified to make certain the statistical results are not dictated by the approach.

MULTIVARIATE DATA PROCESSING: ALIGNMENT

Identified compounds do not go through an alignment process, and therefore, it is very important that identifications are vetted before statistical analysis. Unidentified compounds require alignment. This process is often undervalued but poor alignment can lead to incorrect identifications and therefore poor statistics. Data alignment includes a minimum of two parts. The first part is determining the dot product difference between two extracted spectra. Usually, these spectra have already been filtered to remove chemical noise and so a correlation match factor of 0.3–0.5 is appropriate, based on how much chemical interference is in a particular compound's spectra. The second part is the retention time alignment that needs to be adjusted to allow for the retention time drift in the experiment. The simplest approach is to have a retention time window that allows for experimental drift. This approach works well when there is adequate chromatographic separation and there are no clusters of structurally similar compounds to complicate the alignment process. Fortunately, we can use landmark anchoring to automatically adjust the retention time based on ubiquitous compounds that can be utilized as retention time markers [7]. With accurate mass, it is also possible to identify which ions can be logically formed from the molecular formula of a tentatively identified molecule. On top of using the mass differences, it is possible to create an enhanced match score based on the abundance ratios, isotope spacing, and isotope m/z match. This is automatically done using the molecular formula generator built into Mass Hunter software. By extracting exact mass spectra, it is possible to improve the confidence of identifications by stripping out chemical noise with the same unit mass that would otherwise interfere with the identifications.

One tool to help us judge an experiment's alignment is the mass versus retention time plot, also known as the summary plot. The plot describes the distribution of the aligned and unaligned entities in the experiment (Figure 8.4). The alignment parameters are adjusted while monitoring the summary plot to determine the impact on the distribution of aligned entities in the summary plot. It is also advisable to review the spectra of the aligned entities to verify that a reasonable match factor was applied (Figure 8.5).

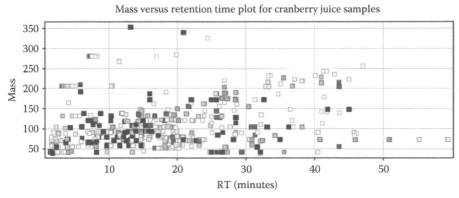

Mass versus retention time plot for cranberry juice samples

FIGURE 8.4 442 unique compounds are identified by chromatographic deconvolution and associated through peak alignment. Note that most of these compounds occur only once or twice and are filtered out by MPP. The mass versus retention time plot shows how many of these 442 compounds were actually found in each sample. The graph shows the mass versus retention time plot. The one-hit-wonders are white, the gray indicates compounds that showed up in half the samples of a particular condition, and the black are in all the samples of a particular condition. Of the 442 entities, 112 are in at least 60% and 77 are in 100% of either condition. Recursion, discussed later, is a useful tool for verifying that deconvolution and alignment worked properly. It is built into software tools like GC/MS ChemStation, GAVIN, and MassHunter ProFinder.

Verifying recursive results for cranberry juice samples

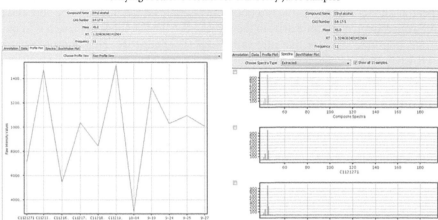

FIGURE 8.5 It is valuable to mine the data to verify that reasonable match factor settings were used for alignment and that the recursion is properly configured. One of the benefits of being able to drill down in an entity list is to see how a compound's concentration changes across the samples (left image). We can also view the spectra to verify that the peak alignment parameters are properly set. We also learn that many things that seem interesting turn out to be statistically insignificant under closer inspection. It turns out that nearly every cranberry juice sample has measurable quantities of ethanol present. Ethanol concentration does not correlate with fermented cranberry juice flavor.

Data should also be normalized in order to remove systematic detector drift that can be caused by matrix effects, changes to the instrument's response, injection variability, and so on. Ideally, internal standards could be applied to normalize the data. However, there are often reasons to normalize all the entities (compounds) to the mean in order to give equal weighting to low and high responding compounds. This plays more of a role when the ultimate goal is to determine causality for a particular flavor characteristic.

Although we did not have accurate mass information for the cranberry juice project, we did have the ability to reprocess the data in order to recover components present in the samples but not necessarily identified due to either low abundance or chemical interferences. This process, known as recursion, does not filter chemical interferences; instead, it verifies a particular ion is found within a specific time interval [8]. Because of this there is a slightly higher risk of a false positive and so the raw data should be checked to assure that the updated identifications are real. In practice, the recursive process is very useful in finding components buried in the noise.

With accurate mass data, such as found with the GC/QTOF, recursion can be done by exporting all the component spectra to the instrument's quantitative software for evaluation. The unidentified compounds are annotated by their retention time and most abundant ion in their spectrum and processed alongside the identified components. In this process, standard quantitation tools can be used to process the data in order to reduce the number of false positives or negatives. This process is fairly manual; however, there are moves to automate this process. MassHunter ProFinder (Agilent Technologies; Santa Clara, CA) is an early example of an automated recursive feature finding workflow.

MULTIVARIATE DATA PROCESSING: DATA FILTERING

Entity filtering can simplify multivariate analysis. The most commonly applied filter is frequency analysis in which only entities (compounds) that were in at least one class (samples with shared characteristics—i.e., control, treated, or blank samples) a certain percentage of the time are retained. A frequency filter of 100% means that the compound must be in all the samples of a particular class. Setting a strong frequency filter greatly reduces the number of possible flavor markers to be evaluated. A strong filter is good when you have a very homogenous data set. If the controls come from all over the world, a strong filter may miss a good marker because of biological variation. This is a reasonable trade-off when the goal is to find an abundant marker that correlates with a desired characteristic. On the other hand, a technique such as recursion or lowering the frequency requirement is beneficial to keeping hard-to-detect compounds from being filtered out. The frequency filter is never reduced to zero as it is necessary to remove compounds that don't show up in more than one sample, commonly referred to as one-hit-wonders.

Two other commonly applied filters are sample variability (either based upon coefficient of variance or standard deviations) and abundance. Typically, these filters are used to reduce the complexity of the data, making it easier to draw statistical conclusions. It should be a warning flag when these filters are needed to separate samples into classes. If this occurs, we need to be vigilant in order to verify that the filters are allowing a significant difference in the samples to be measured instead of biasing the results.

It will always be easier to look for abundant, easy-to-measure compounds that correlate with a certain characteristic as opposed to trying to mine the data further to find all the compounds that correlate with a flavor characteristic of a particular botanical. Median baselining is often used in small studies so that one or two outliers do not distort the results. Outliers have much less effect with larger data sets.

It is possible that causal flavor components are not detectable by the current analytical technique. The polyphenols responsible for olive oil astringency are a good example. It is reported that these water-soluble polyphenols have molecular weights between 500 and 3000 g/mol [9]. Without derivatization, they are not volatile enough to be analyzed by GC/MS. This brings up the possibility of needing multiple analytical techniques to properly characterize a botanical. Fortunately, many of the multivariate statistical analysis tools can combine data from different sources through the Z-transform baseline option. The Z-transform process works by first baselining each entity to the median abundance across all the samples. Then, the standard deviation of the abundance is determined. Finally, each entity in a sample is scaled based on its standard deviation from the mean abundance.

MULTIVARIATE DATA PROCESSING: STATISTICAL ANALYSIS

In this section, we use the olive oil samples as an example but the same principals hold for the cranberry juice or any other type of sample. It no longer matters what technique is used to generate the sample data now that it is aligned, filtered, and normalized. In general, we are following the guided workflow built into mass profiler professional (MPP) but the data quality and the objective of the experiment dictate how samples need to be analyzed.

Principle component analysis (PCA) reduces the differences between samples from a collection of complex variables to principal components. It shows major differences between samples without losing the discriminating power in the data [10]. It is performed via the transformation of measured variables into uncorrelated principal components, each being a linear combination of the original variables. The goal is to identify possible relationships within the classes of data. PCA of the olive oil entities that varied in amount between the sensory test pass and fail samples revealed distinctive grouping of the data into two classes (Figure 8.6). This is usually a good sign that there are obvious statistical differences between the two samples, most of which correlate but some of which are causal.

Analysis of variance (ANOVA) determines what level of variance is accepted as significant for a given entity. ANOVA compares the variance within a class to the variance between classes [11]. Once these variances are understood, it is relatively

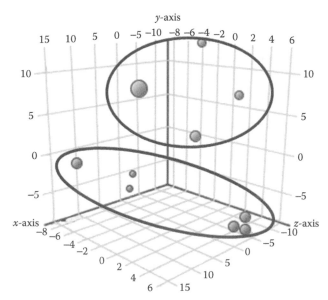

FIGURE 8.6 PCA on conditions shows data clusters. The samples that failed the sensory test are in the top portion of the graph and the ones that passed are in the bottom half of the graph. PCA on conditions allows for the detection of similarity between samples, discriminated by the major trends in the data. Note that there are two distinct phenotypes among the EVOO samples that passed the sensory evaluation. MPP allows for the creation of new entity lists to evaluate the differences between these phenotypes.

straightforward to determine if the null hypothesis is at work here. The null hypothesis is that there are no significant differences between classes. If the standard deviation within the classes is the same as the standard deviation between the classes, then the null hypothesis is proved. You cannot disprove the null hypothesis. You can only give the probability that it is not true. Setting a lower p value provides a greater probability that the null hypothesis is not true and increases the probability that the classes are different. A common threshold is using a probability p value of 0.05. This simply defines that the variance from one sample to another has a 95% or greater probability to be considered significant. By determining statistical significance, we greatly decrease the number of entities or compounds that are considered to correlate with the desired class. There is no fixed rule on what p value should be used and it is often beneficial to reduce the p value if we are clearly filtering out compounds that are already suspected of being causal of a particular characteristic.

Some multivariate statistical tools combine ANOVA and fold change through the use of a volcano plot. This is done by placing the two filters on the axes of a plot. Fold change is typically placed on the x-axis and significance on the y-axis. This filter identifies entities with large abundance differences between pass and fail EVOO samples or tasty and fermented cranberry juice. The volcano plot reveals both up- and downregulated compounds as well as the degree of statistical significance. Fold change only compares *pairs* so a volcano plot is only useful when there are two categories. It does work well with these projects because we are comparing controls to defective samples. In both of these studies, we were interested in compounds that correlated with negative attributes so we were more interested in entities upregulated among the defective samples.

Quite often PCA does not show significant differences between classes. In this case, it makes sense to go back and reevaluate the data. If the sensory panel data are detailed, are there obvious differences in the samples that can confound the data? Are strong characteristics potentially masking more subtle flavors? The PCA data itself will show which samples are dissimilar. This is a good time to use cluster analysis to evaluate some of the differences between samples that belong to the same class. Obvious differences might suggest new ways to group the samples in order to categorize them correctly. Another option is to go back and apply various coefficients of variance filters to look at compounds that are either consistent or inconsistent across a particular category. Often, following these procedures will give new insight into the experiment and a greater appreciation of what can be asked of the data.

Cluster analysis is a powerful method to organize compounds or entities and conditions in the data set into clusters based on the similarity of their abundance profiles. Hierarchical clustering is one of the simplest and most widely used clustering techniques for analysis of mass abundance data. The method follows an agglomerative approach in which the most similar abundance profiles are joined together to form a group. These are further joined in a tree structure, until all data form a single group [12].

One advantage of hierarchical clustering (Figure 8.7) is that it provides entity (compound) as well as sample information in a visual heat map. These data can often be insightful to identify which compounds dominate a particular flavor characteristic. This can be done by determining if any of the compounds are not universally

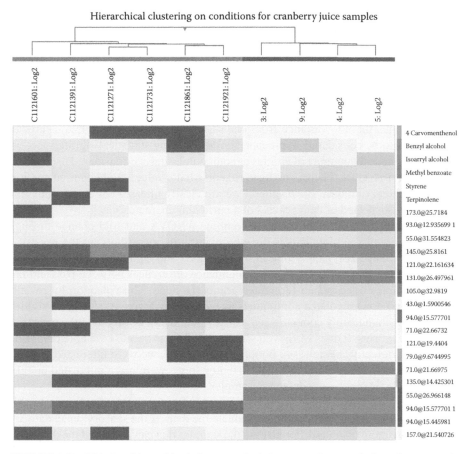

FIGURE 8.7 This is a hierarchical cluster analysis heat map for association of compounds detected in the sample classes. The clustering is shown as a tree diagram (dendrogram) above the heat map. The color range in the map indicates the peak intensity of each compound for each class. Compounds with low intensity are shown in white; those with intermediate intensity are shown in gray, and those with high intensity are shown in black.

present and yet clearly contribute to a flavor characteristic. Specific fusel alcohols may not be present in every fermented or musty cranberry sample, although they clearly contribute to the fermented characteristic. That suggests that they are minor contributors to the fermented or musty characteristic. At this point, we should know which components correlate with a failed sensory test and it is time to apply that information into building a classification model.

There are cases in which either the data are too noisy or there are confounding factors that are keeping the data from parsing along class lines. In these instances, it is still useful to look at fold change data in order to determine what compounds are increasing on average in concentration relative to other compounds. The benefit of expressing the change in concentration as a ratio as opposed to using terms of abundance is that the change is emphasized as opposed to the absolute values. This information, viewed as a whole, is often quite insightful. Is there an increase in the number of fusel alcohols identified in the fermented cranberry juices? Are there any compounds that are upregulated in one class relative to another class? This could be an indication that there are confounding factors throwing off the statistics, factors such as farm, varietal, or storage conditions that are playing a significant role in the flavor profile but are not being identified as a significant variable [13].

DEVELOPING CLASSIFICATION MODELS

The goal of classification is to produce general hypotheses based on a training set of examples that are described by several variables and identified by known labels corresponding to the class information. The task is to learn the mapping from the former to the latter. Numerous techniques based either on statistics or on artificial intelligence have been developed for that purpose [14,15].

MPP automatically builds and tests five different algorithms—decision tree (DT), support vector machines (SVM), naïve Bayes (NB), neural network (NN), and partial least square discriminant analysis (PLSDA). The DT algorithm builds a model based on value tests that branch out from a root node that contains all the samples. Each node is organized by its ability to divide the remaining data into two subsets. The SVM algorithm separates samples into classes by projecting them into higher dimensional space and then determining a separating plane which separates the two classes of points. The NB classifier assumes that the effect of a compound on a given class is independent of the presence or value of other compounds. Although this compound independence assumption is often inaccurate, NB models are not as susceptible to the curse of dimensionality as other techniques and so data sets don't need to expand exponentially with the number of variables [16]. This is an advantage when there are hidden variables in the data. An NN is a system inspired by the structure of biological NNs in which inputs for a node are weighted sums of outputs of other nodes. The weighting of a particular connection is adjusted in the direction of a correct output. Because of this characteristic, it is sometimes capable of resolving classes that linear models cannot. PLSDA uses vector analysis to minimize the number of variables required to explain the separation between the samples and their class assignments. Because of this, it has an intrinsic predicting power and is particularly adapted to situations where there are fewer observations than measured variables [17].

The first step in building the classification model is to train the models with the data, using each of the five model algorithms. The second step is to test each model with unknown sample data. Additional samples that are not used to create the models are used for this purpose. Using these samples, it is straightforward to determine model robustness based on the percent accuracy of the prediction. A common visualization tool applied during this step is a confusion matrix (a table with the true class in rows and the predicted class in columns). By aligning the prediction results in this manner, this matrix shows when a model fails (when it confused classes) (Figure 8.8).

Confusion matrix for cranberry juice samples

Prediction Result	Confusion Matrix

Partial Least Squares Discrimination

	[C] (Predicted)	[F] (Predicted)	[M] (Predicted)	Accuracy
(True) [C]	5	0	0	100.000
(True) [F]	0	4	0	100.000
(True) [M]	0	0	4	100.000
Overall Accuracy				100.000

Support Vector Machine

	[C] (Predicted)	[F] (Predicted)	[M] (Predicted)	Accuracy
(True) [C]	5	0	0	100.000
(True) [F]	0	4	0	100.000
(True) [M]	0	0	4	100.000
Overall Accuracy				100.000

Naïve Bayes

	[C] (Predicted)	[F] (Predicted)	[M] (Predicted)	Accuracy
(True) [C]	5	0	0	100.000
(True) [F]	1	3	0	75.000
(True) [M]	0	2	2	50.000
Overall Accuracy				76.923

Decision Tree

	[C] (Predicted)	[F] (Predicted)	[M] (Predicted)	Accuracy
(True) [C]	4	1	0	80.000
(True) [F]	1	3	0	75.000
(True) [M]	1	1	2	50.000
Overall Accuracy				69.231

Neural Network

	[C] (Predicted)	[F] (Predicted)	[M] (Predicted)	Accuracy
(True) [C]	5	0	0	100.000
(True) [F]	1	3	0	75.000
(True) [M]	0	0	4	100.000
Overall Accuracy				92.308

FIGURE 8.8 The confusion matrix name stems from the fact that the table makes it easy to see if the system is confusing classifications. We can already see that naïve Bayes, decision tree, and neural network are confusing classifications.

An SCP provides a robust way to determine EVOO and cranberry juice quality that can be used in a production QC environment. Small differences between samples can be clearly visualized, using a multivariate analysis of GC/MSD data. One thing to be aware of is the risk of overfitting [18]. This is when a statistical model describes a random error instead of an underlying relationship. It will have a poor predictive performance because it exaggerates minor fluctuations in the data. In order to generate the SCP model with the highest accuracy of prediction, the data quality is crucial. This facilitates construction of the right filtering and prediction model for the samples. An SCP will provide the best results when the sample data are properly filtered. Multiple prediction models allow the evaluation and customization of different prediction models to the analysis. Robust entity lists enable the development of better SCP models which in turn enable improvement of the workflow of quality assurance and quality control of food analysis.

Note that a prediction model switches the analysis from being untargeted to targeted and it is advisable to switch your data mining from untargeted to targeted as well. This could be done through recursion or if no recursion tools are available then create a targeted method using the instrument's quantitation software. Mine the data using that list, import the results into the multivariate statistical package, and rebuild the model. This approach will eliminate *missing values*, the bane of any multivariate analysis. The final method will be of higher quality, and if some of the entities drop out, they should also be removed from the quantification method. That final method is the one best suited for class prediction. The results of which are shown in Tables 8.1 and 8.2.

TABLE 8.1
Cranberry Juice Results

RT (min)	NIST ID	CAS	Odor/Flavor
10.55	Methyl benzoate	93-58-3	F: Phenolic and cherry pit with a camphoraceous nuance
12.918	Ocimenol	5986-38-9	O: Fresh citrus lemon lime cologne sweet mace
14.423	Cherry propanol	1197-01-9	F: Fruity, cherry, sweet, hay-like with cereal and bread-like nuances
15.447, 15.58	p-Menth-1-3n-9-al	29548-14-9	O: Spicy herbal
15.704	(+)-(E) Limonene oxide	6909-30-4	O: Green
21.67	Trimethyl pentanyl diisobutyrate	6846-50-0	
32.992	Benzophenone	119-61-9	O: Balsam rose metallic powdery geranium

Some of the tentatively identified compounds that correlate with a fermented cranberry juice flavor already have a known flavor or odor. In this case, the identifications came from the publicly available online database from The Good Scents Company. Some of these compounds may be responsible for a fermented flavor, while other compounds just correlate with a fermented flavor.

TABLE 8.2
EVOO Results

RT (min)	NIST ID	CAS	Odor/Flavor
20.91	α-Cubebene	17699-14-8	Herbal
27.55	n-Hexadecanoic acid	57-10-3	Faint oily
29.75	Octadecanoic acid, ethyl ester	111-61-5	Waxy

Some of the tentatively identified compounds that correlate with a defective olive oil are tentatively identified from The Good Scents Company's database. It is impressive that multivariate statistics can start with hundreds of components identified in a sample and reduce the data down to a handful of compounds that correlate with a particular classification.

CONCLUSIONS

The power of multivariate software is easy to grasp and it is easy to come up with many interesting experiments. The data quality needs to be high and a good experimental design needs to be implemented in order to get the best possible outcome from a statistical experiment. It might seem that the small data size of the experiments involved here would be problematic; however, by having a statistically relevant but limited sample size, it was easier to evaluate the data and draw meaningful results. In fact, I'd start with a significantly larger data set but would eliminate many samples in the process, to reduce potential confounding factors such as cultivar and country of origin. Once the dimensionality of the data is sufficiently reduced, you can build and test models. By no means are large studies to be avoided but we have to be aware that studies run over weeks or months are invariably introducing new variables, the impact of which need to be determined.

Moreover, it always pays off to manually evaluate the data. It is surprising to see how often a pattern that is visually observable is missed by statistical techniques. It is instructive to go back and look at overlays of TICs and EICs for characteristics fragments (chemical markers) for certain traits. If all the observable patterns are found statistically, then all the data filters are set appropriately and you can have confidence in the results. Otherwise, you need to evaluate the cause of the discrepancies between what is observable and what is found statistically.

All peak finding algorithms are data filters. In some cases, the filtering criteria can be set tightly, and in other cases, the criteria need to be set loosely. In any case, we need to understand how these filters are being applied in order to ascertain if the parameters are set correctly or if we are even using the correct tool. By evaluating the data that is being used for class prediction, we can determine if a missing value is correct, below an intensity threshold, or missed by the peak finding algorithm. One of the best tools developed to deal with identifying false negatives is the recursive workflow. This process mines the original data with a less discriminant filter, typically one based on searching for peak component ions at a defined retention time. This approach does not take noise into account and is therefore more susceptible to false positives, so results need to be confirmed. Fortunately, the latest generation of software tools is building in recursion to minimize both false positive and negative components.

REFERENCES

1. P. Geladi and K. Esbensen. The start and early history of chemometrics: Selected interviews. Part 1. *Journal of Chemometrics* 4, 2005: 337–354.
2. K. Esbensen and P. Geladi. The start and early history of chemometrics: Selected interviews. Part 2. *Journal of Chemometrics* 4, 2005: 389–412.
3. F. David, F. Scanlan, P. Sandra, and M. Szelewski. Analysis of essential oil compounds using retention time locked methods and retention time databases, Agilent Technologies Application Note. 5988-6530EN, May 2002. http://www.chem.agilent.com/Library/applications/5988-6530EN.pdf.
4. S. E. Stein. An integrated method for spectrum extraction and compound identification from gas chromatography/mass spectrometry data. *Journal of the American Society for Mass Spectrometry* 10, 1999: 770–781.
5. S. Baumann and S. Aronova. Olive oil characterization using Agilent GC/Q-TOF MS and mass profiler professional software, Agilent Technologies Application Note 5991-0106EN, March 2012. http://www.chem.agilent.com/Library/applications/5991-0106EN.pdf.
6. R. P. Adams. *Identification of Essential Oil Components by Gas Chromatography—Mass Spectroscopy.* Allured Publishing Corporation: Carol Stream, IL, 2007; pp. 3–4.
7. R. Tautenhahn, G. J. Patti, D. Rinehart, and G. Siuzdak. XCMS online: A web-based platform to process untargeted metabolomic data. *Analytical Chemistry* 84, 2012: 5035–5039.
8. V. Behrends, G. D. Tredwel, and J. G. Bundy. A software compliment to AMDIS for processing GC-MS metabolomics data. *Analytical Biochemistry* 415, 2011: 206–208.
9. E. C. Bate-Smith and T. Swain. "Flavonoid Compounds." In: *Comparative Biochemistry.* H. S. Mason and A. M. Florkin, Eds. New York: Academic Press, 1962; Vol. 3, pp. 755–809.
10. H. Hotelling. Analysis of a complex of statistical variables into principal components. *Journal of Educational Psychology* 24, 1993: 417–441; 498–520.
11. G. M. Clarke and D. Cooke. *A Basic Course in Statistics*; Arnold: London, 2004; pp. 571–599.
12. R. R. Sokal and C. D. Michener. A statistical method for evaluating systematic relationships. *University of Kansas Scientific Bulletin* 38, 1958: 1409–1438.
13. J. E. Wilson. The role of geology, climate, and culture in the making of French wines. Mitchell Beazley: London, 1998; p. 55.
14. J. Boccard, J. L. Veuthey, and S. Rudaz. Knowledge discovery in metabolomics: An overview of MS data handling. *Journal of Separation Science* 33, 2010: 290–304.
15. R. Nisbet, J. Elder, and G. Miner. *Handbook of Statistical Analysis and Data Mining Applications.* Academic Press: Burlington, VT, 2009; pp. 251–256.
16. A. K. Smilde, M. M. W. B. Hendricks, J. A. Westerhuis, and H. C. J. Hoefsloot. "Data Processing in Metabolomics." In: *Metabolomics in Practice: Successful Strategies to Generate and Analyze Metabolic Data.* M. Lämmerhofer and W. Weckwerth, Eds. Weinheim, Germany: Wiley-VCH Verlag & Co., 2013; p. 266.
17. M. Sugimoto, M. Kawakami, M. Robert, T. Soga, and M. Tomita. Bioinformatics tools for mass spectroscopy-based metabolomic data processing and analysis. *Current Bioinformatics* 7, 2012: 100.
18. S. Lawrence, C. L. Giles, and A. C. Tsoi. Lessons in neural network training: Overfitting may be harder than expected. *Proceedings of the 14th National Conference on Artificial Intelligence.* Providence, RI, AAAI, July 29–31, 1997; pp. 540–545.

9 Conformation of Botanical or Bio-Based Materials by Radiocarbon and Stable Isotope Ratio Analysis

Randy Culp

CONTENTS

Introduction..125
Atmospheric ^{14}C Activity...127
Stable Isotopes 130
   Increasing Specificity in Isotopic Measurement 131
   Bio-Based Material Verification..135
Methods: ^{14}C Determinations by AMS and LSC 137
Stable Isotope Ratio Analysis 137
Isotopic Cases of Interest..138
Conclusion 139
Acknowledgments...140
References..140

INTRODUCTION

Isotopic analysis, including both stable isotopes and radiogenic nuclides, had its origin in support of the geological sciences. For decades, the multitude of long-lived, naturally occurring radionuclides and their stable daughter products have been useful as geochronometers and proxies for such processes as continental crust formation and sedimentation rates in the world's oceans [1].

Natural sources of radioactivity come from (1) long-lived radionuclides such as Uranium-238 (^{238}U) and Rubidium-87 (^{87}Rb), (2) decay products of naturally occurring long-lived isotopes of uranium and thorium, and (3) cosmogenic radionuclides such as ^{14}C and tritium (^{3}H). Finally (4) Cosmic radiation composed of nuclear particles and ultraviolet radiation completes the naturally occurring spectrum of radioactivity. Anthropogenic sources of radioactivity include fission reaction by-products and their inherent nuclear waste generated from spent fuel rods and processing materials, medical devices, X-ray and gamma-ray equipment as well as ionizing radiation from ^{3}H containing paints used for luminous dials on clocks and watches and ceramic glazes containing uranium.

Of the many radionuclides, naturally occurring and anthropogenic, ^{14}C and ^{3}H have found usefulness in the authentication of hydrocarbon-based materials with regard to origin or process of formation [2]. ^{3}H, with its relatively short half-life of 12.5 years, has been used less often over the last few decades for authenticity testing due to low abundance. ^{14}C, however, has found great usefulness in authenticating naturalness.

Although ^{14}C can be produced as a by-product in a fission reactor, its abundance occurs naturally through the interaction of cosmic energetic neutrons with nitrogen-14 (^{14}N) in our stratosphere 12–15 km above earth's surface [3]. Similarly, ^{3}H is formed by the interaction of these cosmic neutrons with ^{14}N to form ^{3}H and ^{12}C [4]. The ^{14}C and ^{3}H produced in the upper atmosphere then equilibrate with carbon dioxide (CO_2) and water throughout the atmosphere. CO_2 is incorporated into plants through photosynthesis while ^{3}H equilibrated water is taken up through the soil and plant leaves. For ^{14}C, a steady-state equilibrium is achieved between the ^{14}C production in the atmosphere and its' decay. When a plant or organism dies, no further photosynthesis or uptake of $^{14}CO_2$ occurs. The ^{14}C declines at a specific rate based on its' half-life of 5730 years (Figure 9.1). After death, the remaining ^{14}C represents the time since death occurred of such organism. This is the basis of the radiocarbon dating method.

Although most consumable foods, flavors, and beverages, as well as nonconsumable bio-based products, are derived from recently harvested plants, and this food chain begins with ^{14}C incorporation into plants, these products will exhibit the highest natural level of ^{14}C available at the time of harvest. This is in contrast to fossil fuel-derived products that are lacking or exhibit no ^{14}C activity. This forms the

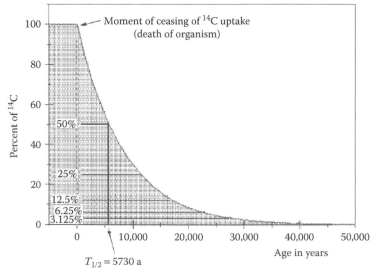

FIGURE 9.1 ^{14}C decay curve portraying decreasing percentage of ^{14}C over time after organism's death. (After S. Bowman, *Radiocarbon Dating, Interpreting the Past* Series. 1990, University of California press.)

basis of radiocarbon authenticity testing of flavoring materials [5–8]. One can then conclude that animals, including humans, will possess the same ^{14}C content as the plants they consume [9,10]. In humans and other animals, exceptions exist in the incorporation of ^{14}C with teeth, bones, and dentine, which have ^{14}C incorporated early in life or over extended periods of time [11–13]. With regard to the ^{14}C concentration or activity, it is also important to recognize that the various biosynthesized hydrocarbons within these recently grown materials exhibit similar or identical ^{14}C activities because they were all produced from an equivalent $^{14}CO_2$. By virtue of its' extremely low abundance, existing in $10^{-10}\%$ relative to the abundance of ^{12}C, the ^{14}C activity of the variety of organic molecules in plants is minimally dependent, albeit indiscernible with current measurement precisions, on the type of organic molecule biosynthesized. One can conclude, within measurement error, ^{14}C is incorporated equally within a specific plant. So whether a regional field of corn produces ethanol for fuel modification or high fructose corn sugar for consumer products, they will exhibit similar ^{14}C activity, if grown in the same year. On a global scale, however, because ^{14}C is produced in the upper atmosphere by interaction with cosmic radiation, which is moderated less by our atmosphere at the earth's poles and higher altitudes, slightly higher ^{14}C activity is found with plants grown at higher latitudes and altitudes.

ATMOSPHERIC ^{14}C ACTIVITY

The global ^{14}C activity has remained relatively stable, over the past tens of thousands of years, based on equilibrium between production and decay. This is at least true within the practical radiocarbon dating range for ^{14}C of approximately 50,000 years. This is based on ^{14}C half-life of 5730 years and 10 half-lives of decay. This stability over time is a major assumption in the field of radiocarbon dating, although a necessary one, to fix the ^{14}C activity at an organism's death. However, there is evidence for excursions from this steady state. Two major perturbations to this steady state have and are occurring currently. First is the nuclear bomb testing effect and second is the combustion of fossil fuels for power generation. During the late 1950s and early 1960s, the aboveground nuclear bomb testing emitted large quantities of ^{14}C and ^{3}H into the atmosphere, as well as many other radioactive isotopes. The level of atmospheric ^{14}C nearly doubled from pre-bomb-testing levels. With the signing of the nuclear test ban treaty, fewer aboveground tests were conducted, and the levels of ^{14}C, present in both northern and southern hemispheres, began to decrease. Through the uptake of CO_2 through photosynthesis and absorption by the world's oceans, the atmospheric ^{14}C activity has been decreasing ever since. Although this spike in the radiocarbon activity has made dating historic artifacts within this time period difficult, this bomb effect has been useful for radiochemists to conclude a so-called vintage age to materials produced from biomass grown during the last 50 years [14–16]. The quantity (annual yield) of nuclear bomb blasts conducted since 1945 is shown in Figure 9.2. The consequential effect to the atmospheric ^{14}C activity since 1940 is shown in Figure 9.3. The second major perturbation is due to the combustion of fossil fuel for power generation. This results in a dilution of atmospheric $^{14}CO_2$ with fossil fuel-derived CO_2, because it contains no $^{14}CO_2$.

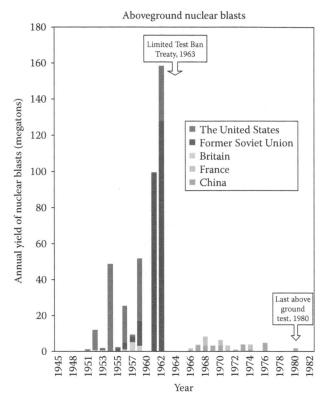

FIGURE 9.2   Annual yield of nuclear blasts versus year. (Data from US EPA.)

This effect reduces the natural level of $^{14}CO_2$ in the atmosphere and changes the ratio of ^{14}C to the heavy and light stable isotopes ^{13}C and ^{12}C [17], also known as the Suess effect named after Hans Suess, the Austrian chemist who noted its effect on radiocarbon dating.

These two modern processes have created a reduction in ^{14}C over the last few decades that must be accounted for in calculating the percentage of modern plant derivation in foods, flavors, and bio-based materials. Because most bio-based products are harvested annually, their respective ^{14}C activity will decrease from previous years to current harvests. Studies have been made by the Center for Applied Isotope Studies to define the ^{14}C activity levels for various products of high biomass concentration such as corn and peanut oils, but also on various flavor compounds of smaller manufactured quantities. Nonetheless, they equally represent the ^{14}C activity for a specific year of harvest. One such study revealed very similar ^{14}C activity levels with decreasing levels over more than 20 years for 10 compounds of interest to the flavor industry. Because other products, such as wine, vinegar, even ethanol for fuel modification, all show similar patterns and ^{14}C activity levels, a reference ^{14}C activity can be derived for use in confirming vintage products and modern harvested materials alike. Figure 9.4 indicates the decreasing ^{14}C activity levels in plant-based products since 1987.

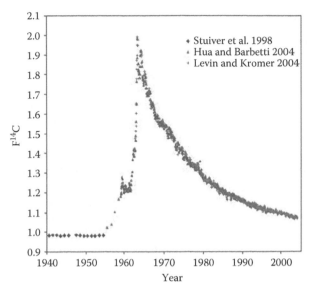

FIGURE 9.3 ^{14}C activity versus year (units F[14]C, fraction modern defined as 1950 ^{14}C activity = 1.0). (Data from M. Stuiver, P.J. Reimer, and T.F. Braziunas, High-precision radiocarbon age calibration for terrestrial and marine samples. *Radiocarbon* 1998, 40: 1127–1151; I. Levin and B. Kromer, The tropospheric 14CO$_2$ level in mid-latitudes of the northern hemisphere (1959–2003). *Radiocarbon* 2004, 46: 1261–1272; Q. Hua and M. Barbetti, Review of Tropospheric bomb 14C data for carbon cycling modeling and age calibration purposes. *Radiocarbon* 2004, 46: 1273–1298.)

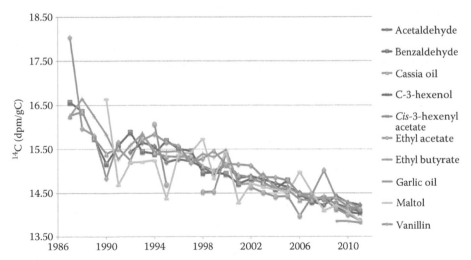

FIGURE 9.4 ^{14}C activity versus year, units in disintegrations per minute per gram carbon (dpm/gC). (Data from R.A. Culp and J.E. Noakes. Two decades of flavor analysis: trends revealed by radiocarbon [14C] and stable isotope [δ13C and δD] analysis. In C.T. Ho, C.J. Mussinan, F. Shahidi, and E.T. Contis, Eds., *Recent Advances in Food and Flavor Chemistry*. Royal Society Chemistry, Cambridge. 2010, pp. 9–27.)

STABLE ISOTOPES

Stable isotopes, those not undergoing radioactive decay, are typically measured and reported as a ratio of the less abundant minor isotope relative to more abundant major isotope, for instance, $^{13}C/^{12}C$ have been used to study processes from discriminating mantle-derived carbon from biogenic carbon [18], fossil-fuel origins [19] to more terrestrial processes such as carbon uptake by the ocean [20] and biomass and most significant photosynthetic processes [21]. Other stable isotope pairs, including hydrogen ($^{2}H/^{1}H$) and oxygen ($^{18}O/^{16}O$), are used to delineate hydrological cycles, geographical provenance through source water determination [22], and paleothermometry [23].

Stable isotopes of the lighter elements are also used in tandem with the ^{14}C content of products to determine their degree of authenticity. Isotopes of the same element differ in the number of neutrons in the nucleus. Existing in much lower abundance, the heavy stable isotope ^{13}C represents approximately 1.11% of all carbon, while ^{12}C makes up nearly 98.89% of all carbon, remembering that ^{14}C makes up a miniscule 10^{-10}% [24]. Their abundance is reflected by their relative stability, and while not undergoing any radioactive decay, the additional neutron in the ^{13}C isotope creates enough mass difference to affect kinetic and equilibrium effects through various physical and chemical processes. These isotope effects or fractionations are the basis for authenticating sources and processes of plant-based products. One of the most common isotope effects, and one used frequently for authentication of plants precursors, is through three unique photosynthetic processes used by plants. Through photosynthesis, plants selectively become enriched in ^{12}C because it requires less energy to acquire the lighter isotope than the heavier ^{13}C. However, depending on atmospheric CO_2 concentration and other factors, plants will use either the Calvin, also known as the C-3, process or the Hatch–Slack, also known as C-4, process by which to fix the available carbon. The C-3 and C-4 represent the 3- and 4-carbon intermediates formed during photosynthesis during these two processes. [25,26]. A third process, or a combination of C-3 and C-4, is referred to as the Crassulecean acid metabolism or CAM pathway for short. The C-3 photosynthetic process fractionates the heavier isotope ^{13}C more than the C-4 pathway. It is common to flowering plants such as apples, grapes, and citrus and grains such as rice, wheat, and barley. In contrast, C-4 plants, which fractionate ^{13}C less, are common to grasses, corn, and cane sugar. Therefore, discrimination between these pathways is useful for determining the addition of corn- or cane-derived sugars to those products reported to be free of such precursors. Figure 9.5 illustrates the range of stable isotope ratios for various classes of carbon containing materials [27]. In addition to the stable isotopes of carbon, the measurement of the stable isotopes of hydrogen is useful in determining the geographical and provenance information useful for authenticating natural materials [28].

Analogous to the three carbon isotopes, hydrogen exists as radioactive tritium (^{3}H or T), and as stable isotopes deuterium (^{2}H or D) and hydrogen (^{1}H). However, unlike photosynthetic pathway dependent, this stable isotope pair is dependent on the hydrologic cycle and the mechanism of evapotranspiration. During evaporation, the lighter isotope or molecule of water evaporates preferentially to the heavier molecule. Conversely, the heavier molecule of water will condense preferentially to the lighter molecule. The result is water masses become progressively lighter in isotopic

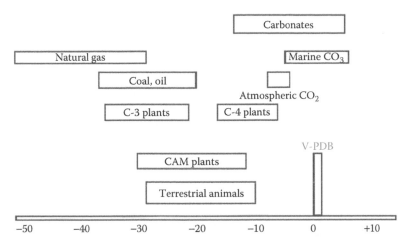

FIGURE 9.5 δ^{13}C value (‰ versus V-PDB). V-PDB, Vienna-PeeDee Belemnite. (Data from F.J. Winkler, Application of natural abundance stable isotope mass spectrometry in food control, In A. Frigerio and H. Milon, Eds., *Chromatography and Mass Spectrometry in Nutrition Science and Food Safety*, Elsevier, Amsterdam, the Netherlands, 1983.)

composition as they move over a landmass. Condensation and resulting groundwater therefore has geographically unique ratios that can be used to assist in product authentication or determining if a process is equivalent to that found in nature. Although the abundance is small for deuterium at 151 parts per million and large for hydrogen at 99.99%, the difference in mass between DH and H_2 results in significant fractionation factors and large deviations in nature [25]. Similar to isotopic ranges for carbon, Figure 9.6 illustrates hydrogen stable isotope ranges of interest [27]. As important for the authentication of bio-based products, but less frequently used, are the stable isotope ratios of oxygen ($^{18}O/^{16}O$) and nitrogen ($^{15}N/^{14}N$) and sulfur ($^{34}S/^{32}S$) for those compounds containing such heteroatoms [29]. Figure 9.7 illustrates the isotope ratio mass spectrometer (IRMS) diagrammatically measuring CO_2. By combining stable isotope ratios of as many elements as possible, along with ^{14}C activity, one is able to create a multidimensional isotopic display of an organic compound, making the possible manipulation of isotopes for the purpose of imitating the real material, a much more daunting task.

INCREASING SPECIFICITY IN ISOTOPIC MEASUREMENT

An advantage of the higher concentrations of stable isotopes, relative to the radioactive isotopes ^{14}C and ^3H, is that within a plant, biosynthesis of different compounds can actually impart a fractionation of the isotopes. This fractionation, although increasing the complexity of interpretation, can reveal valuable information about processes and source materials [30–32]. Table 9.1 illustrates the stable isotope ratio of carbon for a number of compound classes found in plants and their respective range of values from depleted to enriched in ^{13}C relative to the total organic matter value [33]. Separation of compound classes, or even specific compounds adds another dimension to the characterization of materials using isotopic analyses. Compound classes such

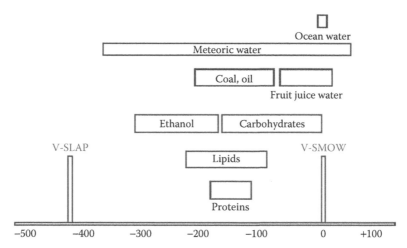

FIGURE 9.6 δD value (‰ versus V-SMOW). V-SMOW, Vienna-Standard Mean Ocean Water. (Data from F.J. Winkler, Application of natural abundance stable isotope mass spectrometry in food control, In A. Frigerio and H. Milon, Eds., *Chromatography and Mass Spectrometry in Nutrition Science and Food Safety*, Elsevier, Amsterdam, the Netherlands, 1983.)

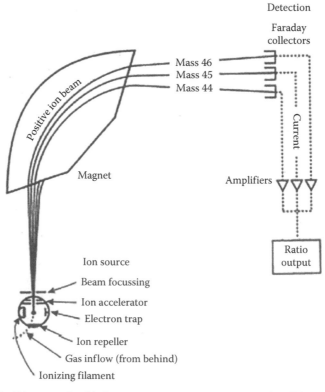

FIGURE 9.7 Diagram of stable isotope mass spectrometer measuring CO_2.

TABLE 9.1

δ^{13}C Value (‰ versus V-PDB) Relative to Organic Compound Class

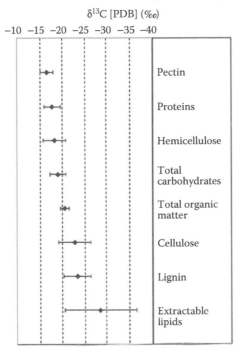

δ^{13}C [PDB] (‰)

Source: E.T. Degens et al., *Deep Sea Res.*, 15, 11–20, 1968.

as proteins, amino acids, and extractable lipids vary in their resistance to degradation or oxidation. Therefore, they can be useful proxies for the chemical processes they undergo whether man-made or natural. The isotopic signature of specific compounds is even more powerful, especially when the potential rises for economically driven fraud in manufacturing food and flavor chemicals of interest on the molecular scale. The use of compound-specific stable isotope analysis is accomplished using sophisticated instrumentation that combines the techniques of gas chromatography (GC) or high-performance liquid chromatography (HPLC) with an IRMS [34–36]. Unlike conventional GCs, the effluent goes into a combustion interface where organic molecules are combusted to CO_2 and enter the IRMS rather than conventional quadrapole mass spectrometer. Figure 9.8 illustrates the principle, and complexity, of this technique diagrammatically. Because so many flavors are complex mixtures of organic compounds, some in major concentration while others at accessory or even trace amounts, the ability to single out these biomarkers of interest for authenticity testing makes the task of adulteration an order of magnitude more difficult to go undetected. In the past, a major component such as benzaldehyde could be manipulated to mimic a natural material, and the flavor would go undetected. Today, accessory compounds such as benzoic acid or parahydroxy-benzaldehyde can reveal adulteration because they reflect a synthetic

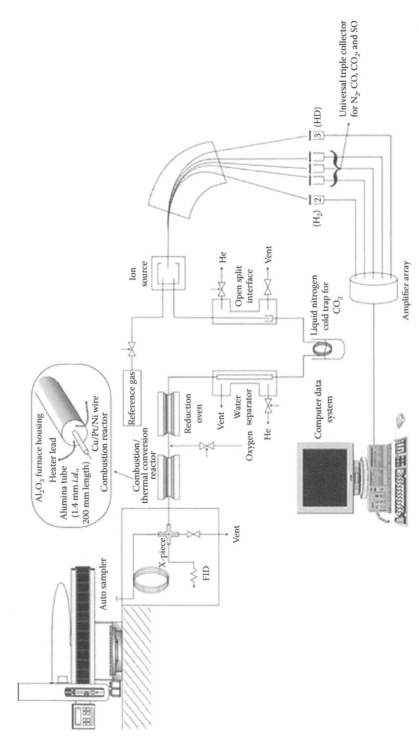

FIGURE 9.8 Diagram of gas chromatograph/combustion reactor/IRMS for compound-specific isotope measurement. (Courtesy of University of Bristol.)

precursor, previously undetected. This new level of specificity in isotopic measurement greatly compliments the conventional GC/MS capability of identification and quantitation of organic compounds in foods and flavors. In addition, another dimension is possible with the other isotope pairs available for measurement in select organic molecules. Nearly, all are hydrocarbons, implying hydrogen-containing molecules (D/H), while most contain oxygen ($^{18}O/^{16}O$) with fewer containing nitrogen ($^{15}N/^{14}N$) and sulfur ($^{34}S/^{32}S$). Each of these isotope ratios can be determined on the same selected compounds of interest where concentration permits, usually at the nanogram level. Multidimensional chemometric analysis can be applied and reveal authenticity with greater accuracy than previously made with bulk isotopic analyses.

Compound-specific analysis is not just limited to the determination of stable isotopes in food and flavor testing. Compound-specific ^{14}C is now used for the same enhanced specificity applied to authenticating accessory and trace components in foods and flavors. Similar to the detection of fossil fuel-derived products in bulk, the absence of ^{14}C in any component in a flavor mixture will indicate a fossil fuel derivation and likely adulteration of the product using ^{14}C labeled compounds. Even when the prominent compound such as cinnamic aldehyde in cassia oil exhibits a plant-based ^{14}C activity, any other component within the mixture displaying the lack of ^{14}C activity will imply a fossil fuel precursor or synthetic process of manufacture. Unlike the GC/IRMS technique described above, one must use a preparative fraction collection system, due to the larger sample requirements of accelerator mass spectrometry (AMS) measurement, to collect enough material for AMS processing. Unlike GC/IRMS requiring only nanogram levels of material, AMS requires a minimum of tens to hundreds of micrograms of carbon. However, the resulting measurement of ^{14}C activity of individual molecular compounds is an extremely powerful tool, although labor intensive.

BIO-BASED MATERIAL VERIFICATION

A new application for the use of ^{14}C determination began in 2004 with the creation of the preferred procurement program. The 2002 Farm Security and Rural Investment Act authorized the development of the United States Federal Biobased Products Preferred Procurement Program (FB4P) requiring preference to the purchase of bio-based products instead of their fossil fuel-based equivalent. From this mandate, the US Department of Agriculture along with Iowa State University begin development of the FB4P program. As part of the development, the Center was asked to help develop methodology to accomplish the goal of confirming natural botanically derived materials from those derived from fossil fuel. Because of our expertise in the field of ^{14}C and stable isotopes, the program was able to come to fruition quickly and successfully. Out of the collaborative effort came the American Society for Testing Materials (ASTM) method D6686, for the determination of bio-based content. Originally, the method used three techniques for measuring the ^{14}C content in products [37]. Method A used an absorbing material (carbosorb) to capture the CO_2 from combustion of a product followed by liquid scintillation counting (LSC). LSC uses specialized chemicals to convert the decay particle energy (beta particle) that occurs when ^{14}C decays to ^{14}N, into light, which is detected by ultrasensitive photomultiplier tubes within the scintillation counter. Method B makes use of

AMS techniques where samples are converted from the combustion product CO_2 into graphite for direct measurement of their ^{14}C content. Very small samples are measured which allows only milligrams of sample to be collected [38,39]. However, heterogeneous samples can present problems due to small sampling size as well as higher potential for contamination. Method C also makes use of LSC but all combustion CO_2 is converted via synthesis into the common solvent benzene. Again, specialized chemicals are added to this benzene, which represents the unknown sample, and it is counted in a scintillation counter. Much higher precision and accuracy is achieved by this method over Method A. In light of the poor performance of Method A, the ASTM has now removed this particular part entirely from ASTM D6866 [40]. A comparison of bio-based measurements by all three methods in ASTM D6866 was made by the Center for Applied isotope Studies and others [41,42].

Details of the methods can be found on the ASTM website with modifications and updates [43,44]. An important update is the reference ^{14}C activity, which is used to determine the 100% bio-based reference. Because of the decreasing atmospheric ^{14}C activity, annual reference ^{14}C activity has been provided by the Center and incorporated into the ASTM method periodically. This adjustment of the reference ^{14}C activity enables reports generated by different laboratories to use the same ^{14}C reference activity for proper comparison.

Unlike food and flavor authenticity, bio-based products are typically nonconsumables and common household or business products. These can range from paints, adhesives, cleaners, aprons, gloves, plastic utensils, tiles, and carpeting. An important commodity tested for bio-based content is fuels and fuel additives. As part of the 2007 Energy Independence and Security Act, the United States requires the use of 36 billion gallons of renewable fuel by 2022 of which 16 billion gallons of this requirement will be derived from cellulosic biofuel material, primarily corn. Ethanol derived from corn is currently added into nearly 90% of all gasoline at a level of 10% with some additions as high as 85% for certain flexi-fuel vehicles. The ASTM D6866-12 method is used for the confirmation of corn, or other plant-derived ethanol, based on the presence of ^{14}C in this fuel. Proper addition of anti-knock ingredients, of which some are partly or wholly bio-based in composition, is another important application of the ASTM method [45,46].

In some cases, it is possible to verify the addition of corn-based ethanol, a C-4 plant material using only stable isotope ratio measurement of $^{13}C/^{12}C$, provided no other source of renewable bio-fuel is used other than that derived from corn. This is because of the unique stable isotope signature for corn, which is considerably distinct from other C-3 plants and fossil fuels. Corn's stable isotope signature is unique in this regard as even a renewable source of ethanol from a C-3 plant such as flaxseed could not be differentiated from fossil fuel $^{13}C/^{12}C$ and thus not be used for authentication by stable isotopes alone. The ASTM D6866-12 method, using ^{14}C determination, eliminates the need for constraining the source of ethanol and is an unambiguous determinant of fossil fuel-derived carbon. A recent study by the Center found a number of reportedly pure, ethanol-free gasoline samples, an important fuel for older engines and especially marine outboard motors, had the typical 10% addition of corn-based ethanol, based on ^{14}C and $^{13}C/^{12}C$ measurements, although they were advertised to be free of added ethanol [47].

METHODS: [14]C DETERMINATIONS BY AMS AND LSC

The AMS method measures ^{14}C activity by direct counting of ^{14}C after separation of interfering atoms. Mass spectrometers accomplish this by separation of ions based on mass and charge. The advantage of an accelerator is removal of interfering molecules that simulate the proper mass and charge but are not the desired ion. An example of the accelerator's capability is the removal of the [13]CH molecule so not to interfere with [14]C. AMS requires only milligram or less quantities of sample so heterogeneity of samples is a problem. Samples are oxidized in either a high temperature furnace with suitable oxidant after evacuation, or using liquid oxidizer such as $KMnO_4$, or acidified using suitable acid to release CO_2. The CO_2 is cryogenically trapped while impurity gases are removed using high-vacuum technology. The purified CO_2 is converted to graphite using an iron catalyst and excess hydrogen. The resulting graphite is pressed into a target holder for placement into the AMS wheel along with standards and background material for calibration and background correction. A suite of unknowns, typically 10–15, is accompanied by 4–6 standards and 2 background graphite samples to complete each group within the target wheel. Typical precision is better than 0.35% using National Institute of Standards and Technology (NIST) standard reference materials. The University of Georgia's Center for Applied Isotope Studies maintains a 500 KV tandem Pelletron accelerator and a 250 KV single source accelerator.

The LSC method measures ^{14}C activity by measuring the decay of ^{14}C to ^{14}N and the release of a high-energy electron or beta particle. As the name implies, samples are converted to a liquid for counting, in this case, one that is advantageous for counting due to high electron and photon transmission and solvating ability. Through benzene synthesis, samples, whether flavors, fuels, or bones, are converted to benzene that acts as sample and solvent for the added scintillator. The decay and beta particle can be seen when it interacts with the fluorescing scintillator and detected with ultra-sensitive photomultiplier tubes. In contrast to AMS samples, LSC requires up to 5 g of carbon greatly reducing the potential for contamination during processing and enabling measurement of heterogeneous samples. Samples are oxidized similarly to AMS in either a high-temperature furnace with suitable oxidant after evacuation, usually pure oxygen, or using liquid oxidizer such as KMnO4, or acidified using suitable acid to release CO_2. The CO_2 is cryogenically trapped while impurity gases are removed using high-vacuum technology. The purified CO_2 is converted to Li_2C_2 and eventually to benzene. NIST standard reference materials and fossil fuel-derived materials are used as calibration standards and background materials, respectively. The Center for Applied Isotope Studies maintains seven Perkin–Elmer and three Quantulus liquid scintillation counters.

STABLE ISOTOPE RATIO ANALYSIS

Stable isotope ratio analysis of carbon ($^{13}C/^{12}C$), oxygen ($^{18}O/^{16}O$), nitrogen ($^{15}N/^{14}N$), and hydrogen (D/H) is measured on produced CO_2, oxygen or carbon monoxide, nitrogen, or hydrogen, respectively, after suitable processing to remove impurity gases. Samples are combusted with suitable oxidizer to form CO_2 and NO_x that is reduced to N_2 for measurement, acidified to release CO_2 for carbonaceous materials or pyrolyzed under high-temperature reduction for CO and H_2 measurement. Calibration is made using NIST,

International Atomic Energy Agency (IAEA), and United States Geological Survey (USGS) stable isotope reference materials and run against a working standard common to each stable isotope laboratory. Results are reported relative to international standards in parts per mil using delta notation (δ) and the heavy isotope such as $\delta^{13}C$ to represent $^{13}C/^{12}C$. The Center for Applied Isotope Studies maintains four stable IRMS, Thermo 253, Delta V plus, Delta XL plus, and 252 capable of running in dual-inlet or continuous flow (He) modes when interfaced to a gas chromatograph or elemental analyzer.

ISOTOPIC CASES OF INTEREST

To demonstrate the utility of isotopic measurements for authenticity testing and confirmation of botanical or bio-based precursors, a few examples of studies developed and completed at the Center for Applied Isotope Studies are listed below.

1. *Anomalous deuterium/hydrogen ratios:* Late in the 1980s, there was concern about the naturalness of cherry and almond extracts and their primary flavoring compound bitter almond oil. Chemically identical to benzaldehyde, bitter almond oil can be derived from the kernels of cherries and almonds. This is not only a laborious extraction process, but involves the formation of amygdalin glucoside with hydrocyanic acid, a rather potent poison. Care must be made in removing this material prior to manufacturing flavors or addition to foods. This high cost of processing may have prompted the use of alternative procedures in manufacturing benzaldehyde. Eventually, less expensive routes of manufacturing, albeit synthetic, were adopted. However, the high price demanding by natural almond oil was enough of an incentive to encourage manipulation of the synthetic materials to mimic natural isotopic signatures. One such process, the catalytic oxidation of toluene, a fossil fuel precursor, was simple enough to detect by its lack of ^{14}C activity. However, when ^{14}C-labeled toluene was added in the correct proportion, which was no simple task, the ^{14}C activity resembled that of a recently grown botanical, such as cherries or almonds. Although measurements revealed some poorly computed additions of ^{14}C to mimic natural levels, leaving some samples with 10 times the expected ^{14}C activity, a clear indication of adulteration with radiolabeled material. However, it was the anomalous D/H ratio that revealed the nefarious manipulations. As a result of the catalytic process of converting toluene to benzaldehyde, a higher ratio of deuterium to hydrogen followed [7,8]. To this day, anomalously high ratios of D/H can indicate catalytic processes not only in benzaldehyde production but other flavors as well.

2. *Vanillin derived from pine trees:* Although many routes of manufacture of vanillin are used today, only authentic vanilla bean-derived vanillin is considered natural by US Food and Drug Administration. Common synthetic forms of vanillin include those made from guaiacol, a pine–tar product; eugenol, found in clove oil; fossil fuel-derived vanillin; and lignin-derived vanillin. The later, lignin is the structural component of plants and found in the terrestrial environment. Woody materials have been used to impart a vanilla aroma to products such as whiskey and wine, since vanillin is one of

many monomers found in wood lignin once the polymeric material is broken down. Although all of these synthetic vanillin precursors, because they are plant derived, except for fossil fuel, have natural ^{14}C activities, lignin-derived vanillin is unique. Lignin vanillin is derived primarily from trees, which take up their ^{14}C over an extended period of time. This fact has proven extremely useful in the study called dendrochronology, the dating of tree-rings, where each annual growth ring reveals the particular year's atmospheric ^{14}C activity level. For recently grown, young trees, the composite ^{14}C activity from center to outside bark may resemble today's ^{14}C activity. However, more mature, larger pulp wood trees where vanillin is commonly derived live much longer and will resemble ^{14}C activities extending further into the past. Old growth trees, those 40 years or older, will indicate the higher but decreasing ^{14}C activity residual from nuclear bomb testing. Therefore, vanillin derived from this source exhibits a much higher ^{14}C activity than other vanillin sources and along with a characteristic stable isotope ratio of carbon and hydrogen is now well described [48]. Initially, the possibility of ^{14}C-labeled material, not out of the realm of possibilities, was suggested for this unknown source until the testing of some pulp wood lignin revealed these unique isotopic signatures.

3. *Bio-based products for fuel production:* Recent modifications to fuel additives and anti-knock ingredients have prompted manufacturers to consider bio-based or renewable materials as fuel additives. ETBE, or ethylene tertiary-butyl ether, has replaced the lead additions to gasoline to prevent engine knocking due to improper combustion for nearly 30 years. ETBE is produced from the addition of ethanol with isobutylene. The ethanol is typically the bio-component in the mixture while isobutylene is fossil fuel derived. This makes the determination of the bio-based additives more difficult when it is not wholly derived from modern botanical precursors. Not only can the proportion of the components making up the ETBE vary, giving inconsistent ^{14}C activities, but also the ETBE concentration within the fuel itself may vary with additions as high as 15% by volume. By using ^{14}C activity calibration curves and calculated carbon concentrations [42], derived from elemental analyses, along with end-member determination of pure ETBE and pure gasoline, the correct proportion of bio-produced ETBE additions could be computed within 2.3% [46]. Both ASTM D6866 methods, AMS and LSC, proved accurate for the proper apportionment of bio-based additives for not only ETBE additives but also ethanol concentrations up to 85% as found in the E85 fuels now on the market. The use of both ^{14}C and the ratios of $^{13}C/^{12}C$ has been instrumental in the accurate measurement of biofuel additions, verification of blending with fossil fuels, and demonstration of the viability and environmental benefits of renewable fuel sources.

CONCLUSION

The use of stable isotopes and radiocarbon for authentication of foods, flavors, bio-based products, and fuels has now become a well-accepted method in analytical chemistry. It works well independently, but even better when coupled with sophisticated analytical techniques such as GC and HPLC. Development of databases of specific compounds and flavoring materials has helped quality control chemists and production managers ensure their products purity, uniformity, and above all integrity as to its source and process of manufacture [49]. The improvement of isotopic resolution, especially in ^{14}C measurement, along with long-term statistics from known sources of materials allows accurate confirmation with more confidence when additions or dilutions with synthetic components have been made. As the number of isotopic tests increase, either from new or improved methods of existing stable isotope ratio determinations, or from characterizing the complex mixture of compounds through GC or HPLC separations, followed by isotopic analysis, the potential for detection of fraudulent material increases dramatically.

ACKNOWLEDGMENTS

The development of isotopic methods and implementation is by no means a simple task performed by a single individual. I acknowledge the hardworking researchers, technicians, and staff at the University of Georgia's Center for Applied Isotope Studies and hope they are content and pleased with the important function they provide the industry.

REFERENCES

1. G. Faure and T.M. Mensing. *Isotopes: Principles and Applications.* 3rd edn., 2005. Wiley, Hoboken, NJ.
2. M.P. Neary, J.D. Spaulding, J.E. Noakes, and R.A. Culp. Tritium analysis of burn-derived water from natural and petroleum-derived products. *J. Agric. Food Chem.* 1997, 45: 2153–2157.
3. W.F. Libby. Atmospheric helium-three and radiocarbon from cosmic radiation. *Phys. Rev.* 1946, 69: 671–672.
4. H. Craig and D. Lal. The production rate of natural tritium. 1961. *Tellus.* 13(1): 85–105.
5. P.G. Hoffman and M. Salb. Radiocarbon (14C) method for authenticating natural cinnamic aldehyde. *J. Assoc. Off. Anal. Chem.* 1980, 63: 1181–1183.
6. B. Byrne, K.J. Wengenroth, and D.A. Krueger. Determination of adulterated natural ethyl butyrate by carbon isotopes. *J. Agric. Food Chem.* 1986, 34: 736–738.
7. M. Butzenlechner, A. Rossmann, and H.L. Schmidt. Assignment of bitter almond oil to natural and synthetic sources by stable isotope ratio analyses. *J. Agric. Food Chem.* 1989, 37: 410.
8. R.A. Culp and J.E. Noakes. Identification of isotopically manipulated cinnamic aldehyde and benzaldehyde. *J. Agric. Food Chem.* 1990, 38: 1249–1255.
9. W.F. Libby. *Radiocarbon Dating.* 1952. University of Chicago Press, Chicago, IL.
10. W.F. Libby. *Radiocarbon Dating.* 1955. 2nd edn. University of Chicago Press, Chicago, IL.
11. R.E.M. Hedges, J.A. Lee-Thorp, and N.C. Tuross. Is tooth enamel carbonate a suitable material for radiocarbon dating. *Radiocarbon* 1995, 37(2): 285–290.
12. B.A. Buchholz and K.L. Spalding. Year of birth determination using radiocarbon dating of dental enamel. *Surf. Interface Anal.* 2009. LLNL-JRNL-411266.
13. K.T. Uno, J. Quade, D.C. Fisher, G. Wittemeyer, I. Douglas-Hamilton, S. Andanje, P. Omondi, M. Litoroh, and T.E. Cerling. Bomb-curve radiocarbon measurement of recent biologic tissues and applications to wildlife forensics and stable isotope (paleo)ecology. 2013. www.pnas.org/cgi/doi/10.1073/pnas.1302226110.
14. R. Nydal and K. Lovseth. Tracing Bomb 14C in the atmosphere 1962–1980. *J. Geophys. Res.* 1983, 88(C6): 3621–3642.

15. K. Dai and C.Y. Fan. Bomb produced 14C content in tree rings grown at different latitudes. *Radiocarbon* 1986, 28: 346–349.

16. Q. Hua, M. Barbetti, and A.Z. Rakowski. Atmospheric radiocarbon for the period 1950–2010. *Radiocarbon* 2013, 55(4): 2059–2072.

17. H.E. Suess. Radiocarbon concentration in modern wood. *Science* 1955, 122: 415–417.

18. J.M. Hayes, I.R. Kaplan, and K.W. Wedeking. Precambrian organic chemistry, preservation of the record. In J.W. Schopf (Ed.), *Origin and Evolution of Earth's Earliest Biosphere*. 1983. pp. 93–134. Princeton University Press, Princeton, NJ.

19. Z. Sofer. Stable carbon isotope compositions of crude oil: applications to source depositional environments and petroleum alteration. *Am. Assoc. Petrol. Geol. Bull.* 1984, 68: pp. 31–49.

20. N.A. Beveridge and N.J. Shackleton. Carbon isotopes in recent foraminifera: A record of anthropogenic CO_2 invasion of the surface ocean. *Earth Planet. Sci. Lett.* 1994, 126: 259–273.

21. H. Craig. The geochemistry of stable carbon isotopes. *Geochem. Cosmochim. Acta.* 1953, 3: 53–93.

22. H. Craig. Isotopic variations in meteoric waters. *Science* 1961, 133: 1702–1703.

23. P.M. Grootes, M. Stuiver, J.W.C. White, S.J. Johnsen, and J. Jouzel. Comparison of oxygen isotope records from the GISP-2 and GRIP Greenland ice cores. *Nature* 1993, 336: 552–554.

24. G. Faure. *Principles of Isotope Geology*. 1st edn. 1977. Wiley, New York.

25. M. Calvin and J.A. Bassham. *The Photosynthesis of Carbon Compounds*. 1962. Benjamin: New York.

26. B.N. Smith and S. Epstein. Two categories of $^{13}C/^{12}C$ ratios for higher plants. *Plant Physiol.* 1971, 47: 380–384.

27. F.J. Winkler. Application of natural abundance stable isotope mass spectrometry in food control. In A. Frigerio and H. Milon (Eds.), *Chromatography and Mass Spectrometry in Nutrition Science and Food Safety*. 1983. Elsevier, Amsterdam, the Netherlands.

28. R.A. Culp and J.E. Noakes. Determination of synthetic components in flavors by deuterium/hydrogen isotopic ratios. *J. Agric. Food Chem.* 1992, 40: 1892–1897.

29. L.W. Doner, O.A. Henry, L.S.L. Sternberg, J.M. Milburn, M.J. DeNiro, and K.B. Hicks. Detecting sugar beet syrups in orange juice by D/H and $^{18}O/^{16}O$ analysis of sucrose. *J. Agric. Food Chem.* 1987, 35: 610–612.

30. J.M. Hayes. Fractionation of carbon and hydrogen isotopes in biosynthetic process. In J.W. Valley and D.R. Cole (Eds.), *Stable Isotope Geochemistry*. 2001, pp. 225–279. Mineralogy Society of America, Blacksburg, VA.

31. P.G. Hoffman and M. Salb. Isolation and stable isotope ratio analysis of vanillin. *J. Agric. Food Chem.* 1979, 27: 352–355.

32. R. Benner, M.L. Fogel, E.K. Sprague, and R.E. Hodson. Depletion of 13C in lignin and its implications for stable carbon isotope studies. *Nature* 1987, 329: 708–710.

33. E.T. Degens, M. Behrendt, B. Gotthardt, and E. Reppmann. Metabolic fractionation of carbon isotopes in marine plankton. II. Data on samples collected off the coasts of Peru and Ecuador. *Deep Sea Res.* 1968, 15: 11–20.

34. R.A. Culp, J.M. Legato, and E. Otero. Carbon isotope composition of selected flavoring compounds for the determination of natural origin by gas chromatography/isotope ratio mass spectrometer. In C. Mussinan and M. Morello (Eds.), *Flavor Analysis: Developments in Isolation and Characterization*. 1997, pp. 260–287. American Chemical Society, Washington, DC.

35. R.F. Dias, K.H. Freeman, and S.G. Franks. Gas chromatography-pyrolysis-isotope ratio mass spectrometry: A new method for investigating intramolecular isotopic variation in low molecular weight organic acids. *Org. Chem.* 2002, 33: 161–168.

36. V.W. Ebongue, B. Geypens, M. Berglund, and P. Taylor. Headspace solid phase micro-extraction-GC/C-IRMS δ13C VPDB measurements of mono-aromatic hydrocarbons using EA-IRMS calibration. *Isotopes Environ. Health Stud.* 2009, 45: 53–58.

37. ASTM. Standard test methods for determining the biobased content of natural range materials using radiocarbon and isotope ratio mass spectrometry analysis. Designation D 6866-06a, ASTM International, West Conshohocken, PA. 2006.

38. A. Cherkinsky, R. Culp, D. Dvoracek, and J. Noakes. Status of the AMS facility at the University of Georgia. *Nucl. Instrum. Methods Phys. Res. B* 2010, 268: 867–870.

39. G.V.R. Prasad, J. Noakes, A. Cherkinsky, R. Culp, and D. Dvoracek. The new 250KV single stage AMS system at CAIS, University of Georgia: Performance comparison with 500KV compact tandem machine. *Radiocarbon* 2013, 55(2/3): 319–324.

40. ASTM. Standard test methods for determining the biobased content of natural range materials using radiocarbon and isotope ratio mass spectrometry analysis. Designation D6866-12, ASTM International, West Conshohocken, PA. 2012.

41. J.E. Noakes, G.A. Norton, R.A. Culp, M. Nigam, and D. Dvoracek. A comparison of analytical methods for the certification of biobased products. In *Advances in Liquid Scintillation Counting, Proceedings of LSC*, Katowice, Poland. S. Chalupnik, F. Schonhofer, J. Noakes (Eds.), Radiocarbon: Tucson, AZ. 2005, pp. 259–271.

42. I.J. Dijs, E. van der Windt, L. Kaihola, and K. van der Borg. Quantitative determination by ^{14}C analysis of the biological component in fuels. *Radiocarbon* 2006, 48(3): 315–323.

43. R.A. Culp and J.E. Noakes. Evaluation of biobased content ASTM method 6866-06A: Improvements revealed by liquid scintillation counting, accelerator mass spectrometry, and stable isotopes for products containing inorganic carbon. In *Advances in Liquid Scintillation Counting, Proceedings of LSC*, Davos, Switzerland. J. Eikenberg, M. Jaggi, H. Beer, and H. Baehrle (Eds.), Radiocarbon: Tucson, AZ. 2008, pp. 269–278.

44. R.A. Culp, J.E. Noakes, D.R. Smith, L. Greenway, F.S. Smith, and B.K. Markowicz. Bio-base product determination by ASTM method D6866-08 using liquid scintillation counting and benzene synthesis: Progress and performance in an expanding market. In *Advances in Liquid Scintillation Counting, Proceedings of LSC*, Paris, France. P. Cassette (Ed.), Radiocarbon: Tucson, AZ. 2010, pp. 23–34.

45. R. Edler. The use of liquid scintillation counting technology for the determination of biogenic materials. In *Advances in Liquid Scintillation Counting, Proceedings of LSC*, Davos, Switzerland. J. Eikenberg, M. Jaggi, H. Beer, and H. Baehrle (Eds.), Radiocarbon: Tucson, AZ. 2008, pp. 261–268.

46. J.E. Noakes, A. Cherkinsky, and R.A. Culp. Comparative radiocarbon analysis for artificial mixes of petroleum and biobased products. In *Advances in Liquid Scintillation Counting, Proceedings of LSC, Paris, France*. P. Cassette (Ed.), Radiocarbon: Tucson, AZ. 2010, pp. 15–21.

47. R. Culp, A. Cherkinsky, and G.V. R. Prasad. Comparison of radiocarbon techniques for the assessment of biobase content in fuels. In LSC2013: Advances in Liquid Scintillation Counting, Barcelona, Spain, *Appl. Rad. Isot.* 2014, 93: 106–109.

48. R. Culp and G.V. R. Prasad. Present-day radiocarbon content of select flavoring compounds reveals vanillin production pathway. *Radiocarbon* 2013, 55(2/3): 1819–1826.

49. R.A. Culp and J.E. Noakes. Two decades of flavor analysis: trends revealed by radiocarbon (14C) and stable isotope (δ13C and δD) analysis. In C.T. Ho, C.J. Mussinan, F. Shahidi, and E.T. Contis (Eds.), *Recent Advances in Food and Flavor Chemistry.* Royal Society Chemistry, Cambridge. 2010, pp. 9–27.

50. M. Stuiver, P.J. Reimer and T.F. Braziunas, High-precision radiocarbon age calibration for terrestrial and marine samples. *Radiocarbon* 1998, 40(3): 1127–1151.

51. I. Levin and B. Kromer, The tropospheric 14CO$_2$ level in mid-latitudes of the northern hemisphere (1959–2003). *Radiocarbon* 2004, 46(3): 1261–1272.

52. Q. Hua and M. Barbetti, Review of Tropospheric bomb 14C data for carbon cycling modeling and age calibration purposes. *Radiocarbon* 2004, 46(3): 1273–1298.

53. S. Bowman, *Radiocarbon Dating, Interpreting the Past Series.* 1990, University of California press.

10 Botanical Isoscapes

Emerging Stable Isotope Tools for Authentication and Geographic Sourcing

Jason B. West

CONTENTS

What Are Stable Isotopes? 143
Carbon Isotope Ratios 145
Nitrogen Isotope Ratios 146
Sulfur Isotope Ratios 146
Hydrogen and Oxygen Isotope Ratios 146
Compound-Specific Isotope Analysis 147
Heavy Isotopes 147
What Are Isoscapes? 148
Detecting Adulteration with Isotopes 148
Isoscapes to Assess Geographic Origin 149
Conclusion 151
References 151

WHAT ARE STABLE ISOTOPES?

Isotopes are elements with unique atomic masses. That is, the isotopes of a given element have the same number of protons, but different numbers of neutrons. Stable isotopes are those isotopes of an element that have not been observed to radioactively decay. Most elements have more than one stable isotope. For example, hydrogen has two isotopes: ^2H and ^1H (often called deuterium and rarely protium, respectively), with the lighter isotope much more abundant than the heavier isotope. The average abundances of the common *light* stable isotopes that are of interest are shown in Table 10.1.

Stable isotope abundances are typically expressed relative to a standard and as ratios of the rare to common isotope in delta (δ) notation:

$$\delta(\text{‰}) = \left(\frac{R_{\text{sample}}}{R_{\text{standard}}} - 1 \right) \times 1000 \qquad (10.1)$$

TABLE 10.1
Abundances of Stable Isotopes Relevant to Botanical Authentication

Element	Isotope	Observed Natural Range (mole fraction)
Hydrogen	^1H	0.999816–0.999974
	^2H	0.000026–000184
Carbon	^{12}C	0.98853–0.99037
	^{13}C	0.00963–0.01147
Nitrogen	^{14}N	0.99579–0.99654
	^{15}N	0.00346–0.00421
Oxygen	^{16}O	0.99738–0.99776
	^{17}O	0.00037–0.00040
	^{18}O	0.00188–0.00222
Sulfur	^{32}S	0.94454–0.95281
	^{33}S	0.00730–0.00793
	^{34}S	0.03976–0.04734
	^{35}S	0.00013–0.00019

Source: M. Berglund and M.E. Wieser, *Pure and Applied Chemistry* 83, 397–410, 2011.

where R is the molar ratio of the heavy to light isotope of the sample and standard, for example:

$$R = \frac{^2H}{^1H} \tag{10.2}$$

The most common method for quantifying naturally occurring stable isotope ratio variation is with isotope ratio mass spectrometry (IRMS). IRMS instruments are gas-source mass spectrometers that precisely quantify mass ratios based on simultaneous measurement of typically two to three masses. Detailed discussion of stable isotope instrumentation, peripherals, sample treatment, and other aspects may be found in other publications and so will not be addressed here (cf. [2,3]). In general, the approaches utilize oxidation/reduction reactions (e.g., for $\delta^{13}C$, $\delta^{15}N$, and $\delta^{34}S$) or high temperature conversion/pyrolysis reactions (e.g., for δ^2H and $\delta^{18}O$) and chromatographic separation of the gases of interest, which are carried into the IRMS via helium carrier gas. Accepted standards used by the international community include V-SMOW (Vienna Standard Mean Ocean Water) for hydrogen and oxygen (δ^2H and $\delta^{18}O$), V-PDB (Vienna Pee Dee Belemnite) for carbon ($\delta^{13}C$), atmospheric air for nitrogen ($\delta^{15}N$), and CDT (Canyon Diablo Troilite) for sulfur ($\delta^{34}S$). These standards can be obtained from the International Atomic Energy Agency (IAEA) and are used to calibrate laboratory *working standards* that are then run with unknowns to permit isotope ratio measurement on a common scale [4,5]. Laser spectrometry is emerging as an additional tool for measuring natural abundance isotope ratios of a variety of materials and promises to be an important component of isotope ratio measurement approaches in the near future [6,7].

The use of stable isotopes in assessing or verifying the authenticity of botanical materials depends on predictable patterns of isotopic variation in plant tissues. Stable isotopes record, in ways that are specific to each isotope, climatic variation, photosynthetic pathway, fertilization, growth environment, and the isotopic composition of plant source water [8]. Thus, there are specific classes of authentication activities for which isotopes may be appropriate and others for which they may not be. Any application of stable isotopes requires an understanding of the underlying drivers of variation. From a fundamental standpoint, plant isotope ratios depend first on the isotopic composition of the source of the element and second on the physical and biochemical isotope effects (fractionation) of absorbing that element and incorporating it in plant tissues through metabolic processes. Thus, it is important to understand how the environment affects the isotope ratios of sources of these elements and then how plant metabolism and other processes result in isotopic fractionations. This understanding may then be applied to a wide range of authentication or geographic origin questions.

CARBON ISOTOPE RATIOS

The carbon isotope ratios of plant tissues are controlled by the isotopic composition of atmospheric CO_2 and the fractionations caused by the three primary pathways of photosynthesis in plants (C_3, C_4, and crassulacean acid metabolism [CAM]). In general, atmospheric CO_2 exhibits limited isotopic variation relative to that caused by physiological processes [9]. A significant exception to this is the use of elevated CO_2 for specialized indoor plant growth. Here the CO_2 may be derived from fossil fuel sources and will therefore have a significantly different isotopic composition compared to atmospheric CO_2, driving large differences in plant $\delta^{13}C$. Plant photosynthesis discriminates against the heavier isotope ($^{13}CO_2$) during CO_2 diffusion and fixation by photosynthetic enzymes [10]. The fixation of CO_2 by Rubisco in C_3 species has a significant competing reaction with oxygen, which results in the net release of CO_2 [11]. The evolution of C_4 photosynthesis that permitted the site of CO_2 fixation and CO_2 reduction to be spatially separated is thought to have been a response to the relatively high ratios of O_2 to CO_2, evolving several times across many plant lineages [12,13]. In C_4 photosynthesis, CO_2 is fixed initially by phosphoenolpyruvate (PEP) carboxylase, which does not have the same oxygenase properties as Rubisco. The initial product of C_4 photosynthesis is a 4-carbon compound that is then decarboxylated, producing an enriched CO_2 concentration that suppresses photorespiration in specialized bundle sheath cells that are spatially isolated from the atmosphere during a secondary fixation by Rubisco. Because CO_2 fixation occurs in a closed environment, very little isotopic discrimination by Rubisco takes place in C_4 photosynthesis, resulting in large differences in the C isotope ratios of C_3 versus C_4 plants. C_4 plants have a $\delta^{13}C$ value of about $-12‰$ compared to typical values of around $-27‰$ for C_3 plants [9]. CAM photosynthesis employs similar enzymes as C_4 photosynthesis, but segregates the carbon and light reactions in time (night versus day), resulting in very high water use efficiencies. CAM species vary considerably in the degree of dependence on CAM photosynthesis and so

CAM species as a group exhibit a wide range of isotope ratios [14,15]. An important source of variation particularly for C_3 plants is the effect of stomatal conductance on CO_2 diffusion and the resulting effective discrimination by Rubisco. In general, lower stomatal conductance (as occurs during water limitation or in response to dry atmospheres) will increase the ^{13}C content of the products of C_3 photosynthesis, a process described in models that relate plant $\delta^{13}C$ to the intercellular concentration of CO_2 (e.g., [10]). Thus, plants grown in relatively dry environments will tend to have higher $\delta^{13}C$ values than those grown in relatively wet environments.

NITROGEN ISOTOPE RATIOS

The N isotopic composition of plant tissues is initially dependent on the isotopic composition of plant source N and the biochemical transformations associated with uptake, assimilation, and N metabolism. The isotopic composition of source N (nitrate, ammonium, organic acids) can vary substantially over a wide range of spatial and temporal scales, and the fractionations associated with uptake and assimilation are not as well constrained, though a large body of literature exists on various aspects [16–19]. Having a specific or narrow set of values, fertilizer application can result in plant N isotope ratios that are distinct from those expected in unfertilized conditions, and it has been argued that this could be used as an indicator of organic cultivation or to identify the use of different kinds of fertilizers (e.g., [20,21]).

SULFUR ISOTOPE RATIOS

Sulfur has been less extensively explored as a tool for authentication application and research, though it has significant potential given the substantial spatial variation expected, especially when comparing certain regions that may differ in sources of plant-available S. Major controls on the spatial variation of sulfur isotopes ($\delta^{34}S$) include proximity to oceanic sources (e.g., sea spray), inputs from fossil fuel combustion, and volcanic emissions [22]. Temporal changes in tree rings have been used to demonstrate recovery of trees from reduced pollution inputs [23]. In general, plant S isotope ratios have been taken to be similar to soil $\delta^{34}S$ (e.g., [24,25]) and so regional variation in $\delta^{34}S$ is expected to drive variation in plant tissues. More recent work, however, has identified substantial fractionations associated with plant metabolism resulting in variation among tissues in $\delta^{34}S$, similar to that observed for N [26–28].

HYDROGEN AND OXYGEN ISOTOPE RATIOS

Plant hydrogen and oxygen isotope ratios are determined initially by the isotopic composition of plant source water (soil water), isotope fractionations associated with transpiration, and biochemical transformations in photosynthesis and other metabolic processes. Source water isotopic variation is caused by climatic patterns associated with precipitation and fractionation associated with evaporation in soils. In general, the liquid to vapor phase transition leads to accumulation of the heavy isotopes ^2H and ^{18}O in the liquid phase, including in surface waters, rain, or water in the leaves of plants [29–31]. Evaporation and condensation involve equilibrium and kinetic phase transitions and transport processes that drive changes in the isotopic composition of the source and product [32–34]. These effects produce the *global meteoric water line* (GMWL): $\delta^2H = \delta^{18}O * 8 + 10$, along which most precipitation and surface waters lie [35], and also cause the isotope ratios of some waters to deviate significantly from the GMWL (e.g., soil water evaporating into a dry atmosphere). With some limited exceptions [36], there is no isotopic fractionation associated with plant absorption of soil water or transport through the plant vascular system. The first major fractionation occurs as liquid water movement slows and goes through the phase transition from liquid to gas and then diffuses out of the leaf as vapor. This process has been modeled based on similar models used for surface waters [37], with a number of modifications for leaf type and other aspects [38–40]. Photosynthesis and subsequent biochemical transformations also have important effects on the isotopic composition of various tissues and compounds in plants [41–44].

COMPOUND-SPECIFIC ISOTOPE ANALYSIS

A subset of stable isotope analysis that either uses offline approaches for compound extraction and purification, or gas or high-performance liquid chromatography prior to pyrolysis or oxidation/reduction reactors offers the opportunity to quantify the isotope ratios of specific compounds of interest [45–48]. These analyses are particularly powerful from the standpoint of reduced uncertainty of sources of isotopic variation inherent to bulk samples that may vary substantially in chemical composition even within tissues. However, given the complexity inherent to any botanical raw material, it is unlikely that the underlying isotope fractionations are understood in detail for even a small subset of compounds that may be of interest, although progress is being made in some areas [49]. Thus, although the approach offers tremendous promise in specific cases and may be combined productively with chemometric approaches [50,51], significant research may be required to identify appropriate compounds of interest and to constrain the expected isotopic variability.

HEAVY ISOTOPES

In addition to the light stable isotopes discussed above, certain isotopes of heavier elements, in particular strontium, have had important application to authentication and geographic origin questions. Strontium is chemically similar to calcium and substitutes for it in plant tissues. ^{87}Sr forms as a beta decay product of ^{87}Rb, which has a long half-life of 4.88×10^{10} years [52], resulting in a wide range of abundances of ^{87}Sr in Earth's minerals based on time of crystallization. All other stable isotopes of strontium are not radiogenic (^{88}Sr, ^{86}Sr, and ^{84}Sr), and because the abundance of ^{86}Sr is similar to that of ^{87}Sr, the stable strontium isotope ratios of materials are reported as ($^{87}Sr/^{86}Sr$). Modern Sr isotope ratios are the result of the ratio of Rb to Sr, the initial $^{87}Sr/^{86}Sr$ of the rock, and time since crystallization [53]. Bedrock $^{87}Sr/^{86}Sr$ can therefore be approximately predicted from rock age and the Rb and Sr concentrations in the rock [54]. There is not likely to be significant isotopic fractionation with uptake given the small relative mass difference between ^{87}Sr and ^{86}Sr, and thus, plant $^{87}Sr/^{86}Sr$ should broadly reflect the $^{87}Sr/^{86}Sr$ of available strontium in soils. This available soil pool of strontium, however, may be significantly different from that of the rock parent material due to atmospheric deposition, Sr cycling by plants, differential weathering, and other processes that move materials across earth's surface [55]. Lead and other heavy isotopes have also been used in select cases, but do not represent a broadly utilized tool and so these will not be discussed here.

WHAT ARE ISOSCAPES?

Isoscapes are isotopic landscapes that represent the spatiotemporal isotope variation of a particular system. The first discussion of isoscapes as such was in the context of spatial variation in plant hydrogen and oxygen isotope ratios [56], but the framework has rapidly expanded to encompass efforts to describe and understand spatiotemporal isotope variation in a wide range of biological and physical systems [57]. An important utility of isoscapes in authentication and origin studies is in defining a spatial domain within which materials may originate in terms of their isotope ratios and then assessing and defining ranges outside of which materials may be identified as nonauthentic. Many of the challenges in using isoscapes for these applications come from a lack of fundamental understanding of the underlying fractionating processes or limited observations to develop the predicted landscapes. Thus, improved isoscape models will come from specific research efforts into the fundamental controls of stable isotope variation and more extensive spatiotemporal data sets. To date, terrestrial isoscapes used in this area are primarily related to water (δ^2H and $\delta^{18}O$) and strontium ($^{87}Sr/^{86}Sr$) because these systems have fairly well-understood underlying processes that drive continuous or discrete spatial variation.

DETECTING ADULTERATION WITH ISOTOPES

A critical component of quality assurance of plant materials is the capacity to assess the presence or absence of adulterants in a product. Stable isotopes have been used in a wide range of applications to detect additives based primarily on a comparison between expected isotope abundances and measured abundances of selected materials. Although fairly limited, there has been some exploration of stable isotopes for assessing the authenticity of supplements and other botanicals currently available (e.g., [58,59]). The use of stable isotopes in the authentication of foods and beverages generally, especially those of high value, provides an opportunity to illustrate the potential for these tools in the authentication of botanical raw materials. Here, two examples will be discussed to highlight potential opportunities in the use of stable isotopes for authenticating botanical raw materials, as well as potential pitfalls.

Honey is produced by bees from the nectar of a wide range of flowers, as well as from honeydew. The origin and authenticity of some honeys are linked to honey's relatively high value, while at the same time cheaper sources of sugar (e.g., high fructose corn syrup) may be added to honey to increase volumes. Because of its high value and the ease with which honey may be adulterated with other sugars, there has been longstanding interest in authentication tools, including several decades of work using stable isotopes [60–65]. It was recognized early that because the nectar sources of bees are plants that use the C_3 photosynthetic pathway, the resulting honey has a C isotope ratio distribution similar to C_3 plants. This allows straightforward detection of the addition of C_4 sugars in particular [60], as well as perhaps other C_3 sugars with different isotope ratios. Geographic variation in the hydrogen isotope ratios of authentic honey was also recognized early, suggesting that δD could be used to authenticate geographic origin [60]. More recent work has confirmed the utility of both C isotopes and compound-specific H isotopes in detecting adulteration and assessing geographic origin [51,65–69]. An interesting related product is so-called royal jelly. Royal jelly is used by nurse bees to produce new queens in a hive [70,71] and is being marketed as a dietary supplement. Because royal jelly is a very low volume product of beehives, feeding supplements and other adulteration of the product or production process may be employed and may decrease value. Here C isotope ratios have been shown to detect the addition of C_4 sugars to royal jelly production, along with the potential in at least one case for N isotopes to also yield additional source information [72–74].

Olive oil is another food item that also has high value based on authentic characteristics of origin and content and so stable isotopes have been explored alongside other methodologies to assess their potential utility in quality assurance and authenticity testing [75–77]. Here isotope ratios combined with other elemental and chemical data have been useful in identifying and verifying authentic olive oils. These two examples provide a potential framework for exploring the use of isotope ratios in the identification of adulteration in botanical materials.

A general comment on the utility of isotopes for the detection of adulteration or substitution of an inferior product in botanical raw materials is perhaps in order. For all products there will be a range of values that represent authentic materials, depending on the full range of soils and climate in which the source plants are grown, as well as the biophysical and biochemical isotope effects (fractionations) associated with a species or variety. For plant species grown in narrow climatic conditions and with perhaps specialized resource use adaptations, or in cases where the likely adulterant is isotopically distinct (e.g., C_4 adulterant added to C_3 material), the use of C and N isotope ratios are likely to represent useful tools in authentication. However, in the case that the likely adulterant or substitute is similar physiologically or perhaps co-occurs with the authentic material, the isotope distributions may overlap significantly. Any overlap of course reduces the utility of isotope data in separating them. It is also worth noting that the ability to detect an adulterant is related to both its degree of isotopic difference and the amount of contaminant. Although these cautions are important, it is likely that there are many cases for which light stable isotopes would be useful in botanical raw material quality assurance.

ISOSCAPES TO ASSESS GEOGRAPHIC ORIGIN

A particularly important aspect of botanical raw material authenticity is geographic origin. Materials grown in a specific region may be more valuable for a variety of reasons, or there may be regulations limiting the use of materials from particular locations. As such, tools for assessing and verifying the region of origin of plant materials are needed. As discussed in the previous section, stable isotopes can provide information on potential contaminants or adulterants in specific cases. Similar information may also be obtained from chemical analyses or the use of genetic tools (see, e.g., recent evidence of significant contamination in [78]). Although chemical and genetic tools are useful for assessing and verifying raw material content, these tools do not generally provide geographic origin information. Here stable isotope analyses coupled to accurate isoscapes describing underlying spatial variation may provide unique information.

There are two primary classes of isoscapes that may be applicable: continuous and discrete. Continuous isoscapes are those that represent spatial isotope variability that is caused by a continuous variable. For example, plant H and O isotope ratios are linked to precipitation isotope ratios through soil water. Because precipitation isotope ratios tend to vary across space via the effects of temperature and storm rainout, plant H and O isotope ratios also tend to vary continuously [79,80]. This may also be the case for sulfur [81] or nitrogen isotopes [82] in specific cases, though the caveats noted above should be recalled. An initial assessment of the degree of variability across geographic regions in plant H and O isotope ratios may be obtained from existing global isoscapes of leaf water [80]. It is important to note that global estimates represent a climatic description of leaf water and likely do not represent what might be expected for individual plant species at smaller spatial scales. Here differences in rooting depth, seasonal growth patterns, leaf morphology, and physiological stress tolerances can be important. Thus, regional isoscapes, perhaps for specific timeframes and developed perhaps for individual species, are likely to be more appropriate for many applications. Discrete isoscapes may occur when discrete processes, for example, those linked to variation in soil characteristics, drive plant isotope ratio variation. Strontium isotopes at the regional to continental scales are a good example of this. Although dust and other inputs are important contributors to available Sr, soil parent material represents an important driver of spatial variation in Sr isotopes and this variation is linked to patterns in bedrock, as previously discussed. Existing Sr isoscapes thus reflect this basic spatial pattern and tend to exhibit a patchwork of variability driven by rock type and geologic age, with new efforts improving the accuracy of Sr isoscapes [83,84]. Both continuous and discrete isoscapes hold promise for inferring geographic origin of botanical materials and indeed may be most useful when used in combination [85,86].

A useful case to illustrate the potential may be found in wine authentication efforts. Wine is a commodity for which content (variety) and origin are important components of its value, and the isotopic composition of wine has been used for decades to assess both aspects. In particular, European wine producers participate in the development of an *authentics* database that is produced annually to establish the range of authentic isotope ratio variation of wines with region of origin designations and regulations [87–90]. Recognizing that spatial variation in wine H and O isotope ratios is linked to grape water, which is itself derived from spatially varying soil water, a model was developed to produce wine $\delta^{18}O$ isoscapes and compared against measured wine $\delta^{18}O$ values [91]. The model performed well and suggests that similar efforts to link underlying processes to plant isotope ratio variation will likely yield robust isoscapes for a wide range of botanical systems.

Another example in this area is work intended to provide information on the geographic origin of illicit substances. Models have been developed for marijuana based on data collected from plants of known geographic origin [92,93]. The results of this work demonstrated that bulk leaf H isotope ratios did provide information on geographic origin, consistent with expectations from an understanding of isotope hydrology and effects associated with plant uptake of soil water and the incorporation of H into plant tissues. Both of these approaches utilized an empirical approach to model generation, primarily due to the limited fundamental understanding of isotopic fractionation for each of the relevant biophysical and biochemical steps to the final bulk product being analyzed. New tools are being developed to allow individual researchers to generate model isoscapes that may be spatially and temporally specific, as well as that incorporate species-specific parameters in generating predicted spatial variation in plant H and O isotope ratios (http://isomap.org). It is expected that continued technological advances in the analytical tools used to quantify isotopic variation, along with expanding research into the underlying mechanistic controls, will continue to improve the specificity and therefore the utility of plant isoscapes for use in a variety of applications to botanical materials.

CONCLUSION

Although stable isotopes have not yet been widely applied to questions of authenticity or geographic origin of botanical materials used for applications to human health and wellbeing, this approach holds significant promise. This is especially true as herbal and other products become more widely used and as a range of relevant industries attempts to improve processes for quality control and assurance. The development of stable isotope tools in other areas, including application to food and beverages, has resulted in a knowledge base and toolset that can easily be transferred to assurance of authenticity and origin of botanical materials.

REFERENCES

1. M. Berglund and M.E. Wieser. Isotopic compositions of the elements 2009 (IUPAC Technical Report). *Pure and Applied Chemistry* 83, 2011: 397–410.
2. P.A. de Groot, ed. *Handbook of Stable Isotope Analytical Techniques*, Vol. 1. Elsevier B. V.: Amsterdam, the Netherlands, 2004; p. 1234.
3. P.A. de Groot, ed. *Handbook of Stable Isotope Analytical Techniques*, Vol. 2. Elsevier B.V.: Amsterdam, the Netherlands, 2009; p. 1398.
4. R.A. Werner and W.A. Brand. Referencing strategies and techniques in stable isotope ratio analysis. *Rapid Communications in Mass Spectrometry* 15, 2001: 501–519.
5. T.B. Coplen. New guidelines for reporting stable hydrogen, carbon, and oxygen isotope-ratio data. *Geochimica et Cosmochimica Acta* 60, 1996: 3359–3360.
6. B. Paldus and A. Kachanov. Spectroscopic Techniques: Cavity-Enhanced Methods. In *Springer Handbook of Atomic, Molecular, and Optical Physics*, G. Drake, ed., Springer: New York, 2006; pp. 633–640.
7. D. Penna, B. Stenni, M. Šanda et al. On the reproducibility and repeatability of laser absorption spectroscopy measurements for δ2H and δ18O isotopic analysis. *Hydrology and Earth System Sciences* 14, 2010: 1551–1566.

8. J.B. West, G.J. Bowen, T.E. Cerling, and J.R. Ehleringer. Stable isotopes as one of nature's ecological recorders. *Trends in Ecology & Evolution* 21, 2006: 408–414.

9. T.E. Dawson, S. Mambelli, A.H. Plamboeck, P.H. Templer, and K.P. Tu. Stable isotopes in plant ecology. *Annual Review of Ecology and Systematics* 33, 2002: 507–559.

10. G.D. Farquhar, J.R. Ehleringer, and K.T. Hubick. Carbon isotope discrimination and photosynthesis. *Annual Review of Plant Physiology and Plant Molecular Biology* 40, 1989: 503–537.

11. J.R. Ehleringer, T.E. Cerling, and M.D. Dearing, eds. *A History of Atmospheric CO_2 and Its Effects on Plants, Animals, and Ecosystems.* Vol. 177. Springer: Berlin, Germany, 2005; p. 534.

12. J.R. Ehleringer, R.F. Sage, L.B. Flanagan, and R.W. Pearcy. Climate change and the evolution of C4 photosynthesis. *Trends in Ecology & Evolution* 6, 1991: 95–99.

13. R.F. Sage. The evolution of C-4 photosynthesis. *New Phytologist* 161, 2004: 341–370.

14. J.A. Raven, C.S. Cockell, and C.L. La Rocha. The evolution of inorganic carbon concentrating mechanisms in photosynthesis. *Philosophical Transactions of the Royal Society B: Biological Sciences* 363, 2008: 2641–2650.

15. K. Winter, J. Aranda, and J.A.M. Holtum. Carbon isotope composition and water-use efficiency in plants with crassulacean acid metabolism. *Functional Plant Biology* 32, 2005: 381–388.

16. P. Hogberg. Tansley review No 95—N-15 natural abundance in soil-plant systems. *New Phytologist* 137, 1997: 179–203.

17. R.D. Evans. Physiological mechanisms influencing plant nitrogen isotope composition. *Trends in Plant Science* 6, 2001: 121–126.

18. E.A. Hobbie and J.V. Colpaert. Nitrogen availability and colonization by mycorrhizal fungi correlate with nitrogen isotope patterns in plants. *New Phytologist* 157, 2003: 115–126.

19. J.M. Craine, A.J. Elmore, M.P.M. Aidar et al. Global patterns of foliar nitrogen isotopes and their relationships with climate, mycorrhizal fungi, foliar nutrient concentrations, and nitrogen availability. *New Phytologist* 183, 2009: 980–992.

20. K.H. Laursen, A. Mihailova, S.D. Kelly et al. Is it really organic? Multi-isotopic analysis as a tool to discriminate between organic and conventional plants. *Food Chemistry.* 141, 2013: 2812–2820.

21. S. Verenitch and A. Mazumder. Carbon and nitrogen isotopic signatures and nitrogen profile to identify adulteration in organic fertilizers *Journal of Agricultural and Food Chemistry* 60, 2012: 8278–8285.

22. B.A. Trust and B. Fry. Stable sulfur isotopes in plants—A review. *Plant, Cell & Environment* 15, 1992: 1105–1110.

23. R.B. Thomas, S.E. Spal, K.R. Smith, and J.B. Nippert. Evidence of recovery of *Juniperus virginiana* trees from sulfur pollution after the Clean Air Act. *Proceedings of the National Academy of Sciences of the United States of America* 110, 2013: 15319–15324.

24. J.M. Monaghan, C.M. Scrimgeour, W.M. Stein, F.J. Zhao, and E.J. Evans. Sulphur accumulation and redistribution in wheat (*Triticum aestivum*): A study using stable sulphur isotope ratios as a tracer system. *Plant, Cell & Environment* 22, 1999: 831–839.

25. S. Rummel, S. Hoelzl, P. Horn, A. Rossmann, and C. Schlicht. The combination of stable isotope abundance ratios of H, C, N and S with Sr-87/Sr-86 for geographical origin assignment of orange juices. *Food Chemistry* 118, 2010: 890–900.

26. G. Tcherkez and I. Tea. S-32/S-34 isotope fractionation in plant sulphur metabolism. *New Phytologist* 200, 2013: 44–53.

27. I. Tea, T. Genter, N. Naulet, M. Lummerzheim, and D. Kleiber. Interaction between nitrogen and sulfur by foliar application and its effects on flour bread-making quality. *Journal of the Science of Food and Agriculture* 87, 2007: 2853–2859.

28. N. Tanz and H.L. Schmidt. Delta S-34-value measurements in food origin assignments and sulfur isotope fractionations in plants and animals. *Journal of Agricultural and Food Chemistry* 58, 2010: 3139–3146.
29. Y. Yurtsever and J.R. Gat. Atmospheric waters. In *Stable Isotope Hydrology: Deuterium and Oxygen-18 in the Water Cycle*, J.R. Gat and R. Gonfiantini, eds. International Atomic Energy Agency: Vienna, Austria, 1981; pp. 103–142.
30. L.B. Flanagan, J.P. Comstock, and J.R. Ehleringer. Comparison of modeled and observed environmental influences on the stable oxygen and hydrogen isotope composition of leaf water in *Phaseolus vulgaris* L. *Plant Physiology* 96, 1991: 588–596.
31. R. Mathieu, D. Pollard, J.E. Cole, J.W.C. White, R.S. Webb, and S.L. Thompson. Simulation of stable water isotope variations by the GENESIS GCM for modern conditions. *Journal of Geophysical Research* 107, 2002: 4037.
32. J.R. Gat. Oxygen and hydrogen isotopes in the hydrologic cycle. *Annual Review of Earth and Planetary Sciences* 24, 1996: 225–262.
33. J.E. Lee and I. Fung. "Amount effect" of water isotopes and quantitative analysis of post-condensation processes. *Hydrological Processes* 22, 2008: 1–8.
34. X.F. Wang and D. Yakir. Using stable isotopes of water in evapotranspiration studies. *Hydrological Processes* 14, 2000: 1407–1421.
35. H. Craig and L. Gordon. Deuterium and oxygen 18 variations in the ocean and marine atmosphere. In *Stable Isotopes in Oceanographic Studies and Paleotemperatures*, E. Tongiorigi, ed., Consiglio Nazionale Delle Ricerche Laboratorio di Geologia Nucleare: Pisa, Italy, 1965; pp. 9–130.
36. P.Z. Ellsworth and D.G. Williams. Hydrogen isotope fractionation during water uptake by woody xerophytes. *Plant and Soil* 291, 2007: 93–107.
37. L.B. Flanagan, J.P. Comstock, and J.R. Ehleringer. Comparison of modeled and observed environmental-influences on the stable oxygen and hydrogen isotope composition of leaf water in *Phaseolus vulgaris* L. *Plant Physiology* 96, 1991: 588–596.
38. M.M. Barbour. Stable oxygen isotope composition of plant tissue: A review. *Functional Plant Biology* 34, 2007: 83–94.
39. A. Kahmen, K. Simonin, K.P. Tu, A. Merchant, A. Callister, R. Siegwolf, T.E. Dawson, and S.K. Arndt. Effects of environmental parameters, leaf physiological properties and leaf water relations on leaf water delta(18)O enrichment in different Eucalyptus species. *Plant, Cell & Environment* 31, 2008: 738–751.
40. L.A. Cernusak, K. Winter, and B.L. Turner. Physiological and isotopic (delta C-13 and delta O-18) responses of three tropical tree species to water and nutrient availability. *Plant, Cell & Environment* 32, 2009: 1441–1455.
41. A. Kahmen, D. Sachse, S.K. Arndt, K.P. Tu, H. Farrington, P.M. Vitousek, and T.E. Dawson. Cellulose delta O-18 is an index of leaf-to-air vapor pressure difference (VPD) in tropical plants. *Proceedings of the National Academy of Sciences of the United States of America* 108, 2011: 1981–1986.
42. J.M. Hayes. Fractionation of carbon and hydrogen isotopes in biosynthetic processes. *Stable Isotope Geochemistry* 43, 2001: 225–277.
43. J.S. Roden and J.R. Ehleringer. Hydrogen and oxygen isotope ratios of tree-ring cellulose for riparian trees grown long-term under hydroponically controlled environments. *Oecologia* 121, 1999: 467–477.
44. M.J. Deniro and S. Epstein. Relationship between the oxygen isotope ratios of terrestrial plant cellulose, carbon-dioxide, and water. *Science* 204, 1979: 51–53.
45. W. Meier-Augenstein. Applied gas chromatography coupled to isotope ratio mass spectrometry. *Journal of Chromatography A* 842, 1999: 351–371.
46. Y. Chikaraishi and H. Naraoka. Compound-specific delta D-delta C-13 analyses of n-alkanes extracted from terrestrial and aquatic plants. *Phytochemistry* 63, 2003: 361–371.

47. J.T.Brenna, T.N. Corso, H.J. Tobias, and R.J. Caimi. High-precision continuous-flow isotope ratio mass spectrometry. *Mass Spectrometry Reviews* 16, 1997: 227–258.
48. E.A. Hobbie and R.A. Werner. Intramolecular, compound-specific, and bulk carbon isotope patterns in C-3 and C-4 plants: A review and synthesis. *New Phytologist* 161, 2004: 371–385.
49. D. Sachse, I. Billault, G.J. Bowen et al. Molecular paleohydrology: Interpreting the hydrogen-isotopic composition of lipid biomarkers from photosynthesizing organisms. *Annual Review of Earth and Planetary Sciences* 40, 2012: 221–249.
50. L.Z. Chen, J. Zhao, Z.H. Ye, and Y.P. Zhong. Determination of adulteration in honey using near-infrared spectroscopy. *Spectroscopy and Spectral Analysis* 28, 2008: 2565–2568.
51. J.M. Camina, R.G. Pellerano, and E.J. Marchevsky. Geographical and botanical classification of honeys and apicultural products by chemometric methods. A review. *Current Analytical Chemistry* 8, 2012: 408–425.
52. R.H. Steiger and E. Jäger. Subcommission on geochronology: Convention on the use of decay constants in geo- and cosmochronology. *Earth and Planetary Science Letters* 36, 1977: 359–362.
53. G. Aberg. The use of natural strontium isotopes as tracers in environmental studies. *Water, Air, & Soil Pollution* 79, 1995: 309–322.
54. B.L. Beard and C.M. Johnson. Strontium isotope composition of skeletal material can determine the birth place and geographic mobility of humans and animals. *Journal of Forensic Sciences* 45, 2000: 1049–1061.
55. J.B. West, J.M. Hurley, F.Ö. Dudas, and J.R. Ehleringer. The stable isotope ratios of marijuana. II. Strontium isotopes relate to geographic origin. *Journal of Forensic Sciences* 54, 2009: 1261–1269.
56. J.B. West, G.J. Bowen, and J.R. Ehleringer. Predicting hydrogen and oxygen stable isotope ratios of plants across terrestrial surfaces: Plant isoscapes. In *American Geophysical Union Fall Meeting*, San Francisco, CA, December 5–9, 2005.
57. J.B. West, G.J. Bowen, T.E. Dawson, and K.P. Tu, eds. *Isoscapes: Understanding Movement, Pattern, and Process on Earth through Isotope Mapping*. Springer Science + Business Media: Dordrecht, the Netherlands, 2010; 487.
58. H. Kroll, J. Friedrich, M. Menzel, and P. Schreier. Carbon and hydrogen stable isotope, ratios of carotenoids and beta-carotene-based dietary supplements. *Journal of Agricultural and Food Chemistry* 56, 2008: 4198–4204.
59. K.M. Herbach, F.C. Stintzing, S. Elss, C. Prestonb, P. Schreierb, and R. Carlea. Isotope ratio mass spectrometrical analysis of betanin and isobetanin isolates for authenticity evaluation of purple pitaya-based products. *Food Chemistry* 99, 2006: 204–209.
60. H. Ziegler, W. Stichler, A. Maurizio, and G. Vorwohl. The use of stable isotopes for the characterization of honeys, their origin and adulteration. *Apidologie* 8, 1977: 337–347.
61. H. Ziegler, A. Maurizio, and W. Stichler. The characterization of honey samples according to their content of pollen and of stable isotopes. *Apidologie* 10, 1979: 301–311.
62. L.R. Croft. Stabe isotope mass-spectrometry in honey analysis. *Trends in Analytical Chemistry* 6, 1987: 206–209.
63. E. Cienfuegos, I. Casar, and P. Morales. Carbon isotopic composition of Mexican honey. *Journal of Apicultural Research* 36, 1997: 169–179.
64. E. Anklam. A review of the analytical methods to determine the geographical and botanical origin of honey. *Food Chemistry* 63, 1998: 549–562.
65. A. Schellenberg, S. Chmielus, C. Schlicht et al. Multielement stable isotope ratios (H, C, N, S) of honey from different European regions. *Food Chemistry* 121, 2010: 770–777.
66. U. Kropf, M. Korošec, J. Bertoncelj, N. Ogrinc, M. Nečemer, P. Kump, and T. Golob. Determination of the geographical origin of Slovenian black locust, lime and chestnut honey. *Food Chemistry* 121, 2010: 839–846.

67. L.A. Chesson, B.J. Tipple, B.R. Erkkila, T.E. Cerling, and J.R. Ehleringer. B-HIVE: Beeswax hydrogen isotopes as validation of environment. Part I: Bulk honey and honeycomb stable isotope analysis. *Food Chemistry* 125, 2011: 576–581.

68. B.J. Tipple, L.A. Chesson, B.R. Erkkila, T.E. Cerling, and J.R. Ehleringer. B-HIVE: Beeswax hydrogen isotopes as validation of environment. Part II. Compound-specific hydrogen isotope analysis. *Food Chemistry* 134, 2012: 494–501.

69. L.A. Chesson, B.J. Tipple, B.R. Erkkila, and J.R. Ehleringer. Hydrogen and oxygen stable isotope analysis of pollen collected from honey. *Grana* 52, 2013: 305–315.

70. A. Buttstedt, R.F.A. Moritz, and S. Erler. More than royal food—Major royal jelly protein genes in sexuals and workers of the honeybee *Apis mellifera. Frontiers in Zoology* 10, 2013: 72.

71. M. Kamakura. Royalactin induces queen differentiation in honeybees. *Nature* 473, 2011: 478–483.

72. A. Stocker, A. Rossmann, A. Kettrup, and E. Bengsch. Detection of royal jelly adulteration using carbon and nitrogen stable isotope ratio analysis. *Rapid Communications in Mass Spectrometry* 20, 2006: 181–184.

73. G. Daniele, M. Wytrychowski, M. Batteau, S. Guibert, and H. Casabianca. Stable isotope ratio measurements of royal jelly samples for controlling production procedures: Impact of sugar feeding. *Rapid Communications in Mass Spectrometry* 25, 2011: 1929–1932.

74. M. Wytrychowski, G. Daniele, and H. Casabianca. Combination of sugar analysis and stable isotope ratio mass spectrometry to detect the use of artificial sugars in royal jelly production. *Analytical and Bioanalytical Chemistry* 403, 2012: 1451–1456.

75. G. Martin. Advances in the authentication of food by SNIF-NMR. In *Magnetic Resonance in Food Science: The Multivariate Challenge*, S.B. Engelsen, P.S. Belton, and H.J. Jakobsen, eds. Royal Society of Chemistry: Cambridge, 2005; pp. 31–38.

76. A. Rossmann. Stable isotope databases for European food products. In *Proceedings of the International Workshop Fingerprinting Methods for the Identification of Timber Origins*, B. Degen, ed. Agriculture and Forestry Research: Bonn, Germany, 2008; pp. 39–46.

77. S.A. Drivelos and C.A. Georgiou. Multi-element and multi-isotope-ratio analysis to determine the geographical origin of foods in the European Union. *Trends in Analytical Chemistry* 40, 2012: 38–51.

78. S.G. Newmaster, M. Grguric, D. Shanmughanandhan, S. Ramalingam, and S. Ragupathy. DNA barcoding detects contamination and substitution in North American herbal products. *BMC Medicine* 11, 2013: 222.

79. G.J. Bowen and J. Revenaugh. Interpolating the isotopic composition of modern meteoric precipitation. *Water Resources Research* 39, 2003: 9.

80. J.B. West, A. Sobek, and J.R. Ehleringer. A simplified GIS approach to modeling global leaf water isoscapes. *PLoS One* 3, 2008: e2447.

81. P. Stack and L. Rock. A delta S-34 isoscape of total sulphur in soils across Northern Ireland. *Applied Geochemistry* 26, 2011: 1478–1487.

82. R. Amundson, A.T. Austin, E.A.G. Schuur, K. Yoo, V. Matzek, C. Kendall, A. Uebersax, D. Brenner, and W.T. Baisden. Global patterns of the isotopic composition of soil and plant nitrogen. *Global Biogeochemical Cycles* 17, 2003: 31.

83. C.P. Bataille, J. Laffoon, and G.J. Bowen. Mapping multiple source effects on the strontium isotopic signatures of ecosystems from the circum-Caribbean region. *Ecosphere* 3, 2012: art118.

84. L.A. Chesson, B.J. Tipple, G.N. Mackey, S.A. Hynek, D.P. Fernandez, and J.R. Ehleringer. Strontium isotopes in tap water from the coterminous USA. *Ecosphere* 3, 2012: art67.

85. J.N. Fenner and C.D. Frost. Modern Wyoming plant and pronghorn isoscapes and their implications for archaeology. *Journal of Geochemical Exploration* 102, 2009: 149–156.

86. L.V. Benson, E.M. Hattori, H.E. Taylor, S.R. Poulson, and E.A. Jolie. Isotope sourcing of prehistoric willow and tule textiles recovered from western Great Basin rock shelters and caves—Proof of concept. *Journal of Archaeological Science* 33, 2006: 1588–1599.

87. O. Breas, F. Reniero, and G. Serrini. Isotope ratio mass-spectrometry—Analysis of wines from different european countries. *Rapid Communications in Mass Spectrometry* 8, 1994: 967–970.

88. M.P. Day, B.L. Zhang, and G.J. Martin. The use of trace-element data to complement stable-isotope methods in the characterization of grape musts. *American Journal of Enology and Viticulture* 45, 1994: 79–85.

89. G.J. Martin, M. Mazure, C. Jouitteau, Y.L. Martin, L. Aguile, and P. Allain. Characterization of the geographic origin of Bordeaux wines by a combined use of isotopic and trace element measurements. *American Journal of Enology and Viticulture* 50, 1999: 409–417.

90. N. Ogrinc, I.J. Kosir, M. Kocjancic, and J. Kidric. Determination of authenticy, regional origin, and vintage of Slovenian wines using a combination of IRMS and SNIF-NMR analyses. *Journal of Agricultural and Food Chemistry* 49, 2001: 1432–1440.

91. J.B. West, J.R. Ehleringer, and T.E. Cerling. Geography and vintage predicted by a novel GIS model of wine delta O-18. *Journal of Agricultural and Food Chemistry* 55, 2007: 7075–7083.

92. J.M. Hurley, J.B. West, and J.R. Ehleringer. Stable isotope models to predict geographic origin and cultivation conditions of marijuana. *Science & Justice* 50, 2010: 86–93.

93. J.M. Hurley, J.B. West, and J.R. Ehleringer. Tracing retail cannabis in the United States: Geographic origin and cultivation patterns. *International Journal of Drug Policy* 21, 2010: 222–228.

11 Nuclear Magnetic Resonance

A Revolutionary Tool for Nutraceutical Analysis

*Kimberly L. Colson, Jimmy Yuk,
and Christian Fischer*

CONTENTS

Introduction 157
 Unique Strengths of NMR 158
 Automation in NMR 162
 Seeing Beauty in a Highly Complex Mixture 163
Targeted Approaches of Qualitative and Quantitative Assessment 163
 Development of an NMR SBASE 166
 Obtaining Quantitative Results 168
Nontargeted NMR Approaches of Qualitative Assessment 170
 Sample Collection and Preparation Considerations for Botanical
 Metabolomics 170
 NMR Acquisition and Processing for Metabolomics Analysis 172
Three Examples: NMR Analysis of Nutraceuticals and Botanicals 173
 Ginseng 173
 Energy Drinks 173
 Undeclared Drug Substance in Regenerect 177
Practical Considerations to Implementing NMR Screening of Nutraceuticals 178
Conclusions 178
References 179

INTRODUCTION

The urgency for enhanced analytical methods for the evaluation of nutraceuticals is summarized well by Kathie Wrick in an excellent review about the impact of regulations in the book entitled *Regulation of Functional Foods and Nutraceuticals* "... a very real problem is the absence of validated analytical methods for use in manufacturing controls and finished product testing to assure that label claims, and therefore customer expectations, are met ... Natural products are some of the most complex matrices found in the world of analytical chemistry. Sometimes they contain thousands of phytochemicals in one plant, which may vary in composition and quantity depending on season and soil ... well designed clinical trials for many herbs and botanicals await standardized methods to validate plant species and analytical methods to establish a quantitative chemical fingerprint for the preparations under study" [1].

Botanical identity and composition analysis are challenging. Typically, identity is accomplished in the field when plant material is harvested and done with highest confidence using voucher specimens and trained experts. Identifying the material taxonomically and avoiding confusion that often results from the use of common names and similar looking plants is essential. Once plant material including leaves, roots, stems, or fruit is pressed, ground, or extracted, the product identity relies on analytical technologies with comparisons to standards. Direct comparisons to standards are challenged by the variability of material that results from different growing conditions, different local cultivars and landraces (LRs), different harvest and processing techniques, and limited availability of applicable standards. Similar complexities to analysis are also observed in dietary supplements, functional foods, and beverages, such as energy drinks (EDs), that are complex mixtures that often contain botanical components.

Traditionally, the application of nuclear magnetic resonance (NMR) spectroscopy to botanical analysis has largely been limited to the study of purified small molecule metabolites. NMR has long been a primary tool for the structure elucidation of natural product metabolites [2–6]. In these cases, the chemical shifts, coupling constants, and coupling patterns are typically analyzed in detail to establish the molecular connectivity of atoms making up a natural product metabolite using a full complement of one-dimensional (1D) and two-dimensional (2D) experiments. Spatial relationships between atoms are also evaluated using NMR to gain insight into stereochemistry and site of substituent attachment. Recent advancements in NMR technology and statistical analysis methods have greatly expanded the use of NMR technology to include applications to complex mixtures. Improvements in linearity, stability, dynamic range, and sensitivity of NMR technology have supported this expansion of the applications of NMR beyond purified materials. Method development for animal metabolomics to study biofluids established a template that is readily adaptable to botanical identity, dietary supplements, and quality assessment of complex mixtures. The use of statistical methods to complex NMR spectra aids the spectral interpretation by focusing on similarities and differences rather than on each NMR signal in the spectrum, as is traditionally done with a purified natural product. Further developments in automation coupled with NMR technology enhancements have expanded the use of NMR spectroscopy beyond the traditional chemical research space and into quality control facilities, botanical analytical and biological laboratories, and production facilities.

Unique Strengths of NMR

Although the traditional analytical techniques for nutraceuticals include high pressure liquid chromatography (HPLC), thin layer chromatography (TLC), gas chromatography–mass spectrometry (GC-MS), and near infrared spectroscopy (NIR), NMR is now emerging as an important technology for these analyses. NMR technology excels in high reproducibility, high compound specificity, and high precision for quantitation. Reproducibility of NMR data assures the consistency of measurements between different laboratories and enables the exchange of NMR spectral databases (SBASEs) and statistical models between facilities. This ability empowers the analyst to make material assessments with relative ease.

Two key properties of NMR make it an inherently reproducible technique including that the (1) intensity of the signal is proportional to the number of atoms giving rise to the signal and (2) NMR frequency of a material is directly proportional to the strength of the magnetic field. Achieving the highest level of reproducibility possible (Figure 11.1) requires attention to standard operating procedures (SOPs) for sample preparation, data acquisition, and data processing. Additionally, proper instrument maintenance and well-designed sampling conditions are essential to realizing the desired reproducibility results.

Also important to the uniqueness of NMR is its inherently high compound specificity enabling the user to identify the components with high accuracy. This characteristic is derived from the fact that each atom that experiences a unique chemical environment displays a unique chemical shift and coupling constant. This allows the identification of structurally related materials with high accuracy including for many stereoisomers and diastereomers where other analytical techniques would fail to resolve the materials. A case in point is quinidine, a pharmaceutical agent used for anti-arrhythmia, and quinine, an alkaloid isolated from the bark of the cinchona tree (Figure 11.2). These two compounds share a common molecular weight of 324.417 g/mol and a common molecular formula of $C_{20}H_{24}N_2O_2$. Although

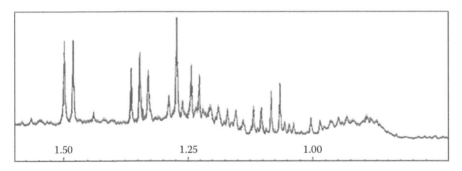

FIGURE 11.1 Reproducibility of NMR as shown by an overlay of 30 replicate NMR spectra of fruit juice including sample preparation acquired on three different Bruker AVIII 400 MHz spectrometers. Samples were prepared by six different people. (Courtesy of Manfred Spraul, Bruker BioSpin GmbH.)

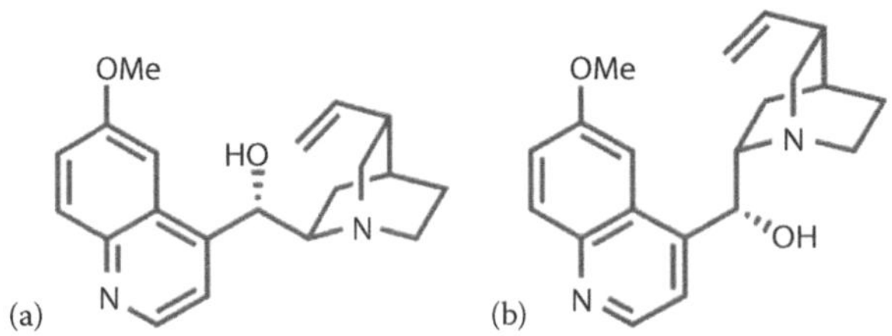

FIGURE 11.2 Chemical structures of (a) quinidine and (b) quinine.

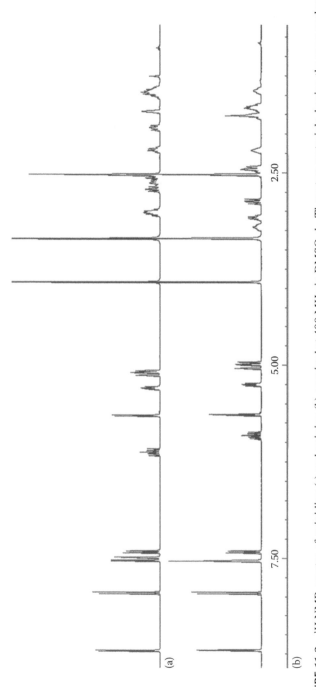

FIGURE 11.3 ¹H NMR spectra of quinidine (a) and quinine (b) acquired at 400 MHz in DMSO-d₆. These two materials, having the same molecular formula and molecular weight, are clearly distinguished using ¹H NMR.

these stereoisomers are easily distinguished using NMR (Figure 11.3), they are indistinguishable using various other analytical techniques such as UV, NIR, or mass spectrometry (MS). When used for screening of nutraceuticals, the high compound specificity of NMR enhances product safety by enabling the detection of adulterants and impurities even of closely related materials.

Quantitation of key components in nutraceuticals is essential to the evaluation of product quality and potential efficacy. Highly quantitative results are obtained with relative ease with NMR. Two standard approaches used for conducting absolute quantitative NMR measurements include using either an internal reference material or an externally calibrated spectrometer [7,8]. When using an internal reference material, a known amount of a compound is added to the sample to be tested. An externally calibrated spectrometer references to a previously acquired NMR spectrum of a known amount of a pure material. The NMR signal is proportional to the number of atoms giving rise to the signal. This principle means that any stable molecule with sharp NMR signals, not necessarily the components to be measured, can be used as a quantitative reference standard for either the internal calibrated or externally calibrated approach. Additionally, the advancement in NMR spectrometer development that enhanced the linearity of an NMR spectrometer has enabled the use of single-point calibrations and has made an externally calibrated spectrometer preferred (Figure 11.4). The external calibration holds as long as the software and hardware of the spectrometer remains unchanged thus making it possible to acquire data for months without recalibration on most systems; especially in the case of nutraceuticals, where complex mixtures are involved (Figure 11.5), externally calibrated spectrometers are beneficial as a result of reducing overlap and sample stability issues. Mixtures having many metabolites are more apt to experience sample stability issues than NMR samples having a single component. Therefore, it is best not to add an additional compound to the mixture if possible.

In a recent NMR study by Hicks et al. [9], factors contributing to potential interference in obtaining quantitative results were studied on a botanical product

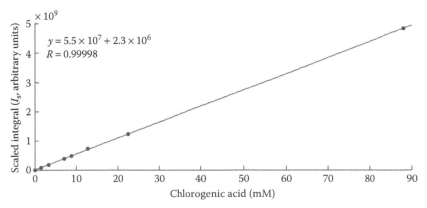

FIGURE 11.4 Linearity of NMR as demonstrated with a calibration curve of known chlorogenic acid standards (mM) measured on a five decimal place gravimetric scale, prepared in DMSO-d_6 against the scaled integral (I_s) in arbitrary units of the NMR signal at 6.8 ppm.

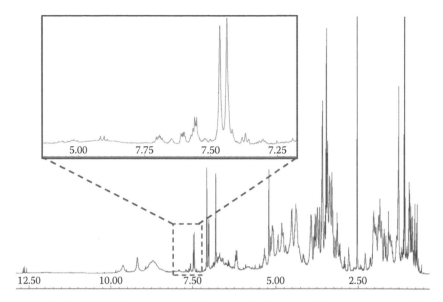

FIGURE 11.5 ¹H NMR spectrum of blueberry leaf extract (*Vaccinium angustifolium*) dissolved in DMSO-d_6 showing the typical complexity of the NMR spectrum of a dried crude ethanol extract (90%) from *Vaccinium* spp. Data acquired on a Bruker 600 MHz AVIII NMR spectrometer.

(blueberry leaf, *Vaccinium angustifolium*). It was determined that there was no significant interference from experimental sources such as different field strengths, extraction method, mesh size, gravimetric scale precision, NMR spectroscopy tube type, pulse program, amount of starting dry material, or day-to-day operation [9]. Attention to instrument optimization and experiment choice also plays a significant role in the level of quantification accuracy achievable [8,10]

Combining the distinctive features of NMR including (1) reproducibility, (2) compound specificity, and (3) high precision of quantitation with its ability to be heavily automated explains why NMR is now being utilized as an effective new tool for the study of nutraceuticals [11,12]. Global supply chains, regulations, and the desire for growth in the nutraceutical industry will all contribute to the continued growth of this emerging new analytical technology for nutraceutical analysis. NMR contributes a robust method for identity, strength, composition, and purity determination to enhance the product quality. This thorough knowledge of the components and composition aids in the understanding of the biological effects and proper usage of these beneficial materials as clinical trial results become available.

Automation in NMR

NMR is capable of being highly automated such that a non-NMR experienced user can operate an NMR spectrometer and obtain analyzed results. Functions such as tune, match, lock, shim, pulse calibration, phase, baseline correct, and Fourier transform are readily automated on any modern NMR. Additionally, NMR analysis may be automated either by using specifically designed software or with the use of automation scripts. To achieve fully automated acquisition, processing, and analysis, the NMR spectrometer must be initially optimized for automated use. Moreover, the acquisition and analysis method must be established by an NMR knowledgeable individual. In the pages of this chapter, we focus on describing the steps that would be taken to conduct a manual analysis of a nutraceutical so to make apparent the steps required in a fully automated nutraceutical evaluation.

SEEING BEAUTY IN A HIGHLY COMPLEX MIXTURE

Traditionally, NMR spectroscopists studying natural products have sought highly purified samples. These pure materials are used to study the molecular structure in detail (primary, secondary, and tertiary) and dynamics of compounds ranging from very simple compounds to very complex materials. In the early 1990s, studies on human body fluids became common and this began a new era of looking at highly complex mixtures. The field of metabolomics, which is the study of small molecule metabolites produced by an organism, is now well established, and the application of this approach to natural product material or other complex mixtures is expanding. The small molecule metabolites studied in biofluids and natural product extracts not only consist of amino acid, sugars, short-chain fatty acids, and vitamins common in many organisms, but also contain many unique bioactive compounds. There are many advantages in utilizing metabolomics in understanding and evaluating the composition of botanical extracts. One major advantage is that it allows the investigation of the biochemistry of the botanical material and gives insight into the plants unique chemical composition. Furthermore, studying these complex mixtures provides insight into the sample history as well as the botanical identity. Purification steps may be reduced or eliminated in the study of botanicals when taking a metabolomics approach. Sample information may be gained very rapidly, through the use of both targeted and nontargeted methods. With the potential of using these methods in complete automation, this can reduce the amount of labor required for material assessment once an evaluation method is established. The depth of information that can be obtained from the NMR spectrum of a crude botanical extract is considerable. Results in our facility have revealed that NMR has the ability to distinguish species [13,14], detect diseased plant material [15], distinguish the LR where a plant was grown [16], and even discriminate harvest dates of plants [14]. Our results are far from unique as many others have found a vast wealth of knowledge that may be garnered from NMR spectra [17,18].

TARGETED APPROACHES OF QUALITATIVE AND QUANTITATIVE ASSESSMENT

Analysis of nutraceuticals involves targeted and/or nontargeted approaches. Targeted approaches involve evaluating a sample for key metabolites to provide a qualitative and often a quantitative assessment of the material. For example, the popular botanical material *Aloe vera* is commonly evaluated using a qualitative targeted approach

[19,20] where the key metabolites, acetylated polysaccharides, glucose, and malic acid, assist in identifying the sample as *Aloe vera* and the degradation products acetic acid, lactic acid, succinic acid, fumaric acid, and formic acid indicate the product quality. Additionally, maltodextrin serves as a marker to indicate the intentional adulteration of *Aloe vera* that may have been used to artificially elevate the desirable acetylated polysaccharides. Qualitative assessment by NMR is greatly enhanced through the use of an NMR SBASE. Combining the reproducibility of NMR with the high compound specificity empowers the SBASE to be a highly valuable tool for laboratories utilizing NMR in studies of nutraceuticals. An NMR SBASE entry of a specific metabolite typically contains a *cleaned* NMR spectrum devoid of signals from solvent, reference standards, impurities, and noise. The entry may be of a pure material or a commonly used mixture. Utilizing appropriate SOPs, SBASEs can be developed to be used in the identification of key components with relative ease. Comparison of line positions, J-couplings, and the line shape of the individual resonances from the nutraceutical to the pure component enables the analyst to match the resonances and determine the presence of components as seen on *Aloe vera* in Figure 11.6. While some non-NMR users would consider the complexity of an NMR

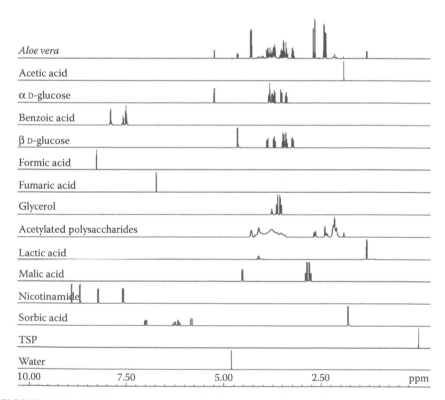

FIGURE 11.6 NMR spectrum of dried *Aloe vera* leaf material as compared to NMR SBASE entries used for automated analysis of *Aloe vera* that include common components, degradation products, and NMR reference standards. Data for all the samples were acquired at 600 MHz in 150 mM phosphate buffer at pH 7.4.

spectrum for a specific compound in a mixture to be a nuisance, this complexity brings assurance that the identity of the component is properly assigned. With careful attention to experimental conditions when designing the SOP for the material, the NMR user may reduce the chemical shift variations between the metabolites in a pure material, and hence the SBASE, and the chemical shift within the nutraceutical helps to make identification easier. To aid in the interpretation of highly overlapped regions of the NMR spectrum in a nutraceutical, a 2D J-resolved experiment may be utilized (Figure 11.7). This experiment resolves the chemical shift against the coupling constant. Therefore, two atoms having the same chemical shift are often easily distinguished by the coupling pattern. Coupling constants, being defined by nearby neighbors and the molecule's geometry, are typically different for different atoms. In molecules with a small number of protons or a small number of proton–proton couplings, the identification of the component is more complex and often requires experiments utilizing heteronuclear information. The 2D ^{1}H,^{13}C-HSQC (heteronuclear single quantum coherence spectroscopy) experiment may be employed in such cases. This experiment also permits good distinction between closely related structures such as actein and 26-deoxyactein found in black cohosh (Figure 11.8).

The targeted approach of nutraceutical evaluation by NMR provides identity and composition analysis of the material through identity of specific components. Quantitative evaluation of the degradation products of *Aloe vera* and maltodextrin provides a valuable product assessment. Information on the nutraceutical's strength may also be ascertained from the crude material through quantification if the active

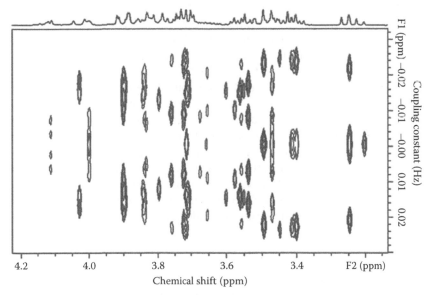

FIGURE 11.7 Expansion of the 2D J-resolved spectrum of *Aloe vera* showing the heavily overlapped sugar region at 600 MHz. The 2D J-resolved spectrum shows a specific proton signal as a function of chemical shift (*x*-axis) and coupling constant (*y*-axis). The pattern for each proton in this spectrum may be used for identification and quantification of signals that are difficult to evaluate directly from the 1D ^{1}H spectrum.

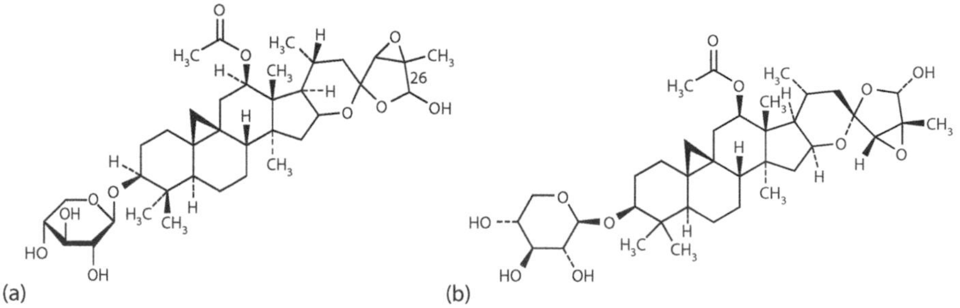

FIGURE 11.8 Chemical structures of (a) actein and (b) 26-deoxyactein found in black cohosh.

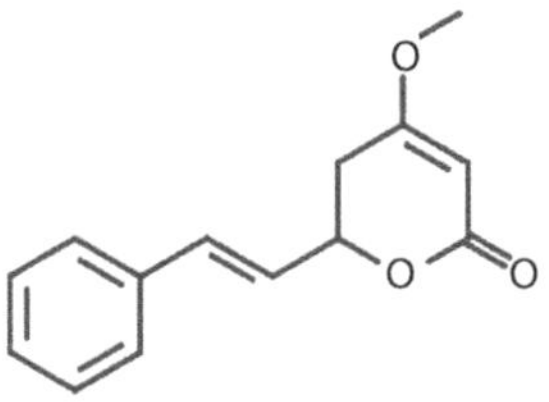

FIGURE 11.9 Chemical structure of kavain, a main component of *Kava kava*.

component has reasonably resolved NMR signals and the metabolites are available in adequate quantity. Kavain (Figure 11.9), the key component in *Kava kava* (*Piper methysticum*) [21], is easily observed in a crude *Kava kava* sample by spectral comparison as shown in Figure 11.10. Kavain is considered the active component of this material and therefore the strength may be evaluated based on the amount of kavain present. A study of three different lots showed considerable variation in the amount of kavain: 8.0 mM (lot 1), 2.7 mM (lot 2), and 8.3 mM (lot 3) [22]. As in the case of kavain in *Kava kava*, a single metabolite or a couple of metabolites often dominated the NMR spectrum and are easily identified and quantified.

Purity of a complex mixture is difficult to assess based on targeted analysis methods alone unless expected impurities and adulterants are well known. If known, these impurities are searched for using an NMR SBASE. If they are present in observable quantities, then they may be identified and quantified from a single NMR spectrum without further purification of the sample. A nontargeted approach, if available, is a better tactic for determining purity because it can be used for the detection of both predicted and not predicted impurities and adulterants.

DEVELOPMENT OF AN NMR SBASE

The development of an NMR SBASE is best preceded by careful consideration of the ways the SBASE may be used in the future. Could it be used for multiple nutraceuticals? Would it be important to distinguish a particular metabolite if it were overlapped in ^{1}H but not in ^{13}C? What solvent would be the most versatile for products that will be evaluated? Are other easily detected nuclei such as phosphorus or fluorine present that may be useful in identification and/or quantification of the

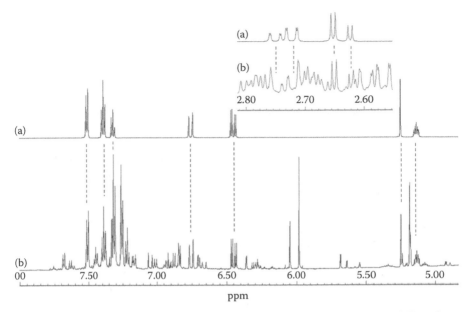

FIGURE 11.10   (a) NMR SBASE entry of kavain and (b) NMR spectrum of *Kava kava*. NMR signals for kavain in the *Kava kava* sample at 7.51, 7.39, 6.75, and 6.44 ppm are adequately resolved and present in significant quantity for quantification. The concentration of kavain using these signals was determined by NMR to be 8.0 mM.

molecule? SBASE entries may be pure materials or mixtures. General guidelines for SBASE development are as follows:

1. Develop an SOP for the SBASE development. Ideally this SOP is similar to the method used for the actual nutraceutical sample that will be measured. Using common SOP increases the quality of matches from analyte to SBASE entry.
2. As part of the SOP, consider current and future needs with regard to the temperature and field strength of analysis. Both of these factors have significant influence on the spectra and thus on the utility of the SBASE.
3. As part of your SOP, determine a naming structure for your SBASE entries. This will assist in manual and automated searches against your SBASE. For example, an SBASE directory for 1,1-dichloroethane may include a tag called *1dnoig-d2o-600* where the experiment is a 1D Nuclear Overhauser Spectroscopy experiment with inverse gated decoupling *1dnoig*, the solvent is D_2O *d2o*, and the data are a 600 MHz *600* spectrum. Searches for SBASE entries for particular experiments, solvents, and field strengths are easy using such nomenclature structure.
4. Use a solvent where chemical shift changes are expected to be minimal. If D_2O is used, consider using a buffer to reduce chemical shift changes observed from sample to sample. A buffer commonly used in our facility with botanicals measured in D_2O is a 150 mM KH_2PO_4 buffer (pH 7.4, 200 µM NaN_3, 0.01%

trimethylsilyl propanoic acid [TSP]). Buffers should be chosen to give optimal stability of components and chemical shifts. Highly volatile solvents may result in chemical shift changes as a result of solvent evaporation and are best avoided.

5. To maximize the utility of your SBASE, use a solvent in which current and future samples are expected to be soluble. Test as many samples as possible before deciding on the solvent to use. We often use dimethyl sulfoxide (DMSO-d_6) as a solvent because various materials are readily solubilized. Additionally, observing the exchangeable protons may be advantageous for distinguishing metabolites.

6. Run spectra as soon as possible after sample preparation. Chemical shift changes occur with sample degradation, evaporation, and binding of components to the NMR tube. It is recommended to acquire the NMR spectrum shortly after the SBASE sample is prepared.

7. Evaluate the structure of the material being added to the SBASE entry carefully by assigning all peaks and validating coupling coherences. Purchased samples are not always structurally correct. The benefit obtained from the NMR SBASE is only as good as the quality of the SBASE.

8. Acquire as many spectra as possible when you have the sample available. An SBASE entry only needs to be generated once for a particular compound or material. While the experiments needed depend on the specific application, some common experiments beneficial to collect for an SBASE include 1D ^{1}H NOESY; 2D J-resolved; 1D CPMG (Carr–Purcell–Meiboom–Gill); ^{1}H correlated spectroscopy (COSY); 1D ^{13}C, distortionless enhancement by polarisation transfer with retention of quanternaries (DEPTq); and ^{1}H,^{13}C-HSQC.

9. Keep metadata for your SBASE entries. Metadata include information such as suppliers and known ingredients that contribute to harvestable information about the sample.

OBTAINING QUANTITATIVE RESULTS

Attention to several key factors improves the accuracy of quantitative results by NMR [8,10,23]. Quantitative comparisons across spectra are enhanced by utilizing SOPs for sample preparation, NMR acquisition, data processing, and analysis. NMR factors critical to obtaining high accuracy of quantification are easily addressed and include the following:

1. *Flat baselines:* Baseline distortions may lead to substantial errors in quantification. However, by choosing the right experiment baseline distortions may be greatly reduced. A 1D ^{1}H NOESY experiment has been widely accepted by the metabolomics community for its ability to provide a relatively flat baseline. Instrument optimizations including temperature stability and parameter optimization greatly enhance the flatness of the baseline of the NMR spectrum.

2. *Pulse width calibration on each sample:* A 90° pulse width varies from sample to sample particularly when changes in ionic strength occur. To quantify a sample relative to an external standard, it is necessary to know the flip angle used with each sample.

3. *An adequate relaxation delay:* At least $5 \times T1$ is needed when a 90° observation pulse is used for accurate quantification. Alternatively, if a reduced relaxation time is used, then calculations are required to adjust the quantification result as a function of the relaxation delay.
4. *Signal-to-noise:* A signal-to-noise for the resonance being evaluated of greater than 1000:1 is often suggested to give highly accurate quantitative results.
5. *Experiment choice:* Experiments differ in the level of quantification that each can deliver over the spectral range measured. When using an experiment that has not be tested previously, it is advisable to conduct spiking experiments to evaluate the precision of a given experiment.
6. *Accurate phase correction:* Phase of the NMR spectrum is affected by many factors including lock parameters and core instrument delay parameters.

Quantification of an individual compound can either be done by region integration or line shape fitting. Region integration is performed if the signals are isolated with no background signal. Background signal may result from artifacts of the instrumentation or genuine material signal resulting from large molecules such as large proanthocyanidins or cellular material. In case of overlapping signals or signals on top of a background, line shape fitting is the method of choice (Figure 11.11). Line shape fitting adjusts the individual signal parameter in an iterative process. It is required to include the overlapping peaks in the fit. The line shape can be either calculated from chemical shifts and coupling constants [23], and may be directly obtained from a SBASE entry, or a set of peaks describing the multiplet to fit. The simpler the coupling pattern, such as the anomeric proton of a sugar molecule which typically is a doublet line shape pattern, the more precise the quantification may be. Higher-order multiplets contain many unresolved signals, and this makes it difficult to quantify them. It is required to determine chemical shifts and coupling constants from the reference spectrum. Higher-order multiplets occur when the chemical shift difference is relatively small compared to the coupling between these peaks. Using higher magnetic field strength could potentially simplify the multiplet to a pattern more easily quantified. Fortunately, most metabolites have some NMR signals that have simple patterns in less congested regions of the spectrum in a nutraceutical. Quantification on a few or even a single resonance of a metabolite is adequate.

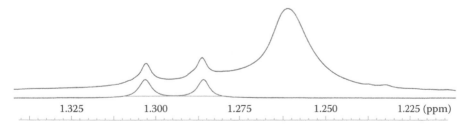

FIGURE 11.11 Quantification of lactic acid in blueberry (*Vaccinium angustifolium*) leaf extract (top line) using the methyl doublet. Quantification was performed using an iterative line shape fit of the doublet peaks (bottom line) and other peaks in region.

NONTARGETED NMR APPROACHES
OF QUALITATIVE ASSESSMENT

Nontargeted NMR approaches to nutraceutical analysis allows a rapid assessment of material and can provide extensive information on the sample. Information pertaining to material identification, purity, adulterants, plant health state, growing location and/or processing location, processing method, and formulations are some of the information accessible from a nontargeted assessment of a nutraceutical. Accordingly, botanical origin, identity, purity, and composition of a nutraceutical product may be evaluated using the combined use of statistical methods and quantification strategies. Typical nontargeted approaches include both univariate and multivariate chemometric methods. Methodologies utilizing NMR-based metabolomics are well summarized in an excellent review by D.G. Cox et al. [24]. Using an NMR-based metabolomics approach, Robert Verpoorte and coworkers gained insight into the metabolic discrimination of *Ilex* and *Verbascum* L. (*mulleins*) species for chemotaxonomic classification [17,19].

A typical objective for the application of nontargeted analysis is pattern recognition. Pattern recognition has two analysis styles: clustering (grouping) and classification of samples [24]. Pattern recognition requires a statistical analysis of the data set. A typical approach is dimension reduction: projecting the high-dimensional data on to lower-dimensional space while retaining as much as possible of the variance of the data set [25–28]. Most often a principal component analysis (PCA) is performed. PCA transforms the data set into a new coordinate system; the new principal components (PC) axes are uncorrelated and which are sorted by variance: the first PC has the highest variance, higher PCs explain small variances. To reduce the dimensionality, the first few PCs are typically used. The rest are typically regarded as noise and ignored. This is an unsupervised method because the membership of individual samples of a data set is not known to the PCA algorithm. Prior to the statistical analysis, data are centered and scaled [27]. An overview about standard scaling methods can be found in the book by D.E. Axelson [28].

Classification or outlier detection is based on distance measures in the reduced data space. SIMCA (soft independent modeling of class analogies) is based on PCA analysis and often used for classification. First, the SIMCA model needs to be defined (1) by the selection of the reduced model space and (2) by the definition of maximum allowed distances. A sample is tested by being projected into the reduced model space. Two distances are taken into account: distance to the model (the residual standard deviation) and the distance to the model center (within reduced model space) [26,27,29]. Reproducibility of the measurement, as afforded by using SOPs, is very important for such an analysis. Otherwise, without having reproducible results, additional variance is added to the data set which can easily mask important trends [27].

SAMPLE COLLECTION AND PREPARATION CONSIDERATIONS
FOR BOTANICAL METABOLOMICS

To gain such a wealth of information about a sample, some investment is required including (1) gaining representative samples; (2) metadata that define these samples of a given type; and (3) a standard procedure for sample collection, processing, and data acquisition. Many experimental variables need to be carefully considered before pursuing a metabolomics approach for quality control and product assessment. From sample preparation, extraction, and storage, a standard procedure is highly recommended in order to obtain comparable and reproducible results. The sample selection and preparation steps are the initial considerations that need to be chosen with care. It is important to recognize that the collection of the botanical material can have significant influence over the resulting analytical data as factors such as sampling time, season, origin, attitude, land conditions, and plant part may result in variation in metabolites and quantity of metabolites observed. For example, Bavaresco et al. [30] collected three varieties of grapes, *V. vinifera* L. "Barbera," "Croatina," and "Malvasia di Candia aromatica," from North-West Italy at four different elevations (150, 240, 320, and 420 m above sea level) and detected fluctuations in the concentrations of stilbenes (*trans*-resveratrol, *trans*-piceid, and *cis*-piceid). From the results, the altitude had a key role in the overall stilbene concentrations as they all increased in all grape varieties as elevation raised to 320 m but decreased at 420 m. With well-documented metadata in this case, NMR screening may be used for determining the best altitude for growing the grape varieties for optimal stilbene concentration or determining which variety of grape has the largest variation in stilbene concentration at specific altitudes.

Samples to be evaluated may be direct from a botanical or an already processed product. Standard procedures for each case are required to ensure reproducibility and enable product comparisons. Already processed product often requires only very simple sample preparation procedures. There is no need for a large quantity of material for NMR-based metabolomics. Approximately 10–50 mg is adequate for a ^{1}H NMR spectrum. For a sample direct from a plant, preparation and extraction are key steps to maximize the quantity and range of metabolites for analysis as is often desired when performing species identification by NMR. Grinding at low temperatures and sonication of the botanical material in the presence of a solvent have been common procedures to ensure efficient breakage of the cells to obtain the highest quantity of metabolites [31]. However, due to the high diversity of polarities and pH ranges, it is difficult to extract all metabolites with a single solvent extraction system. For example, alkaloids are soluble mainly in nonpolar solvents at higher pHs while they are soluble in aqueous solvents at lower pHs [32]. Brown et al. [33] did a comparison of various single solvent extraction systems such as acetonitrile, benzene, chloroform, methanol, DMSO, and D_2O for NMR-based metabolomics of a biological material. From the results, methanol, DMSO, and D_2O extracted higher concentrations of metabolites than acetonitrile, benzene, and chloroform. A combination of the solvents has been utilized as well with a mixture of CD_3OD and $KH_2PO_4 D_2O$ buffer (to adjust for chemical shift fluctuations due to pH) and was successful in extracting out many classes of compounds such as amino acids, organic acids, short-chain fatty acids, sugars, phenolics, and terpenoids [34,35]. Varying mixed solvents of methanol have also been used [32,36–39] to gain a complement of desired metabolites from specific botanical. After preparation and extraction of the samples, it is recommended to store the samples at −80°C until NMR samples are prepared in order to minimize any biochemical reactions or potential degradation [40]. Regardless of the extraction method or sample preparation method used, it is essential to standardize on one method for nontargeted analysis. An example of an extraction procedure, courtesy of John T. Arnason [16] of the University of Ottawa, is as follows. Ground ginseng root was extracted by sonication three times for 25 min each time at room temperature (25°C) in 70% methanol, using 10 mL twice and 4 mL for the third extraction. The phases were centrifuged using a 5810 R centrifuge at 2,220 g and the supernatants were pooled and brought to 25 mL in a volumetric flask. Extracts were dried using a centrifugal evaporator and lyophilized using a Super Modulyo freeze dryer. The samples were then prepared for NMR analysis.

Example: NMR sample preparation procedure for store purchased Ginseng Extract powder product or a dried extract of ginseng is as follows:

1. Preparation of NMR solvent containing reference material.
   a. Add 157 uL 300 mM deuterated DSS (4,4-dimethyl-4-silapentane-1-sulfonic acid) to 100 mL of deuterated solvent (DMSO-d_6).
   b. DSS is used for a line shape reference and, in part, a chemical shift reference.
2. Prepare sample on the day of NMR data acquisition.
   a. Weigh 25 mg of ginseng sample.
   b. Dissolve ginseng sample in 1 mL solvent containing referencing material.
   c. Vortex for ~30 s to 1 min and centrifuge (≥13,000 rpm) for 10 min.
   d. Transfer 600 uL of solution into an NMR tube.
   e. Transfer NMR sample to acquisition queue or store sample in freezer until data acquisition is possible.

NMR ACQUISITION AND PROCESSING FOR METABOLOMICS ANALYSIS

A 1D ^{1}H NMR spectrum of a botanical material can give a plethora of chemical information due to its high sensitivity and its common occurrence in many organic molecules. With prior knowledge of the chemical shifts and coupling constants, as used in an NMR SBASE, many metabolites detected in the NMR spectrum can be easily identified. However, before a high-quality NMR spectrum can be obtained, it is necessary to ensure that the acquisition and processing procedures are optimized so the results are reproducible. Some acquisition parameters that should be checked are pulse calibration of the nuclei of interest, number of scans, temperature, and recycle delays. To understand more about optimizing the various acquisition parameters, two excellent reviews are recommended from Pauli et al. [23], Simpson et al. [41], and Burton et al. [8]. Pulse programs also must be checked for reproducibility. The standard experiments that often make up a metabolomics study include the 1D ^{1}H NOESY and 1D ^{1}H CPMG spectroscopy for a total acquisition time of about 20 min. Inverse C-13 gating is commonly utilized with these experiments to eliminate the C-13 coupling signals. The experiments utilized for a nontargeted approach are by no means limited to proton and 1D experiments only. The 2D J-resolved spectroscopy experiment and ^{13}C NMR are heavily used. The 2D ^{1}H, ^{13}C-HSQC experiments are also being explored [42–44].

THREE EXAMPLES: NMR ANALYSIS OF NUTRACEUTICALS AND BOTANICALS

GINSENG

Cultivated North American ginseng (*Panax quinquefolius*) is a valuable medicinal crop and a common dietary supplement in the United States [45], Canada, and Asia. Major interest has been focused on developing techniques to distinguish various ginseng LRs and ginseng species (*P. quinquefolius* vs. *P. ginseng*) to assess quality and to protect intellectual property. NMR-based chemometric methods have been successful in distinguishing between different Ontario LRs of ginseng and also ginseng species such as *P. ginseng* [16]. In Figure 11.12a, PCA was used to differentiate between each ginseng LR with LR 2 and 5 being the most significant from the rest of the group. This was similar in Figure 11.12b, as the North American ginseng LRs were clearly separated from the Asian ginseng. To provide a means for quality control and authenticity, an extension of the previous study would be to develop a SIMCA model using PCA for classification. In Figure 11.13, a SIMCA model of LR 5 using 22 samples was generated and used in automation to screen and compare various types of ginseng. With minimum and maximum confidence intervals set for each classification model, ginseng samples that are close in profile will be considered in the model and authentic while ginseng that are in different LRs or species will be considered an outlier.

Similarly, as an application of NMR as a quality control tool, a producer of a nutraceutical would generate a nontargeted method for their product using acceptable samples. Once a method is generated, material validation is very rapid, only taking about 15 min and can be done in complete automation. If the newly produced sample is consistent with the parameters set in the nontargeted method (no unknowns/adulterants/impurities detected and below or above acceptable concentration limits), then the producer has high confidence that the sample conforms to the identity, purity, composition, and strength of the defined product as detectable by NMR.

ENERGY DRINKS

EDs are complex mixtures containing a large variety of compounds such as sugars (sucrose, glucose, etc.), artificial sugars (sucralose, acesulfame K, erythritol, etc.), vitamin B's (niacin, ninositol, riboflavin, etc.), stimulants (caffeine, taurine, gluconolactone, and L-cartinine), preservatives (benzoic acid, sorbic acid, and citric acid), and botanical materials (ginseng, gurana, yerba mate, and milk thistle extracts). NMR spectroscopy can be a powerful tool for quality control and authentication of EDs with many components being identified using an NMR SBASE. A targeted automated NMR method for EDs can be applied using a qualitative and quantitative approach. In Figure 11.14, an ED (undisclosed brand) was screened for potential components using an NMR ED SBASE. Using the ^1H chemical shift information from each compound, identification and quantification information could be obtained. A pass/fail report can then be generated after specific factors are matched (i.e., no unknowns/adulterants/impurities are found and compounds are above/below a certain

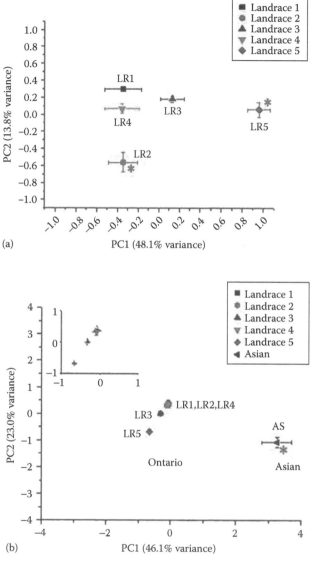

FIGURE 11.12 Mean principal component analysis (PCA) scores plot of ^{1}H NMR spectra of: (a) five Ontario ginseng landraces and (b) the five Ontario ginseng landraces (*P. quinquefolius*) compared with Asian (AS) ginseng (*P. ginseng*). Each point represents the mean principal component score for each landrace and the error bars represent the standard error of the mean. The key indicates the ginseng landrace or specie for each point. The asterisk represents the landrace or specie that was significantly different from the rest of the groups using Tukey's multiple comparison test. (Replicated from J. Yuk et al., *Analytical and Bioanalytical Chemistry*, 405, 4499–4509, 2013.)

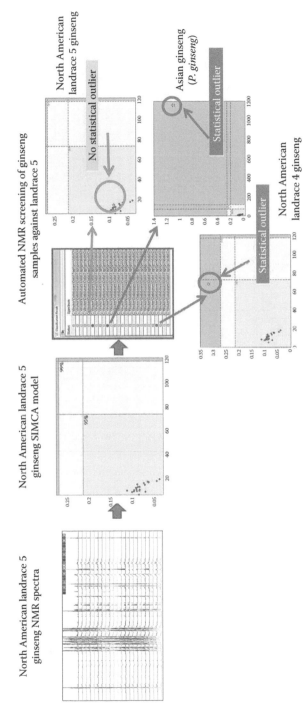

FIGURE 11.13 A schematic of an automated NMR screening approach of ginseng samples using a North American ginseng landrace 5 (*P. quinque-folius*) SIMCA model. The SIMCA model is generated using all the ginseng landrace 5 NMR spectra. Ginseng samples that fit the model are labeled in the model (North American landrace 5 ginseng) while any samples that lie outside the 95%–99% confidence interval of the model are labeled as a statistical outlier. The North American ginseng landrace 5 SIMCA model was specific enough to differentiate between ginseng species (Asian ginseng *P. ginseng*) and landraces of the same specie (North American landrace 4 ginseng). The automated NMR SIMCA screening method was performed using Bruker Assure-Raw Material Screening software v.1.5.

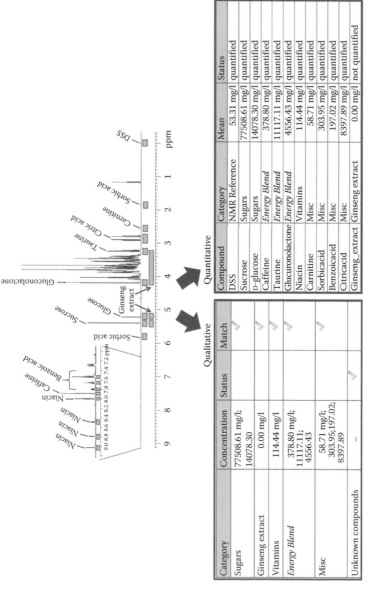

ED NMR spectrum with individual components identified

Qualitative

Category	Concentration	Status	Match
Sugars	77508.61 mg/l; 14078.30		✓
Ginseng extract	0.00 mg/l		✓
Vitamins	114.44 mg/l		✓
Energy Blend	378.80 mg/l; 11117.11; 4556.43		✓
Misc	58.71 mg/l; 303.95;197.02; 8397.89		✓
Unknown compounds	–		✓

Quantitative

Compound	Category	Mean	Status
DSS	NMR Reference	53.31 mg/l	quantified
Sucrose	Sugars	77508.61 mg/l	quantified
D-glucose	Sugars	14078.30 mg/l	quantified
Caffeine	Energy Blend	378.80 mg/l	quantified
Taurine	Energy Blend	11117.11 mg/l	quantified
Glucuronolactone	Energy Blend	4556.43 mg/l	quantified
Niacin	Vitamins	114.44 mg/l	quantified
Carnitine	Misc	58.71 mg/l	quantified
Sorbicacid	Misc	303.95 mg/l	quantified
Benzoicacid	Misc	197.02 mg/l	quantified
Citricacid	Misc	8397.89 mg/l	quantified
Ginseng_extract	Ginseng extract	0.00 mg/l	not quantified

FIGURE 11.14 A schematic of an automated NMR targeted screening method of ED. Two simultaneous approaches are taken with a qualitative and quantitative direction. In the qualitative approach, ED compounds are first matched in the ED spectrum and any unknowns are checked. The quantitative approach is then applied using the matched compounds and concentration values are calculated using an external quantification standard from a calibrated NMR spectrometer. The automated NMR targeted screening method was performed using Bruker Assure-Raw Material Screening software v.1.5.

concentration limit). An NMR targeted method for complex mixtures such as EDs can be a powerful approach in utilizing the high reproducibility and high-throughput capability of NMR.

UNDECLARED DRUG SUBSTANCE IN REGENERECT

In many cases, it will not be possible to gain enough samples to define a model for a specific material. It may also not be possible to generate an NMR SBASE of all possible suspected components. For example, adulteration of herbal product formulations with weight loss or sexual enhancement undeclared drug derivatives is common. Gaining enough nonadulterated product to generate a relevant model is not possible. Also, access to the drug or drug derivatives is restricted. However, NMR analysis is still very useful. In the case of the dietary supplement called Regenerect, a product that was voluntarily recalled in April 2011 as a result of an FDA lab analysis detecting sulfoaildenafil [46], NMR easily identified peaks resulting from the undeclared drug substance (Figure 11.15). Adulterated samples often contain the undeclared drug at a therapeutic dose level. In the case of Viagra® (sildenafil citrate), the therapeutic dose is about 50 mg. In the Regenerect sample, the relative concentration of the sildenafil analog, sulfoaildenafil, is very high relative to the botanical material and is therefore readily observable. Three of the declared natural extract components including silkworm, white willow bark, and oyster extract are present in the material at low concentrations relative to the undeclared drug substance.

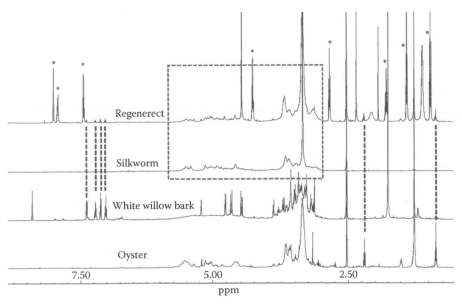

FIGURE 11.15 NMR spectrum of nutraceutical Regenerect lot 100521 that, according to the supplement facts, contains extracts of silkworm, white willow bark, and oyster. In addition, large quantities of a material (indicated with asterisk) is present that was identified by the FDA as sulfoaildenafil in 2011.

PRACTICAL CONSIDERATIONS TO IMPLEMENTING NMR SCREENING OF NUTRACEUTICALS

The method of analysis of any given material must be fit for purpose. In some cases, an NMR method will be adopted to take advantage of the high compound specificity, reproducibility, and/or high quantification accuracy. Conveniently, an NMR spectrometer is a versatile instrument that can be easily used for many different types of analysis without the need to change hardware or clean the system between materials. As a result, NMR spectrometers are rarely dedicated to specific analysis and therefore may switch to a different material just as easily as running the same material repeatedly. This versatility combined with a rapid analysis (often 10–20 min per sample) makes an NMR spectrometer an affordable addition to many analytical laboratories. Disposable costs are low, typically being under $5 per sample for deuterated solvent and an NMR tube. The cost per sample screened over a 10-year lifetime of a 400 MHz instrument may be as low as $1 per sample, excluding the cryogen maintenance cost that varies depending on negotiated contracts and location. Laboratories utilizing staff not trained in NMR spectroscopy often utilize cryogen service contracts to fill the magnet with cryogens as needed. Using this approach, lower cryogen maintenance costs may be achieved for sites with small cryogen demand through agreements with large-scale users of cryogens.

Space requirements of NMR spectrometers have been reduced tremendously through the standard use of shielded magnets. The minimum footprint required for a 400 MHz NMR is about 5 feet by 7 feet with access to fill cryogens required. The 400 MHz magnets now have the 5 Gauss line, considered the primary safe distance for metal objects in the vicinity of the magnet, at the edge of the magnet. The greatly improved safety and smaller space requirement makes siting an NMR in many analytical labs possible. For screening purposes, a 400 MHz NMR system is typically adequate for nutraceutical analysis. NMR systems for both screening and research use, such as the structure elucidation of bioactive components, are typically 600 MHz systems.

CONCLUSIONS

Recent advancements in NMR technology have revolutionized its capabilities and applications. The resulting improved safety, lowered costs, and improved automation make this highly sophisticated technology now accessible to the general analytical laboratories that are traditionally not versed in NMR spectroscopy. With strengths in reproducibility, compound specificity, and absolute quantification, NMR provides a valuable tool for nutraceutical quality and authenticity analysis. Targeted and nontargeted screening approaches, which may be conducted simultaneously from a single NMR spectrum, can provide valuable insight into the identity, purity, strength, and composition of the nutraceutical product. Use of NMR SBASEs and statistical models make analysis of these highly complex mixtures possible even in full automation. Additionally, the sharing of NMR SBASEs and models with users at other sites and on different instruments of the same field strength empowers users to screen materials with cross-lab consistency and benefit from development efforts of other laboratories.

With increasing pressures from regulation and global trade for improved quality control to provide better safety and an understanding of product efficacy, NMR provides a timely addition to the nutraceutical analysis toolbox. Added safety will benefit many, from the consumer who desires a consistent and efficacious product to the producers and suppliers who wish to grow a long-standing profitable business. With an increasing number of people maintaining or improving their health with high-quality nutraceuticals, our society benefits in immeasurable ways.

REFERENCES

1. K.L. Wrick. "The Impact of Regulations on the Business of Nutraceuticals in the United States: Yesterday, Today and Tomorrow." In *Regulation of Functional Foods and Nutraceuticals: A Global Perspective.* C.M. Hasler, Ed., Blackwell Publishing: Ames, IA, 2005, pp. 16–19.
2. G.A. Morris. Modern NMR techniques for structure elucidation. *Magnetic Resonance in Chemistry* 24(5), 1986: 371–403.
3. L.M. Jackman and S. Sternhell. *Applications of Nuclear Magnetic Resonance Spectroscopy in Organic Chemistry.* 2nd Edition; Pergamon Press: New York, 1969.
4. G.E. Martin and A.S. Zektzer. *Two-Dimensional NMR for Establishing Molecular Connectivity.* Wiley-VCH: New York, 1988.
5. R.C. Breton and W.F. Reynolds. Using NMR to identify and characterize natural products. *Natural Products Reports* 30(4), 2013: 501–524.
6. N. Bross-Walch, T. Kühn, D. Moskau, and O. Zerbe. Strategies and tools for structure determination of natural products using modern methods of NMR spectroscopy. *Chemistry & Biodiversity* 2(2), 2005: 147–177.
7. G. Wider and L. Dreier. Measuring protein concentrations by NMR spectroscopy. *Journal of the American Chemical Society* 128 (8), 2006: 2571–2576.
8. I.W. Burton, M.A. Quilliam, and J.A. Walter. Quantitative 1H NMR with external standards: Use in preparation of calibration solutions for algal toxins and other natural products. *Analytical Chemistry* 77, 2005: 3123–3131.
9. J.M. Hicks, A. Muhammad, J. Ferrier, A. Saleem, A. Cuerrier, J.T. Arnason, and K.L. Colson. Quantification of chlorogenic acid and hyperoside directly from crude blueberry (*Vaccinium angustifolium*) leaf extract by NMR spectroscopy analysis: Single-laboratory validation. *Journal of AOAC International* 95, 2012: 1406–1411.
10. E. Saude, C. Slupsky, and B.D. Sykes. Optimization of NMR analysis of biological fluids for quantitative accuracy. *Metabolomics* 2(3), 2006: 113–123.
11. M. Spraul, B. Schütz, P. Rinke, S. Koswig, E. Humpfer, H. Schäfer, M. Mörtter, F. Fang, U.C. Marx, and A. Minoja. NMR-Based Multi Parametric Quality Control of Fruit Juices: SGF Profiling. *Nutrients* 1, 2009: 148–155.
12. D.W. Lachenmeier, E. Humpfer, F. Fang, B. Schuetz, P.Dvortsak, C. Sproll, and M. Spraul. NMR-Spectroscopy for Nontargeted Screening and Simultaneous Quantification of Health-Relevant Compounds in Foods: The Example of Melamine. *Journal of Agricultural Food Chemistry* 57(16), 2009: 7194–7199.
13. M.A. Markus, S.M. Luchsinger, J. Yuk, J. Ferrier, J.M. Hicks, K.B. Killday, C.W. Kirby et al. Distinguishing *Vaccinium* species by chemical fingerprinting based on NMR spectra. Validated with spectra collected in different laboratories. *Planta Medica*, 80 (08/09), 2014: 732–739.

14. J. Ferrier, M.A. Markus, J.M. Hicks, M.J. Balick, A. Cuerrier, J.T. Arnason, and K.L. Colson. Subgeneric identification with 42 species of *Vaccinium* L. *(Ericaceae)* from wild and botanical gardens using nuclear magnetic resonance (1H NMR) spectroscopy, manuscript in preparation.

15. C. Neto, E. Yiantsidis, A. Dovell, F. Caruso, S. Luchsinger, and K.L. Colson. 1H-NMR analysis of cranberry leaves (*Vaccinium macrocarpon*) to detect disease-related stress and variation in metabolites among cultivars, manuscript in preparation.

16. J. Yuk, K.L. McIntyre, C. Fischer, J. Hicks, K.L. Colson, E. Lui, D. Brown, and J.T. Arnason. Distinguishing Ontario ginseng landraces and ginseng species using NMR-based metabolomics. *Analytical and Bioanalytical Chemistry* 405, 2013: 4499–4509.

17. H.K. Kim, K.S. Saifullah, E.G. Wilson, S.D. Kricun, A. Meissner, S. Goraler, A.M. Deelder, Y.H. Choi, and R. Verpoorte. Metabolic classification of South American Ilex species by NMR-based metabolomics. *Phytochemistry* 71(7), 2010: 773–784.

18. M.I. Georgiev, K. Ali, K. Alipieva, R. Verpoorte, and Y.H. Choi. Metabolic differentiations and classification of *Verbascum* species by NMR-based metabolomics. *Phytochemistry* 72(16), 2011: 2045–2051.

19. J. Edwards. In *Aloe Vera Leaf, Aloe Vera Leaf Juice, Aloe Vera Inner Leaf Juice: Standards of Identity, Analysis, and Quality Control.* R. Upton, Ed., American Herbal Pharmacopoeia®, Scotts Valley, CA, 2012; pp. 33–42.

20. P. Jiao, Q. Jia, G. Randel, B. Diehl, S. Weaver, and G. Milligan. Quantitative 1H-NMR spectrometry method for quality control of aloe vera products. *Journal of AOAC International* 93(3), 2010: 842–848.

21. H. Sauer and R. Hänsel. Kava lactones and flavonoids from a Piper species endemic to Papua, New Guinea (in German). *Planta Medica* 15(4), 1967: 443–458.

22. K.L. Colson, J. Hicks, J.L. Sullivan, J.L. Yang, and C. Okunji. Kava Extract Evaluation: Comparison of NMR and HPLC Results. *American Society of Pharmacognosy Annual Conference*, poster presentation, July 10–14, St. Petersburg, FL, 2010.

23. G.F. Pauli, B.U. Jaki, and D.C. Lankin. Quantitative 1H NMR: Development and potential of a method for natural products analysis. *Journal of Natural Products* 68(1), 2005: 133–149.

24. D.G. Cox, J. Oh, A. Keasling, K. Colson, and M. Hamann. *The Utility of Metabolomics in Natural Product and Biomarker Characterization.* BBAGEN (in press).

25. I.T. Jollife. *Principal Component Analysis.* 2nd Edition. Springer: New York, 2002.

26. B.G.M. Vandeginste and S.C. Rutan. *Handbook of Chemometrics and Qualimetrics: Part B.* Elsevier: Amsterdam, the Netherlands, 1998.

27. K. Esbensen, S. Schoenkopf, and T. Midtgaard. *Multivariate Analysis in Practice.* CAMO: Trondheim, Norway, 1996.

28. D.E. Axelson. *Data Preprocessing for Chemometric and Metabonomic Analysis.* MRi Consulting: Chicago, IL, 2010.

29. M. Otto. *Chemometrics.* Wiley: New York, 1999.

30. L. Bavaresco, S. Pezzutto, M. Gatti, and F. Mattivi. Role of the variety and some environmental factors on grape stilbenes. *Vitis: Journal of Grapevine Research* 46 (2), 2007: 57–61.

31. H.M. Heyman and J.J.M. Meyer. NMR-based metabolomics as a quality control tool for herbal products. *South African Journal of Botany* 82, 2012: 21–32.

32. F. van der Kooy, F. Maltese, Y.H. Choi, H.K. Kim, and R. Verpoorte. Quality control of herbal material and Phytopharmaceuticals with MS and NMR based metabolic fingerprinting. *Planta Medica* 75(7), 2009: 763–775.

33. S.A.E. Brown, A.J. Simpson, and M.J. Simpson. Evaluation of sample preparation methods for nuclear magnetic resonance metabolic profiling studies with *Eisenia fetida*. *Environmental Toxicology and Chemistry* 27(4), 2008: 828–836.

34. H.K. Kim, Y.H. Choi, C. Erkelens, A.W.M. Lefeber, and R. Verpoorte. Metabolic fingerprinting of *Ephedra* species using 1H-NMR spectroscopy and principal component analysis. *Chemical & Pharmaceutical Bulletin* 53(1), 2005: 105–109.

35. O. Hendrawati, O. Hendrawati, Q. Yao, H.K. Kim, H.J.M. Linthorst, C. Erkelens, A.W.M. Lefeber, Y.H. Choi, and R. Verpoorte. Metabolic differentiation of *Arabidopsis* treated with methyl jasmonate using nuclear magnetic resonance spectroscopy. *Plant Science* 170(6), 2006: 1118–1124.

36. G. Le Gall, I.J. Colquhoun, and M. Defernez. Metabolite profiling using 1H NMR spectroscopy for quality assessment of green tea, *Camellia sinensis* (L.). *Journal of Agricultural and Food Chemistry* 52(4), 2004: 692–700.

37. J.L. Ward, C. Harris, J. Lewis, and M.H. Beale. Assessment of 1H NMR spectroscopy and multivariate analysis as a technique for metabolite fingerprinting of *Arabidopsis thaliana*. *Phytochemistry* 62(6), 2003: 949–957.

38. H. Wu, A.D. Southam, A. Hines, and M.R. Viant. High-throughput tissue extraction protocol for NMR- and MS-based metabolomics. *Analytical Biochemistry* 372(2), 2008: 204–212.

39. L. Suhartono, F. Van Iren, W. de Winter, S. Roytrakul, Y.H. Choi, and R. Verpoorte. Metabolic comparison of cryopreserved and normal cells from *Tabernaemontana divaricata* suspension cultures. *Plant Cell, Tissue and Organ Culture* 83(1), 2005: 59–66.

40. K.-H. Ott and N. Aranibar. "Nuclear Magnetic Resonance Metabolomics: Methods for Drug Discovery and Development." In *Metabolomics*. W. Weckwerth, Ed., Humana Press: Totowa, NJ, 2007; p. 247–271.

41. A.J. Simpson, D.J. McNally, and M.J. Simpson. NMR spectroscopy in environmental research: From molecular interactions to global processes. *Progress in Nuclear Magnetic Resonance Spectroscopy* 58(3/4), 2011: 97–175.

42. Q.N. Van, H.J. Issaq, Q. Jiang, Q. Li, G.M. Muschik, T.J. Waybright, H. Lou, M. Dean, J. Uitto, and T. Veenstra. Comparison of 1D and 2D NMR spectroscopy for metabolic profiling. *Journal of Proteome Research* 7, 2008: 630–639.

43. S. Masoum, D.J.-R. Bouveresse, J. Vercauteren, M. Jalali-Heravi, and D.N. Rutledge. Discrimination of wines based on 2D NMR spectra using learning vector quantization neural networks and partial least squares discriminant analysis. *Analytica Chimica Acta* 558(1/2), 2006: 144–149.

44. J. Yuk, J.R. McKelvie, M.J. Simpson, M. Spraul, and A.J. Simpson. Comparison of 1-D and 2-D NMR techniques for screening earthworm responses to sub-lethal endosulfan exposure. *Environmental Chemistry* 7(6), 2010: 524–536.

45. K. McIntyre, A. Luu, C. Sun, D. Brown, E.M.K. Lui, and J.T. Arnason. Ginsenoside variation within and between Ontario ginseng landraces: Relating phytochemistry to biological activity. In *The Biological Activity of Phytochemicals*. D.R. Gang, Ed., Springer: New York, 2011; pp. 97–107.

46. FDA Notice. Regenerect: Recall—Undeclared drug ingredient. http://www.fda .gov/Safety/MedWatch/SafetyInformation/SafetyAlertsforHumanMedicalProducts/ ucm253416.htm, Posted 05/02/2011.

12 Quality and Authenticity of Complex Natural Mixtures

Analysis of Honey Using NMR Spectroscopy and Statistics

István Pelczer

CONTENTS

Introduction 184
Discussion 185
    Advantages of NMR Spectroscopy 185
    ^{1}H Detection by NMR 186
    ^{13}C Detection by NMR 186
    Statistical Analysis 191
Component ID and Bookkeeping 193
Experimental 193
    Sample Preparation 193
    NMR Data Acquisition and Processing 193
Analysis of qC-NMR Spectra 194
    Identification of ^{13}C Satellites 194
Summary and Conclusions 196
Acknowledgments 196
References 196

Dedication

This chapter is dedicated to my stepmother, Mrs. István Pelczer Sr., a Master of the Folk Art of the Hungarian State, who is making the beautiful and outstanding Hungarian honey-breads (Figure 12.1) by the thousands, all by fantasy, no two alike.

FIGURE 12.1 Decorative Easter eggs made of Hungarian honey-bread by Dr. István Pelczer Sr., with self-designed Hungarian folk motifs.

INTRODUCTION

Honey is an essential nutrient and ancient source of sugar with unique biological effects, a complex mixture of biological origin. Analytical evaluation of honey can reveal its origin, its quality, and whether if it is real honey or manipulated/adulterated. Honey is one of the food products, which is often adulterated and/or treated in ways, which is not acceptable or is outright dangerous. With the proliferation of pesticides and other industrial pollutants, bees can be adversely affected either directly [1] or indirectly [2] and could introduce contamination to the products of the hive. Therefore, detailed analysis and quality control using highly robust, quantitative, and informative analytical techniques are of high value. At the same time, this analysis provides more knowledge about the enigmatic life and biology of bees as well.

A variety of methods can be used for analysis of honey; most of them rely on some kind of separation or fractionation of the original material. Chromatography and mass spectrometry (MS), also in combination, can be used at very high sensitivity and are indispensable for identification and targeted analysis of trace materials [3,4]. MS also can identify the origin of selected ingredients by tracing isotope ratios, similar to the SNIF-NMR (site-specific natural isotope fractionation nuclear magnetic resonance) analysis [5]. NMR can also be quite successful for targeted analysis and quantification of selected components, such as amino acids [6]. Detailed proteomic analysis may have difficulties too [7], although very recent results show promise to identify genetic fingerprint of the plants where the pollen came from [8]. Integrated analytical approach is probably the best way to accomplish comprehensive assessment of honey composition and characterization, mostly in a specifically targeted fashion [9,10]. NMR spectroscopy, as it will be discussed later, offers the only avenue, which can take honey as a whole and provides a targeted or nontargeted, holistic, quantitative assessment of most ingredients.

DISCUSSION

NMR spectroscopy is one of the two major analytical methods, next to MS (which is most often preceded by some chromatography separation), to study metabolites and metabolic mixtures [11,12]. NMR spectroscopy has special benefits for such studies due to its quantitative nature, high dynamic range, robustness, reproducibility, and easy sample preparation [11]. These metabolic studies usually aim at diagnostic statistical analysis of large number of samples also in combination with component identification and quantification. Most of these metabolic analyses apply ^{1}H-NMR spectroscopy, which offers the highest sensitivity. However, with relatively recent introduction of cryoprobes [13–15], which can be optimized for ^{13}C detection, natural abundance ^{13}C-NMR has become a very competitive alternative. Statistical approaches have not yet been developed or presented for this area, but complex natural mixtures can be efficiently analyzed using direct detection ^{13}C-NMR methods with special benefits, such as high dispersion of chemical shifts (which are closely correlated with the molecular structure), singlet nature of the resonances, very little overlap of signals, easy peak identification, and deconvolution in data processing.

Detailed analysis by NMR can avoid any separation and uses samples in their native condition. A variety of sugars, small molecules, such as free amino acids, and proteins compose honey. The dominant sugar components usually have highly overlapping ^{1}H-NMR spectra; therefore, qualitative and quantitative assessment of the complex mix of mono- and oligosaccharides in honey can be especially challenging. Quantitative, cryoprobe-assisted ^{13}C-NMR (qC-NMR) is best suited for this purpose due to the high level of dispersion and specificity of carbon chemical shifts. General advantages of NMR spectroscopy are summarized below.

Advantages of NMR Spectroscopy

The main advantages of NMR spectroscopy can be briefly summarized as follows:

- Quantitative by nature, highly reproducible, robust, easy to automate.
- Not selective, it is unbiased (does not rely on functional groups, extinction coefficients, ionization, fragmentation properties, etc.).
- Nondestructive, the sample can be easily recovered.
- The sample often can be used in its native condition (easy sample preparation).
- Highly cost-effective (typical average cost per sample is approximately $10).
- Sensitivity is highly improved by cryoprobe technology.
- It can be used for microsamples (few microliters).

The main disadvantage of NMR is its relative low sensitivity compared to chromatography, MS, or optical methods, which is largely compensated today by the introduction of crypoprobes and microsample NMR. There is a relatively high capital investment required for a competitive and capable NMR instrument, but usually the *per-sample* expense is low and is in the regime of a few dollars. High level of automation makes long-term continuous operation available.

NMR analysis of honey can include ^{1}H-NMR, ^{13}C-NMR, and ^{31}P-NMR (which has not been much informative in our experience and will not be discussed further here). Representative ^{1}H- and ^{13}C-NMR spectra are shown in Figure 12.2.

^{1}H DETECTION BY NMR

The advantages and disadvantages of ^{1}H-NMR analysis are as follows. It offers the highest sensitivity; therefore, it is best for the identification of low-concentration components. It has, however, relatively poor dispersion, and often there are many overlapping multiplets; the sugar region is especially overcrowded (Figure 12.3). Therefore, small contributing components are impossible to analyze without using complicated multidimensional NMR methods. Radiation damping may be a serious issue, especially when using high-sensitivity, high-Q probeheads, such as cryoprobes. In order to avoid radiation damping and simply to remove the overwhelming water signal, solvent suppression is usually necessary. Suppression of the water signal comes with the penalty of losing or altering some of the valuable region of the neighborhood resonances, however.

The sugar resonances are obviously dominant features in the ^{1}H-NMR spectrum and can obscure other, small components of fingerprint value. One can consider suppressing these sugar resonances by experimental means using band-selective pulses, for example. The alternative strategy is, however, keeping all signals (but the water) and to do selective suppression later, in data processing. The first option offers best sensitivity for the small components while paying double penalty; some added technical complications and losing valuable information (except if running two parallel experiments with and without suppressing the sugar region, respectively). However, with modern instruments with high sensitivity and high dynamic range, it is better to keep all data staying with simpler experimental conditions (water signal suppression only) and without loss of any information. Separation of information from the sugar region and from the rest of the ^{1}H-NMR spectrum can be done later, using tailored data processing. An example for this is presented later in the section about statistical analysis.

^{1}H-NMR spectra may require specific attention to data processing details (phase correction, baseline correction, peak alignment), fine adjustments often to be done manually. At the same time, a significant advantage is that the methodology and protocols for statistical analysis of ^{1}H-NMR spectra (multivariate statistics, statistical total correlation spectroscopy [STOCSY] analysis, etc.) are well established and well developed [16–18].

^{13}C DETECTION BY NMR

In our practice, we prefer to use ^{13}C detection taking advantage of optimized cryoprobes coupled with high-quality data processing and analysis tools to elucidate the sugar composition of samples *in situ*. A pioneering, early ^{13}C-NMR study of

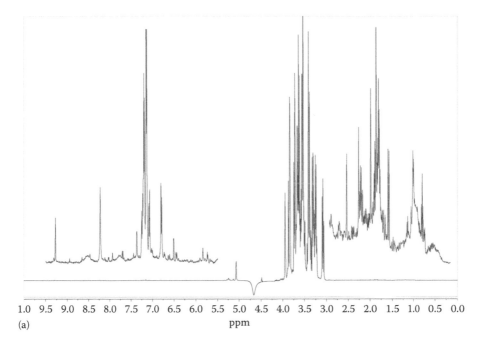

(a)

ppm

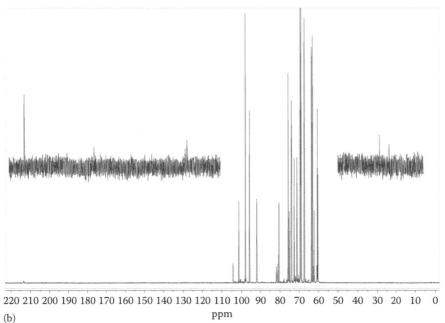

(b)

ppm

FIGURE 12.2 Representative ¹H- and ¹³C-NMR spectra of honey samples (Manuka) (inserts show segments with high vertical expansion).

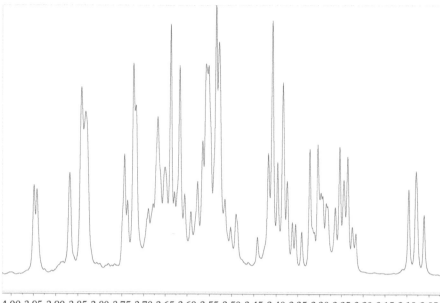

4.00 3.95 3.90 3.85 3.80 3.75 3.70 3.65 3.60 3.55 3.50 3.45 3.40 3.35 3.30 3.25 3.20 3.15 3.10 3.05

ppm

FIGURE 12.3 ¹H-NMR of the sugar region in a typical honey sample (Manuka). Small components are clearly overwhelmed by the dominant sugar resonances.

carbohydrates in honey used a conventional room-temperature probe [19]. Recently, high-sensitivity qC-NMR [13–15] has become a very competitive tool for analysis of complex mixtures, such as natural products, biofluids, or food/beverage samples. Direct-detection ¹³C-NMR has been shown to be useful for structural biology (using fully ¹³C-labeled proteins) [20,21], especially with cryoprobe sensitivity enhancement [13,22,23], and also for other complex mixtures at natural abundance [14], including metabolic mixtures [24]. Possibility of quantitative analysis of biological mixtures in their native condition is a special benefit provided by NMR spectroscopy. In comparison with ¹H-NMR, ¹³C-NMR has the extra advantages of better dispersion, singlet nature of all peaks, and no need for solvent suppression, while it offers more characteristic, more reliable correlation between actual chemical shifts and structure. As the acquisition time can be extended for ¹³C detection, the spectral resolution can be as high as the natural line width allows. This is a significant advantage, especially considering the rare occurrence of any serious overlap in ¹³C except symmetry. Spectral deconvolution/curve fitting methods can be quite effective to further resolve these spectra, as they have to deal with simple singlets. Very complex mixtures, such as the sugar region of honey, can then be analyzed in detail assessing the components with help of database information, prediction tools, and well-designed software routines. In case of high dynamic range spectra, the intensity of the ¹³C satellites of the high-concentration components may be comparable with the resonances of other ingredients of low concentration. This needs special attention, yet provides extra structural information.

Regardless of the possibly hours-long ^{13}C data acquisition, the result is a highly information-rich spectrum that carries all ^{13}C-related information. It is well resolved, is well dispersed, easily covers a high dynamic range of ca. 4–5 magnitudes, and can be made fully quantitative provided enough material (not necessarily more than 10–20 mg for small molecule mixtures) is available. With the ^{13}C-detection optimized cryoprobes, a couple of hours or so data acquisition usually provides excellent results. When detecting ^{13}C, no water suppression is necessary at all. This is another significant advantage as no section of the spectrum is lost, the full range of the receiver gain can be utilized, and no radiation damping or other instability effects should be considered during the acquisition.

The ^{13}C-NMR spectrum is very well dispersed over about 200 ppm, and the signals are all singlets when proton decoupling is applied (with the relatively rare exception of the presence of other heteronuclear coupling). The ^{13}C-NMR chemical shift is less prone to various environmental effects, such as solvation, salt effects, and interaction with other components, temperature change, or other through-space shielding effects, situated one layer deeper in the molecular structure than the protons. Quaternary carbons are quite often part of the molecular skeleton and report about important structural features, which might not be recognized only from the ^{1}H-NMR spectra. Therefore, it is highly desirable to learn the information carried by the carbon atoms, possibly using direct detection methods.

A representative ^{13}C-NMR spectrum, expanded to the sugar region and with high vertical expansion to highlight small components, is shown in Figure 12.4, with the associated APT (attached proton test) spectrum [25] stacked above it. Next to the dominant sugar components, which make ^{1}H-NMR very difficult to evaluate, many small-quantity ingredients (in the regime of 0.1%, or less, of the concentration of the main components) can also be identified and are readily available for analysis. All these components will provide a characteristic fingerprint for to quickly and easily characterize the brand, the harvest time, and will be the avenue to identify tampering with the process or the product itself. Figure 12.5 shows a comparison of six commercial honey samples highlighting a very small segment of the APT spectrum. Subtle but characteristic differences are quite clearly visible.

It is often argued that *inverse* detected correlation methods are superior and provide the same information what could be extracted from direct ^{13}C-detection 1D or nD spectra. There are impressive examples of such applications demonstrating the capacity of such approaches for component identification and quantitative analysis of metabolic mixtures, such as cell extracts [26,27] also using nonlinear sampling to reduce overall acquisition time [28]. With full acknowledgment of the power of these approaches, direct ^{13}C detection alternatives remain complementary and competitive. Sampling over the very large frequency regime for ^{13}C in the incremented dimension with sufficient resolution is still quite time consuming, especially when very closely positioned signals need to be resolved. The inverse correlation methods rely on actual coupling constant values, which may vary and may be unknown in advance, especially in case of long-range connectivities, which make quantitative analysis more difficult [29]. Samples exhibiting extra line broadening represent a special case, which uniquely benefit of direct ^{13}C-detection. Decoupling over the wide range of ^{13}C frequencies,

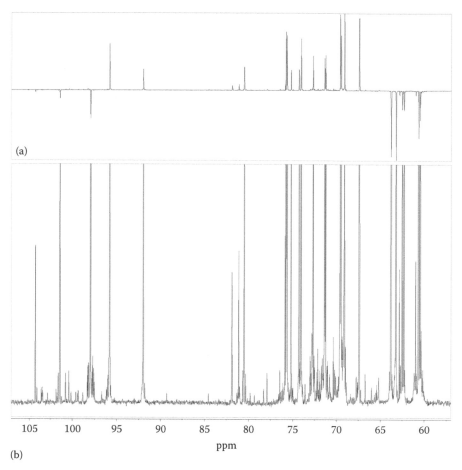

FIGURE 12.4 The sugar region of the quantitative ^{13}C-NMR spectrum of late harvest Japanese Bamboo honey with high vertical expansion (b) and the corresponding APT spectrum (a). In the APT spectrum, CH_2 and quaternary carbon resonances are phased negative, while CH resonances are positive. Many components of small relative concentration are well visible.

related sample heating, and solvent suppression in the ^{1}H domain are also matter of consideration.

In summary, direct detection, cryoprobe-assisted ^{13}C-NMR provides the following advantages:

- Good sensitivity with cryoprobe, best for sugar components
- Great dispersion, minimal overlap
- No radiation damping, solvent suppression is not necessary
- Sugar region is especially information-rich, fingerprint value
- Data processing is straightforward
- Statistical analysis is *in the making* and will also be beneficial
- Spreadsheet-based analysis (see also later) is practically a given

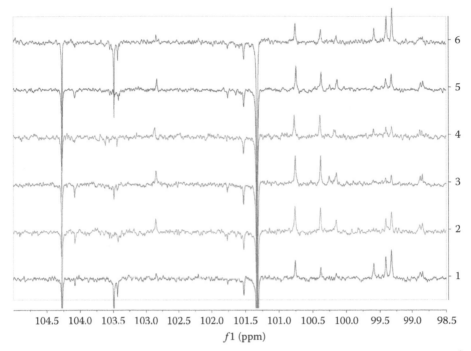

104.5 104.0 103.5 103.0 102.5 102.0 101.5 101.0 100.5 100.0 99.5 99.0 98.5

*f*1 (ppm)

FIGURE 12.5   Small section of the attached proton test (APT) ^{13}C-NMR spectra of selected commercial honey samples. CH resonances are phased to the positive side of the baseline while quaternary carbon resonances point to the negative side. Small, but characteristic differences can help to identify the various brands and origin. These differences are practically not observable at all in the ^{1}H spectrum due to the presence of other, overwhelming large peaks and extensive overlap.

STATISTICAL ANALYSIS

Statistical analysis of honey spectra has some precedence in the literature [30–32] and can be done in various ways using the same data, keeping or removing the dominant sugar resonances. Chemometrics [16] and multivariate discriminant analysis [17] are probably most helpful to identify various brands or different batches of samples. With a large enough reference database, new samples can be blindly sorted out falling into one or the another statistical category—as it was demonstrated very attractively recently for metabolic patterns of various tumor tissues upon simple MS analysis of the smoke while cutting with electric knife [33]. It is easy to remove the sugar resonances/region from this analysis; one has only to pay attention to appropriate normalization of the set of spectra [34] with or without these resonances. As an illustration, Figure 12.6 compares orthogonal partial least squares discriminant analysis (O-PLS-DA) coefficient plots [17] of commercial clover and wildflower honey (commonly sold in a small plastic *bear* container) with all sugar resonances kept (top) or suppressed (bottom), respectively. In this analysis, negative intensities highlight components, which are in excess in the *clover* samples, while positive intensities belong to components, which are relatively more abundant in the

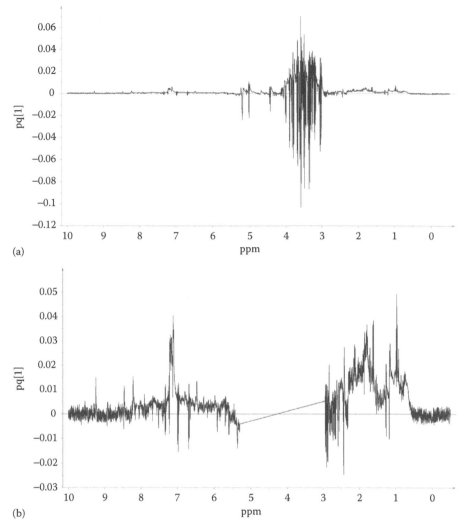

FIGURE 12.6 Illustrative O-PLS-DA coefficient plots (SIMCA v.13.0, Umetrics, Umeå, Sweden) of clover and wildflower (*bear*) honey samples. In (a), the sugar resonances are also part of the analysis, while in (b), this section has been suppressed in data processing. Many small components are well visible in this case.

bear samples. Clearly, the *clover* honey has more sugars in this comparison, while the *bear* honey samples have more proteins, respectively (better emphasized in the *sugar-free* statistics). There are well-visible resonances of several small components, which have characteristically differential concentrations in the two sample sets.

Fine details of the honey composition provide valuable fingerprint information in ^{1}H-NMR, especially when the low-concentration components are concerned. For the analysis of the sugar components, ^{13}C-NMR is more suitable, as discussed above. It is far from customary at this time, but ^{13}C-NMR spectra can be analyzed statistically, too.

COMPONENT ID AND BOOKKEEPING

^{13}C-NMR offers more characteristic/reliable correlation between chemical shifts and structure and thus is more useful than ^{1}H-NMR for identifying components. qC-NMR provides tremendous amount of information in a single spectrum both about the closely related sugar components and other ingredients of smaller quantity. Computer-assisted database building and interactive data analysis help to identify large number of components and provide fingerprint information.

Identification and quantification of the components can utilize extensive database information and spectrum prediction tools (e.g., Bio-Rad Laboratories, www.knowitallu.com; Chenomx, Edmonton, Alberta, Canada, www.chenomx.com; and the Madison Metabolomics Consortium Database (MMCD) portal, http://mmcd.nmrfam.wisc.edu; [35]). It is critical to have capability to adjust the fine details of individual peak positions for good match and then possibly subtract the identified component from the spectrum for successive quantitative analysis of the mixture. Chenomx's NMR Suite offers well-designed interactive tools for such manipulations while comparing the mixture spectrum to individual items from the database either designed by Chenomx or built by the user. Other major software manufacturers are in the process of introducing various incarnations of similar flexibility as well. It is a novel approach to extract only the digital identification of each resonance from the time-domain signal (position, integral, half-height width, and phase) and fully rely on spreadsheet-based information from that on, instead of conventional processing of the spectrum [36]. This concept of spreadsheet-based data management can be implemented using frequency-based analysis by precise deconvolution, too.

qC-NMR spectra carry a tremendous amount of information which has to be assessed and stored in an organized fashion, preferably in a user-defined database. Passing large amount of information into such a database can be a tedious process. We have designed a Java-based interface, Spectrum Builder, to pass any ^{13}C-NMR peak-list directly to Chenomix's user-defined database in a user-friendly, high-throughput fashion [37].

EXPERIMENTAL

SAMPLE PREPARATION

Samples have been prepared fresh and submitted to NMR analysis immediately without extra conditioning. Honey samples have been mixed with deionized water in about 1:5 ratio and homogenized using a stirrer in order to reduce viscosity of the sample. For deuterium lock, ca. 25 µL D_2O was added.

NMR DATA ACQUISITION AND PROCESSING

All experiments were run on Bruker Avance-III spectrometers (500 MHz ^{1}H frequency) at 295 K controlled temperature either using a ^{13}C-detection optimized dual C/H cryoprobe (DCH, ^{13}C-^{1}H//^{2}H) or a cryo-QNP probehead (^{1}H/^{31}P,^{13}C,^{15}N//^{2}H), products of Bruker-Biospin, Billerica, MA. The ^{1}H-NMR spectra were acquired suppressing the water signal using the excitation sculpting (ES) method [38]. ^{13}C APT spectra were also collected regularly [25]. Typical ^{13}C-NMR acquisition times were between half an hour to a couple of hours (for quantitative ^{13}C experiments with half a minute recycle delay).

Data processing was conducted using MNova up to v.9.0 (MestreLab Research S.L., Santiago de Compostela, Spain). We use an in-house written added interface module [37] to feed ^{13}C data efficiently into a user-defined database for large-scale analysis in Chenomx NMR Suite (Chenomx, Edmonton, Alberta, Canada).

ANALYSIS OF qC-NMR SPECTRA

There are three major phases of detailed analysis, all assisted by software routines, which are as follows:

- Data processing
- Identification of ^{13}C satellites
- Component ID and bookkeeping

The best way to identify and integrate individual resonances is applying deconvolution or curve fitting methods. This can be done in the frequency domain or using time domain Bayesian analysis [36].

IDENTIFICATION OF ^{13}C SATELLITES

In high dynamic range spectra, ^{13}C satellites may be comparable with resonances of low-concentration components; in this case, their specific identification is essential. The satellites will also provide additional structural information, such as $^{1}J_{C,C}$ and isotope shift values.

Identification of the ^{13}C satellites can be accomplished by experimental means taking spectra at two different magnetic fields, when the satellites will remain at the same position relative to the peak of origin. One can consider running an APT or J-modulated experiment tailored to the expected $^{1}J_{C,C}$ values; each will offer characteristic phase modulation of the satellites. Alternatively, a software routine can be designed to find these satellites, saving experiment time. A possible simple algorithm is shown below:

---

1. Find a peak above a selected threshold
2. Calculate 0.55% intensity
3. Estimate $^{1}J_{CC}$ based on chemical shift
4. Find corresponding doublet (consider isotope shift)
5. Allow manual verification/intervention
6. Cluster main peak with satellites
7. More satellites? (up to three)   Yes   (go to 3)
8. No
9. Move to next peak   (go to 1)

In our practice, we apply the global spectrum deconvolution (GSD) algorithm of MestreNova (MestreLab Research, Santiago de Compostela, Spain, www.mestrelab .com) to do highly reliable curve-fit analysis of the frequency domain data (Figure 12.7a).

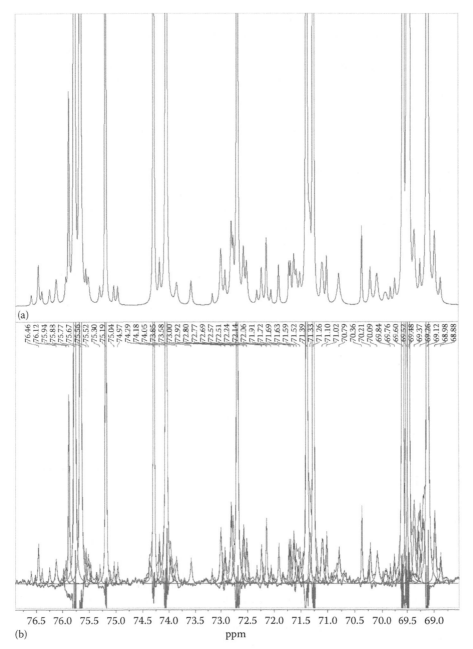

(a)

(b)

76.5 76.0 75.5 75.0 74.5 74.0 73.5 73.0 72.5 72.0 71.5 71.0 70.5 70.0 69.5 69.0

ppm

FIGURE 12.7 Curve fit (b) of a small section of the quantitative ^{13}C-NMR spectrum of a typical honey sample (Manuka), using the global spectral deconvolution tool in MNova (v.9.0), and the corresponding synthetic spectrum (a).

The result may require/benefit of further manual inspection and, if necessary, correction/editing of the peak list. Eventually, the resonances are all represented digitally in the peak table and a synthetic spectrum can also be generated for visualization purposes (Figure 12.7b). The peak table, which can be annotated in MNova itself, can also be exported to offline software tools (such as Excel) for further analysis and bookkeeping or to create a coherent user-defined database (such as in Chenomx; www.chenomx.com) for interactive or automated analysis of new spectra. An in-house written interface [37] makes this latter process more efficient and user-friendly.

SUMMARY AND CONCLUSIONS

NMR spectroscopy is a highly efficient and informative tool for analysis of honey samples in their native conditions for the purpose of characterization and quality control. ^{1}H-NMR is best suited for assessment of small components and for statistical classification. Direct-detection, cryoprobe-enhanced qC-NMR spectroscopy is a powerful approach for detailed analysis of the complex sugar mixture in honey. This analysis benefits of advantageous technical conditions for data acquisition, high dispersion, and little overlap on the spectrum and the highly characteristic nature of the ^{13}C chemical shifts. Component identification and bookkeeping can be highly assisted by available databases and sophisticated software tools.

ACKNOWLEDGMENTS

Joseph G. Lelinho (www.hilltophoneyllc.com) has been providing many well-characterized and well-documented honey samples, as well as lots of education on bees and honey. Students of the NMR Practicum 2009 summer program, Andrew Choi, Suhyun Kim, and Ayaan McKenzie, at Department of Chemistry, Princeton University, contributed with NMR analysis of commercial honey samples. Christina Ha (University of Michigan) has done extensive work on a larger set of honey samples while participating in the NSF-founded REU program in 2011.

REFERENCES

1. M. J. Palmer, C. Moffat, N. Saranzewa, J. Harvey, G. A. Wright, and C. N. Connolly. Cholinergic pesticides cause mushroom body neuronal inactivation in honeybees. *Nature Communications*, 2013:1634, DOI:10.1038/ncomms2648.
2. R. D. Gilring, I. Lusebrink, E. Farthing, T. A. Newman, and G. M. Poppy. Diesel exhaust rapidly degrades floral odours used by honeybees. *Scientific Reports* 3, 2013:2779, DOI:10.1038/srep02779.
3. M. Lee. (Ed.) *Mass Spectrometry Handbook.* Wiley, New York, 2012.
4. C. Cordella, I. Moussa, A.-C. Martel, N. Sbirrazzuoli, and L. Lizzani-Cuvelier. Recent developments in food characterization and adulteration detection: technique oriented perspectives. *Journal of Agricultural and Food Chemistry* 50, 2002:1751–1764.
5. P. Lindner, E. Bermann, and B. Gamarnik. Characterization of citrus honey by deuterium NMR. *Journal of Agricultural and Food Chemistry* 44, 1996:139–140.
6. P. Sandusky and D. Raftery. Use of selective TOCSY NMR experiments for quantifying minor components in complex mixtures: Application to the metabonomics of amino acids in honey. *Analytical Chemistry* 77, 2005:2455–2463.

7. F. Di Girolamo, A. D'Amato, and P. G. Righetti. Assessment of the floral origin of honey via proteomic tools. *Journal of Proteomics* 75, 2012:3688–3693.
8. I. Pelczer, J. G. Lelinho, and D. H. Perlman. Integrated analysis of honey by NMR and MS for characterization of the composition and proteomic/genomic determination of its origin. In *Magnetic Resonance in Food Science* series, Royal Society of Chemistry (in preparation).
9. P. Tuchado, I. Martos, L. Bortolotti, A. G. Sabatini, F. Ferreres, and F. A. Tomas-Barberan. Use of quinoline alkaloids as markers of the floral origin of chestnut honey. *Journal of Agricultural and Food Chemistry* 57, 2009:5680–5686.
10. J. A. Donarski, S. A. Stephen, M. Harrison, M. Driffield, and A. J. Charlton. Identification of botanical biomarkers found in Corsican honey. *Food Chemistry* 118, 2010:987–994.
11. J. C. Lindon and J. K. Nicholson. Spectroscopic and statistical techniques for information recovery in metabonomics and metabolomics. *Annual Review of Analytical Chemistry* 2008:1:2.1–2.25.
12. E. M. Lenz and I. D. Wilson. Analytical strategies in metabonomics. *Journal of Proteome Research* 6, 2007:443–458.
13. N. Shimba, H. Kovacs, A. S. Stern, A. M. Nomura, I. Shimada I, J. C. Hoch, C. S. Craik, and V. Dötsch. Optimization of ^{13}C direct detection NMR methods. *Journal of Biomolecular NMR* 30, 2004:175–179.
14. Z. Zhou, R. Kümmerle, J. C. Stevens, D. Redwine, Y. He, X. Qiu, R. Cong, J. Klosin, N. Montañez, and G. Roof. ^{13}C NMR of polyolefins with a new high temperature 10 mm cryoprobe. *Journal of Magnetic Resonance* 200, 2005:328–333.
15. H. Kovacs, D. Moskau, and M. Spraul. Cryogenically cooled probes—A leap in NMR technology. *Progress in NMR Spectroscopy* 46, 2005:131–155.
16. J. Trygg, E. Holmes, and T. Lundstedt. Chemometrics in metabonomics. *Journal of Proteome Research* 6, 2007:469–479.
17. J. M. Fonville, S. E. Richards, R. H. Barton, C. L. Boulange, T. M. D. Ebbels, J. K. Nicholson, E. Holmes, and M.-E. Dumas. The evolution of partial least squares models and related chemometric approaches in metabonomics and metabolic phenotyping. *Journal of Chemometrics* 24, 2010:636–649.
18. S. L. Robinette, J. C. Lindon, and J. K. Nicholson. *Analytical Chemistry* 85, 2013:1282–1289.
19. V. Mazzoni, P. Bradesi, F. Tomi, and J. Casanova. Direct qualitative and quantitative analysis carbohydrate mixtures using ^{13}C NMR spectroscopy: Application to honey. *Magnetic Resonance in Chemistry* 35, 1997:S81–S90.
20. B. H. Oh, W. M. Westler, P. Darba, and J. L. Markley. Protein carbon-13 spin systems by a single two-dimensional nuclear magnetic resonance experiment. *Science* 240, 1988:908–911.
21. T. E. Machonkin, W. M. Westler, and J. L. Markley. Paramagnetic NMR spectroscopy and density functional calculations in the analysis of the geometric and electronic structures of iron-sulfur proteins. *Inorganic Chemistry* 44, 2005:779–797.
22. I. Bertini, C. Luchinat, G. Parigi, and R. Pieratteli. NMR spectroscopy of paramagnetic metalloproteins. *ChemBioChem* 6, 2005:1536–1549.
23. K. Takeuchi, D. P. Frueh, S. G. Hyberts, Z.-Y. J. Sun, and G. Wagner. High-resolution 3D CANCA NMR experiments for complete mainchain assignments using Cα direct detection. *Journal of the American Chemical Society* 132, 2005:2945–2951.
24. H. C. Keun, O. Beckonert, J. L. Griffin, C. Richter, D. Moskau, J. C. Lindon, and J. K. Nicholson. Cryogenic probe ^{13}C NMR spectroscopy of urine for metabonomic studies. *Analytical Chemistry* 74, 2002:4588–4593.
25. S. L. Patt and J. N. Shoolery. Attached proton test for carbon-13 NMR. *Journal of Magnetic Resonance* 46, 1982:5–539.

26. I. A. Lewis, S. C. Schommer, B. Hodis, K. A. Robb, M. Tonelli, W. M. Westler, M. R. Sussman, and J. L. Markley. Method for determining molar concentrations of metabolites in complex solutions from two-dimensional ^{1}H-^{13}C NMR spectra. *Analytical Chemistry* 79, 2007:9385–9390.

27. I. A. Lewis, R. H. Karsten, M. E. Norton, M. Tonelli, W. M. Westler, and J. L. Markley. NMR method for measuring carbon-13 isotopic enrichment of metabolites in complex solutions. *Analytical Chemistry* 82, 2010:4558–4563.

28. S. G. Hyberts, G. J. Heffron, N. G. Tarragona, K. Solanky, K. A. Edmonds, H. Luithardt, J. Fejzo et al. Utrahigh-resolution ^{1}H-^{13}C HSQC spectra of metabolite mixtures using nonlinear sampling and forward maximum entropy reconstruction. *Journal of the American Chemical Society* 129, 2007:5108–5116.

29. M. Lolli, D. Bertelli, M. Plessi, A. G. Sabatini, and C. Restani. Classification of Italian honeys by 2D HR-NMR. *Journal of Agricultural and Food Chemistry* 56, 2008:1298–1304.

30. J. F. Cotte, H. Casabiance, S. Chradon, J. Lheritier, and M. F. Grenier-Loustalot. Application of carbohydrate analysis to verify honey authenticity. *Journal of Chromatography A* 1021, 2003:145–155.

31. C. Cordella, J. S. L. T. Militão, M.-C. Clément, P. Drajnudel, and D. Cabrol-Bass. Detection and quantification of honey adulteration via direct incorporation of sugar syrups or bee-feeding: Preliminary study using high-performance anion exchange chromatography with pulsed amperometric detection (HPAEC-PAD) and chemometrics. *Analytica Chimica Acta* 531, 2005:239–248.

32. J. A. Donarski, S. A. Jones, and A. J. Charlton. Application of cryoprobe ^{1}H Nuclear magnetic resonance spectroscopy and multivariate analysis for the verification of Corsican honey. *Journal of Agricultural and Food Chemistry* 56, 2008:5451–5456.

33. J. Balog, L. Sasi-Szabó, J. Kinross, M. R. Lewis, L. J. Muirhead, K. Veselkov, R. Mirnezami et al. Intraoperative tissue identification using rapid evaporative ionization mass spectrometry. *Science Translational Medicine* 5, 2013:194ra93, DOI:10.1126/scitranslmed.3005623.

34. F. Dieterle, A. Ross, G. Schlotterbeck, and H. Senn. Probabilistic quotient normalization as robust method to account for dilution of complex biological mixtures. Application in ^{1}H NMR metabonomics. *Analytical Chemistry* 78, 2006:4281–4290.

35. Q. Cui, I. A. Lewis, A. D. Hegeman, M. E. Anderson, J. Li, C. F. Schulte, W. M. Westler, H. R. Eghalbina, M. R. Sussman, and J. L. Markley. Metabolite identification via the Madison Metabolomics Consortium Database. *Nature Biotechnology* 26, 2008:162–164.

36. K. Krishnamurthy. CRAFT (complete reduction to amplitude-frequency table)—Robust and time-efficient Bayesian approach for quantitative mixture analysis by NMR. *Magnetic Resonance in Chemistry*, 2013, DOI:10.1002/mrc.4022.

37. E. Dellapenna. *Spectrum Builder: A New Tool for NMR Database Construction.* Sr. Thesis, Department of Chemistry, Princeton University, Princeton, NJ, 2012.

38. T. L. Hwang and A. J. Shaka. Water suppression that works. Excitation sculpting using arbitrary waveforms and pulsed field gradients. *Journal of Magnetic Resonance Series A* 112, 1995:275–279.

13 Application of CRAFT to Data Reduction of NMR Spectra of Botanical Samples

Krish Krishnamurthy and David J. Russell

CONTENTS

Introduction 199
Complete Reduction to Amplitude Frequency Table 202
Application of CRAFT to Botanical Samples 204
Conclusion 214
References 214

INTRODUCTION

By its very nature, the study of botanical samples is the study of mixtures. When dealing with a complex mixture, the primary characteristics of interest are the identity and concentration estimation of the individual constituents of the mixture. Analytical technology has progressed to the point where large sets of complex samples can be rapidly and automatically analyzed to provide this information [1]. Online mass spectrometric detection of the analytes resulting from various chromatographic separation techniques is probably the most common means currently employed for botanical profiling [2]. These methods are predicated on physical separation of the mixture into individual compounds, a requirement that is sometimes difficult to achieve. Furthermore, the response factor for detection of the individual compounds can vary by as much as 10,000, or more. This creates the situation where quantitation of any individual component requires the availability of a pure standard reference sample of that component.

Although most often thought of as solely a tool for structural analysis, NMR is increasingly recognized as a critical quantitative tool in many applications. As an analytical technique for the study of botanicals, NMR has several appealing characteristics:

1. The NMR experiment is quantitative, exhibiting a molar absorptivity response factor of 1 for any given nuclei. This means that the concentration of any resonance, including those of unknown compounds, can be accurately measured without the need for any type of calibration or reference

standards. Practically, all compounds of botanical origin contain both hydrogen and carbon atoms, and thus, NMR represents a universal detector for these compounds.

2. The NMR experiment can be conducted on almost any botanical sample, from extracts of solid samples to juices to neat samples of liquids. Sample preparation is simple, often requiring just addition of a small aliquot of buffer, making this tedious task easy to automate.

3. The NMR experiment can be conducted and analyzed without the need to physically separate the botanical sample into its individual components. This greatly increases the throughput of method.

4. Where the identity of some components might be in question, NMR can often provide definitive structural information. Should the quality of the data preclude a complete structural analysis, even partial NMR analysis can be used to segregate resonances from unknown compounds into classes such as fats, carbohydrates, and amino acids, which is sometimes enough to meaningfully complete the analysis at-hand.

The NMR spectra of botanical samples are typically quite complex and contain many resonances that overlap when displayed in the frequency domain. NMR spectroscopy provides a variety of experimental techniques to extract further structural information from spectra. Techniques such as DOSY [3] and LC-NMR [4] are often used to separate signals based on their physical properties (diffusion in the former and retention time in the later). Although these approaches are valuable, they require a dedicated spectroscopist to achieve the best spectral quality, and they do not lend themselves easily to applications where many samples require analysis.

In high-throughput scenarios, where a large number of similar spectra from a single study need to be analyzed (such as in metabolomics studies or food science analysis), use of spectral databases is almost universal to deconvolute the peaks of interest and extract quantitative information. This necessitates significant prior knowledge, whether in the form of the database or in the form of operator interaction, or both. These operations are not automated. Minute differences in the sample preparation or just the natural variation in sample-to-sample consistency causes the NMR frequencies recorded for individual resonances to shift slightly, and these perturbations are not uniform for each resonance in the sample. As such, time-consuming and tedious spectral matching where a human operator must inspect every data set and empirically match those data to database reference spectra is the normal workflow used for these types of studies. This can add as much as an hour of time to the data reduction step for every sample in a given study.

In addition to the problem with the time required for each analysis, spectral deconvolution techniques are traditionally applied to data in the frequency domain. The accuracy of the results obtained from frequency domain deconvolution is inherently limited. Each resonance in a data set is defined by four attributes: frequency, decay rate, amplitude, and phase. For frequency domain spectra, quantification of a peak is accomplished using either peak height or integration of the total area under the curve. In either case, the resulting value is a function of decay rate, amplitude, and phase.

Traditional NMR-based quantification methods were based on the measurement of the area under the curve. When each resonance of interest is well resolved from all other resonances, this method can be very accurate. *Well resolved*, however, needs to be defined for each resonance as a function of the line width at half-height observed for each peak. Mathematical analysis of a Lorentzian decay function indicates that significant intensity for each resonance extends at least +/−33 times the half-height for each peak, and some researchers have suggested that an even wider integral area should be considered for optimal results [5].

In the absence of well-resolved peaks, the only tool available for data analysis is comparison of peak heights for individual resonances. Unfortunately, peak heights are a poor indicator of the total area of each peak [6].

Consider the three overlaid spectra displayed in Figure 13.1. The peak heights observed for each line are clearly different, meaning that in a traditional analysis the concentration returned for each sample would reflect the observed differences in peak height. These spectra were generated, however, from a single data set. They graphically demonstrate the dependence on peak height when peaks of identical concentration are recorded with a difference of a single data point of resolution in the line width. The line width observed for each resonance in a research sample is typically limited by the quality of the magnetic field in which the sample is measured, known as *shimming*. Differences in the magnetic field across the volume of the sample that are greater than ~1 part per billion can affect the resulting line shape, and this much variation can be easily observed even when a single sample is shimmed and recorded multiple times.

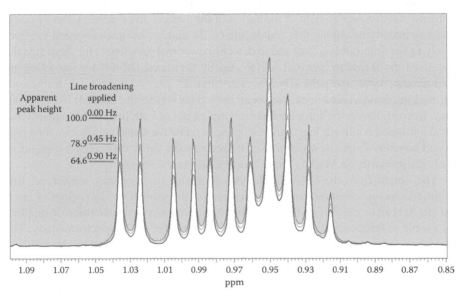

FIGURE 13.1 Methyl region of the NMR spectrum recorded on a natural product sample showing the effect of line width on sample height. All three spectra are processed from the same FID so the area under each resonance is identical, while the peak heights are significantly different, which can lead to significant errors when using peak height for quantitation.

COMPLETE REDUCTION TO AMPLITUDE FREQUENCY TABLE

Time-domain analysis (i.e., extraction of frequency, amplitudes, line width, and phase parameters) of NMR data is another approach for deconvolution. One such time-domain deconvolution approach is the analysis of one-dimensional or one-dimensional arrayed FIDs using Bayesian probability theory [7]. Griffin and coworkers [8] demonstrated the application of a Bayesian deconvolution approach as an alternative to the traditional binning approach for NMR metabolomics analysis. Key benefits of extracting quantitative information from the FID is that the amplitude of an extracted signal is independent of its relative phase with respect to other signals in the FID, and it is unaffected by minor errors in shimming the magnetic field. One key practical bottleneck in FID analysis is that it is neither visual nor intuitive (i.e., outside the typical operator's *comfort zone*). Another bottleneck of varying severity is the increased computation speed with increasing complexity of the FID. We present an algorithm that facilitates the conversion of the FID into a frequency, amplitude, line width, and phase table in an automated and highly time-efficient fashion. The output of CRAFT (complete reduction to amplitude frequency table) [9] can be used, for example, for data mining of quantitative information using fingerprint chemical shifts of compounds of interest and/or statistical analysis of modulation of chemical quantity in a botanical sample.

The CRAFT workflow is shown in Figure 13.2. It uses a two-step approach to time-domain spectral analysis. Step one, the FID is digitally filtered and downsampled to several sub-FIDs, and step two, these sub-FIDs are modeled as sums of decaying sinusoids using Bayesian probability theory [7]. In the examples presented here, exponentially decaying signal models are used throughout, but an exponential decay is by no means a requirement or limitation of the method. In principle, CRAFT could use any underlying engine that can decide on the number of signals needed to effectively fit the time-domain data and to determine optimal parameters for these signals. We used the Bayesian approach developed by Bretthorst [10–14] for the following two reasons. First, and generally, Bayesian statistics provide an effective framework for making these kinds of decisions (i.e., how many signals are needed) in a rigorous way. Second, Bretthorst's implementation makes use of both linear-least squares and gradient-based nonlinear least squares fitting to refine the signal parameters for a proposed number of signals, which makes it exceptionally faster than related approaches that rely primarily on Markov chain Monte Carlo calculations.

The output from the CRAFT analysis is a table of frequency, amplitude, line width (decay rate constant), and phase for each of the sub-FIDs. The regions of interest (ROIs) can be defined by the user with the spectrum as a visual guide or supplied as a table of frequency limits based on prior knowledge of the spectrum/study. The digital filter bandwidth and the downsampling factor are controlled by the width of the associated ROI [15]. To further accommodate for potential influence of transition band characteristics of the digital filters as well as to cover the bulk of the intensities in the central, nondownweighted region of the ROI, the filter bandwidth is wider (15%–20%) than the chosen ROI. Consequently, in contiguously defined ROIs, neighbors will invariably have overlapped regions. The CRAFT output tables from each of the sub-FIDs (i.e., subtables) are then combined appropriately, taking into account

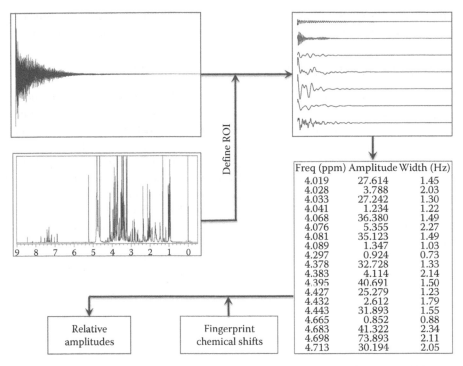

Freq (ppm)	Amplitude	Width (Hz)
4.019	27.614	1.45
4.028	3.788	2.03
4.033	27.242	1.30
4.041	1.234	1.22
4.068	36.380	1.49
4.076	5.355	2.27
4.081	35.123	1.49
4.089	1.347	1.03
4.297	0.924	0.73
4.378	32.728	1.33
4.383	4.114	2.14
4.395	40.691	1.50
4.427	25.279	1.23
4.432	2.612	1.79
4.443	31.893	1.55
4.665	0.852	0.88
4.683	41.322	2.34
4.698	73.893	2.11
4.713	30.194	2.05

FIGURE 13.2 Diagram showing the CRAFT workflow. ROI are selected, the FIDs are downsampled and processed, then the results table can be aligned, matched against fingerprint files, and output to various spreadsheet formats as desired.

the overlapped neighboring ROIs, into a single table or set of tables (each representing one FID) for further data mining.

The fingerprint chemical shift of a chemical is a simple line list of frequencies (in parts per million [ppm]) +/– a segment width. A segment width would be a narrow window (typically 2 or 3 Hz) of frequency band. The data mining from the CRAFT would be done within these frequency constraints (i.e., chemical shift +/– segment width/2). All models in the CRAFT within this window of frequencies are co-added to generate the output model for the fingerprint. In this approach, if one is interested in changes in relative concentration of a chemical (as in a metabolomics study), a complete list of NMR chemical shift of the compound of interest is not needed, but only those of unambiguous resonances. A visual assessment of the quality of the CRAFT analysis is typically achieved by simulating a synthetic FID from the CRAFT and reviewing Fourier transformed trace of the residual (difference) between the experimental FID and the synthetic FID. Thus, an automatic CRAFT analysis sequence would typically include the following steps: (1) Digital filtration and downsampling the ROI; (2) Bayesian analysis of the resultant sub-FIDs to generate tables of frequency/amplitude/decay rate constant/phase of all the exponentially decaying models; (3) combination of all CRAFTs from individual sub-FIDs, taking into account potential overlap in the ROI definition; and finally, (4) simulation of synthetic FIDs for user evaluation. The combined use of frequency windowing using downsampling technique with highly time-efficient strategy and algorithms developed by Bretthorst results in significantly fast and robust CRAFT analysis tool for spectra (FID) of very complex mixtures.

APPLICATION OF CRAFT TO BOTANICAL SAMPLES

Under the Dietary Supplement Health and Education Act of 1994 guidelines for dietary supplements published by the federal Food and Drug Administration (FDA), manufacturers of dietary supplements must provide verification that all products meet established identity, purity, strength, and composition requirements. The guidelines do not stipulate which analytical method(s) are required to comply with dietary supplement cGMP's. Instead, the FDA requires manufacturers to conduct analytical tests such that each botanical component can be authenticated using a scientifically valid method.

As an example of how NMR spectroscopy and CRAFT could be used to provide the data required for compliance with the FDA guidelines, we undertook a study of over-the-counter soybean-based dietary supplements.

Soybean supplements were chosen due to the wealth of previous knowledge published on these types of compounds. Soy supplements are marketed commercially as "A natural way to replenish the aging body's declining estrogen levels, ... relieve menopausal symptoms, such as hot flashes, as well as decrease the risk of heart disease and osteoporosis, without promoting breast cancer" [16].

The key components of soybean extracts that have been implicated in their biological activity are known as the isoflavones. Isoflavonoids are a subclass of flavonoid phenolic compounds having the basic C_6—C_3—C_6 structure shown in Figure 13.3. Genistin and daidzin are the major isoflavonoid glycosides found in soybeans (*Glycine max*) and kudzu (*Pueraria lobata*). Genistein and daidzein are the aglycone analogs of genistin and daidzin, respectively. Genistein is not naturally found at high abundance in soybeans, but it can be formed when harsh processing conditions are used.

The key spectroscopic feature that distinguishes isoflavonoids from other flavonoids is the isolated proton signal arising from the hydrogen atom at the 2 position.

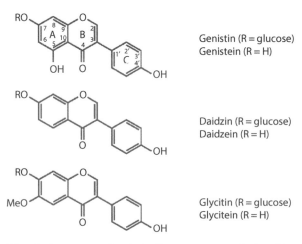

Genistin (R = glucose)
Genistein (R = H)

Daidzin (R = glucose)
Daidzein (R = H)

Glycitin (R = glucose)
Glycitein (R = H)

FIGURE 13.3 Chemical structures for six isoflavones commonly found in soybean.

This proton displays a singlet resonance in the NMR spectrum between 8.0 and 8.5 ppm. Isoflavonoids commonly have a hydroxyl group located at the 5', 7', and 4' positions. In addition, glycosylation, when present, usually occurs at the 7' position (as in genistin) or the 4' position.

For this study, three different soybean dietary supplements were purchased from a local health food store. According to the product label, one product contained only soybean extract: soybean supplement A, containing 80 mg of soy isoflavones per capsule. The other two products contained soybean extract plus other components: soybean supplement B, containing soybean, wild yam, black cohosh, Dong Quai, and additional herbs; and soybean supplement C, containing 50 mg of soy isoflavones plus black cohosh root extract, wild yam, sage leaf, chasteberry, vervain, astragalus, and motherwort.

One capsule of each dietary supplement was extracted with dimethyl sulfoxide-d$_6$ (DMSO-d$_6$). After extraction, the samples were filtered and transferred to NMR tubes for analysis. Fifteen replicates for each different dietary supplement were prepared, for a total of 45 samples. NMR spectra were then collected using an Agilent 500 MHz spectrometer equipped with an OneNMR Probe. Proton, Carbon, and PureShift 1D (PS1D) [17,18] NMR spectra were collected on each sample.

Proton spectra are the most common NMR data used for analysis of botanical samples. A spectrum that is representative of the proton NMR data collected for this study is shown in Figure 13.4. This experiment has very high sensitivity, allowing

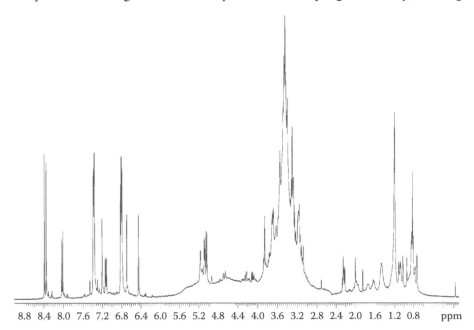

8.8 8.4 8.0 7.6 7.2 6.8 6.4 6.0 5.6 5.2 4.8 4.4 4.0 3.6 3.2 2.8 2.4 2.0 1.6 1.2 0.8   ppm

FIGURE 13.4 Representative proton NMR spectrum collected for the soybean dietary supplements in this study. The high frequency region of the spectrum is dominated by the aromatic resonances of the isoflavone components, the middle of the spectrum contains primarily the carbohydrate and residual water signals, and the low frequency region is made up of fatty acid and amino acid aliphatic resonances.

good signal-to-noise spectra to be acquired after just a minute or two of data acquisition. One of the biggest drawbacks to using proton data for sample screening or library comparisons lies in the frequency dependence of the resulting spectra. The majority of the individual resonances in these spectra are homonuclear coupled to other resonances, giving the classic multiplet type of patterns that encode so much valuable structural information. The center frequency of these patterns corresponds to the chemical shift of the resonance, reported as a ppm based on the field of the magnet used to collect each spectrum; chemical shifts are invariant with respect to the size of the magnetic field. The distance between the individual lines in one resonance pattern is measured in Hz and is also a fixed value, as implied by the term *coupling constant* used to describe this characteristic. Unfortunately, the number of Hz per ppm is not fixed and varies with the field strength. Therefore, a spectrum observed for a particular sample at one frequency will not match the spectrum recorded for the same sample at any other frequency (see Figure 13.5). This aspect of proton NMR spectroscopy significantly complicates the automated comparison of these spectra to a spectral database, necessitating the use of same field strength for the analysis as that used to generate the database.

In those cases where decreased spectral density is desirable, carbon NMR is the most common experiment to which researchers turn. A spectrum that is representative of the carbon NMR data collected for this study is shown in Figure 13.6.

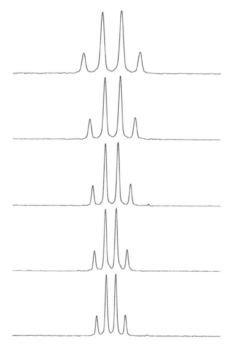

FIGURE 13.5 A stacked plot of NMR spectra collected on the same sample at 400, 500, 600, 700, and 800 MHz, from top to bottom. The chemical shift region is plotted identically for each spectrum. These data clearly show the increasing dispersion observed at higher frequency.

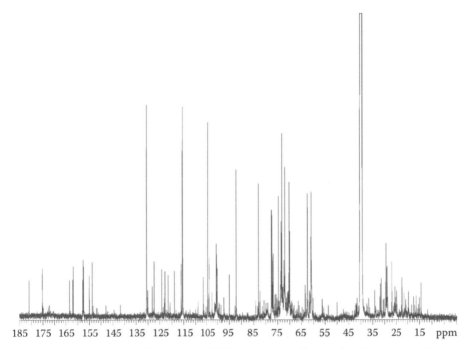

185 175 165 155 145 135 125 115 105 95 85 75 65 55 45 35 25 15 ppm

FIGURE 13.6 Representative carbon NMR spectrum collected for the soybean dietary supplements in this study.

The spectral width is much wider than that of a proton spectrum and the resonances are typically decoupled, meaning that the signals are each represented by a single line. This greatly simplifies the interpretational aspect of the experiment and removes the field-based nature of library searching. Unfortunately, the sensitivity of this experiment is about 400 times lower than the proton experiment conducted on the same sample. Even for a botanical extract where the availability of raw material is not limited, data acquisition times of 10–30 min are typical.

A comparatively new experiment that combines the higher sensitivity of a proton-detected NMR experiment with the field independence and simplicity of the carbon experiment is the PS1D experiment. This is functionally a broadband proton-decoupled proton experiment, yielding a single line for each proton resonance in a sample. Data can be collected quantitatively. A spectrum that is representative of the PS1D NMR data collected for this study is shown in Figure 13.7.

The first step of CRAFT analysis for each experiment type involves selecting the regions of the spectra where resonances of interest are found. For each experiment type described above, a representative spectrum was used to define unique ROI (23 for proton, 12 for carbon, and 11 for PS1D). The ROIs are represented as boxes overlaid on top of each of the representative spectra in Figures 13.8 through 13.10. These ROI from each FID were digitally filtered and downsampled to generate one sub-FID each. For the convenience of this discussion, we define *sub-FIDs* as partial, downsampled time-domain data that represent and contain only the frequencies within the bandwidth of the ROI. Bayesian analysis was performed

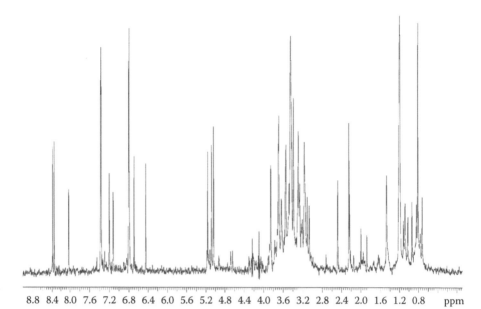

FIGURE 13.7 Representative pure shift 1D NMR spectrum collected for the soybean dietary supplements in this study. This experiment returns a broadband proton-decoupled proton spectrum, giving decreased overlap of signals as compared to a traditional proton spectrum.

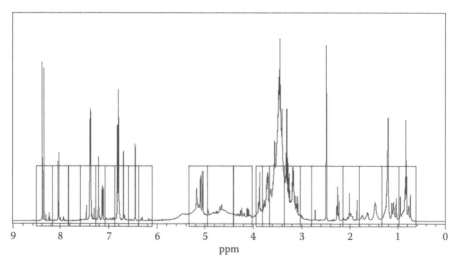

FIGURE 13.8 Proton spectrum of a soybean supplement showing the 23 ROIs used for CRAFT analysis.

on each of these sub-FIDs to generate tables of frequency, amplitude, decay rate constant, and phase of all the exponentially decaying models and combined to generate one CRAFT per FID. Simulations of synthetic FIDs for each of the models in the CRAFT allow the user to quickly and visually evaluate the effectiveness of the deconvolution process.

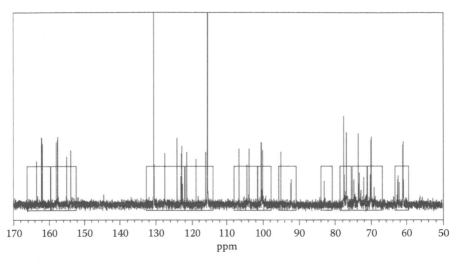

FIGURE 13.9   Carbon spectrum of a soybean supplement showing the 12 ROIs used for CRAFT analysis.

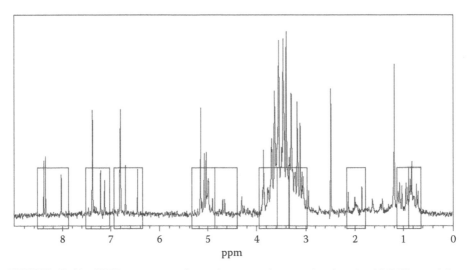

FIGURE 13.10   PS1D spectrum of a soybean supplement showing the 11 ROIs used for CRAFT analysis.

   Visual confirmation of the CRAFT results is most easily accomplished by the creation of a CRAFT stacked plot. As shown in Figure 13.11, the CRAFT models can be used to synthesize an FID for any section of the spectrum that was originally captured for analysis. In Figure 13.11b three traces are displayed. On the bottom is the spectrum after Fourier transformation of the experimental FID, in the middle is the spectrum obtained by Fourier transformation of the FID synthesized from the CRAFT result table, and the trace at the top is the spectrum obtained by Fourier transformation of the difference of the two FIDs. Figure 13.11a is a display showing the individual models used to construct the CRAFT FID.

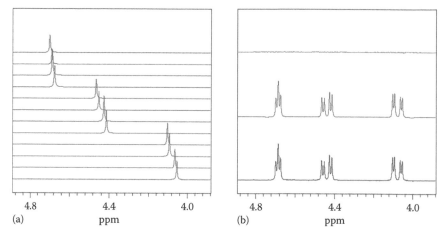

FIGURE 13.11   CRAFT stacked plot showing the aromatic region of a soybean dietary supplement NMR spectrum. Panel (b) shows the experimental, CRAFT, and residual spectra (bottom to top, respectively), while panel (a) the individual CRAFT models.

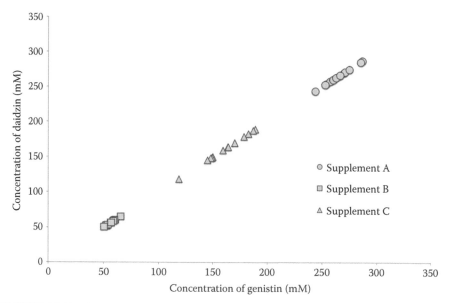

FIGURE 13.12   Bivariate analysis of the 45 soy dietary supplement samples showing the relationship between the measured concentrations of daizin and genestin in each sample.

Once the spectra have been reduced through CRAFT analysis, the results can be used as the basis for a wide range of analyses. For those data sets where only a restricted number of variables are being considered, simple visualization of the relationship between those variables is often sufficient to form a conclusion. An example of this approach is shown in the simple bivariate analysis displayed in Figure 13.12. Even using a simple method such as this, discrimination between the three different types of samples is obvious. This type of analysis would be useful for monitoring the consistency of a given soybean supplement during processing and manufacturing or as a tool to monitor the production of isoflavonoids when soybean plants are grown under various conditions.

As an example of more complete targeted analysis, the concentration of the individual isoflavones can be used to demonstrate potency for both the individual compounds and total isoflavone dose in the sample, as well as to determine content uniformity across each type of supplement and across the entire sample set.

The measured quantity of the individual isoflavones is shown in Figure 13.13. The information conveyed after the spectra are converted to a spreadsheet is powerful. Content uniformity for the supplement C sample set is clearly lower than that observed for the other two products, and the differences in total isoflavone dose from each product are obvious.

Although visual inspection of the results in a bar graph does convey a useful representation of the data set, more complex statistical analysis methods are available for higher dimensionality data. Principal component analysis (PCA) is a common method used to quickly investigate variance in very large data sets and is supported in CRAFT. In essence, PCA allows visualization of the quantitative difference in a collection of multivariate data. PCA analysis is the method of choice for exploratory studies where analysis of the entire complement of a complex mixture is used to compare samples. It allows a collection of related spectra to be characterized according to the degree of similarity between them. As such, it provides a rapid and effective way to identify classes of spectra and to relate these classes to properties of interest.

The trends in CRAFT multivariate data from the soybean dietary supplements can be interrogated from multiple orientations using PCA.

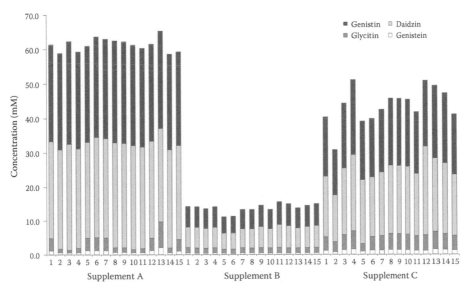

FIGURE 13.13   Comparison of the concentration of individual isoflavones and the total isoflavone dose across the 45 soybean dietary supplement samples. The differences in total dose and content uniformity are readily apparent.

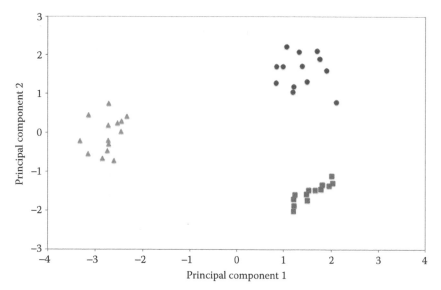

FIGURE 13.14 PCA analysis of the four primary isoflavones identified in the dietary supplement sample set. The concentration of each component was extracted from the CRAFT analysis of the aromatic region of the spectra. This demonstrates the clear discrimination of each sample type from the others based on a targeted analysis.

One such example is the PCA representation obtained using concentrations of the four isoflavone components of each sample as the sole input. As shown in Figure 13.14, the three products are well segregated from each other, yet variability within some clusters is pronounced. This is an orthogonal way of visualizing the same similarity and uniformity information displayed in the bar chart (Figure 13.13), only the amount of variability between the sample types and within the clusters is captured in a more obvious way using this representation.

For many botanical investigations, the first questions that researchers want to ask revolve around authenticity and origin. In these situations, identification of individual components is less important that identifying the gross differences between a given set of samples. By indiscriminately capturing all the peaks observed in the carbon NMR spectra, a holistic PCA plot can be rendered that reports on the gross similarity between each of the three products, including results from the fatty acid and oil components, carbohydrate materials, and amino acids. As shown in Figure 13.15, when the entire carbon spectrum is used as the input for PCA analysis, the three sample types are well separated. This type of analysis requires no interpretation of individual lines or resonances and can be conducted in an automated fashion by novice users.

The nature of carbon NMR data collected under normal conditions is to provide a single line for each chemically distinct carbon atom in the sample. Because these chemical shift measurements are frequency independent, they can readily be used as the basis for a database. Unfortunately, the low sensitivity of this experiment has significantly limited its application to botanical analysis. The proton experiment is much more sensitive, but the fact that the resonance multiplet patterns change with the magnetic field strength of the instrument used to collect the data means that both

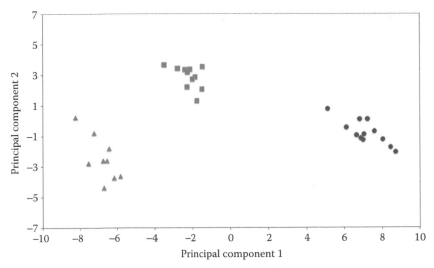

FIGURE 13.15 PCA analysis of the result table created by CRAFTing the entire carbon spectrum of each sample in the study. This demonstrates the clear discrimination of each sample type from the others based on untargeted analysis using carbon NMR.

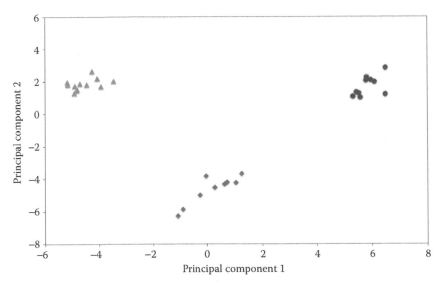

FIGURE 13.16 PCA analysis using an untargeted investigation of the full PS1D data sets collected for the soybean dietary supplements.

creation of a database and comparison to a database are more complicated. The PS1D experiment overcomes these issues and offers a comparatively sensitive proton-detected spectrum without multiplet patterns that has the frequency-independence of carbon NMR. Figure 13.16 shows the PCA result obtained by an untargeted analysis of the PS1D spectra collected in this study. Once again, discrimination and variation between the three products is obvious.

CONCLUSION

NMR spectroscopy provides numerous advantages to the analysis of dietary supplements, including acquisition of a chemical fingerprint, availability of quantitative information for a number of different compounds in a single sample, and structural verification data via one simple experiment. This information can be used for raw material quality control, for process and final product control, authenticity verification investigations, and adulterant screening. As we have seen, there are several options for the type of experiment that can be recorded, and the analysis of the data can be either untargeted, to broadly gather information about the complete sample matrix, or targeted, to analyze for specific known compounds.

As these activities have historically been based on less-than-optimal frequency domain quantification techniques, there has been limited adoption for NMR methods in this arena due the time-intensive nature of spectral matching and the error inherent in that method. CRAFT allows fully automated conversion of the raw NMR spectra into a data-mining friendly spreadsheet format with very high fidelity and accuracy. This advancement allows rapid screening methods, both targeted and untargeted, to be implemented easily and to be employed effectively in high-throughput environments.

REFERENCES

1. R.L. Prior and G. Cao, *J. AOAC Int.* 83(4), 2000: 950–956.
2. J. Sherma, *J. AOAC Int.* 86(5), 2003: 873–881.
3. K.F. Morris and C.S. Johnson, *J. Am. Chem. Soc.* 114, 1992: 3139–3141.
4. K.I. Burton, J.R. Everett, M.J. Newman, F.S. Pullen, D.S. Richards, and A.G. Swanson, *J. Pharm. Biomed. Anal.* 15, 1997: 1903–1912.
5. L. Griffiths and A.M. Irving, *Analyst* 123, 1998: 1061–1068.
6. P.A. Hays and R.A. Thomson, *Magn. Reson. Chem.* 47, 2009: 819–824.
7. G.L. Bretthorst, *An Introduction to Parameter Estimation Using Bayesian Probability Theory: Maximum Entropy and Bayesian Methods*, Kluwer Academic Publishers, Dordrecht, the Netherlands, 1989.
8. D.V. Rubtsov, C. Waterman, R.A. Currie, C. Waterfield, J.D. Salazar, J. Wright, and J. L. Griffin, *Anal. Chem.* 82, 2010: 4479–4485.
9. K. Krishnamurthy, *Poster presented at the 54th Experimental NMR Conference*, Asilomar, CA, 2013.
10. G.L. Bretthorst, *J. Magn. Reson.* 88, 1990: 533–551.
11. G.L. Bretthorst, *J. Magn. Reson.* 88, 1990: 552–570.
12. G.L. Bretthorst, *J. Magn. Reson.* 88, 1990: 571–595.
13. G.L. Bretthorst, *J. Magn. Reson.* 93, 1991: 369–394.
14. G.L. Bretthorst, *J. Magn. Reson.* 98, 1992: 501–523.
15. Krishnamurthy, K., *Magn. Reson. Chem.*, 2013, doi:10.1002/mrc.4022.
16. University of California, Berkley, CA, Wellness webpage, http://www.wellnessletter.com/ucberkeley/dietary-supplements/soy-supplements/#.
17. K. Zangger and H. Sterk, *J. Magn. Reson.* 124, 1997: 486.
18. J. A. Aguilar, S. Faulkner, M. Nilsson, and G. A. Morris, *Angew. Chem. Int. Ed.* 49, 2010: 3901–3903.

A Fit-for-Purpose Approach

Cynthia Kradjel

CONTENTS

Introduction 215
Basic Principles and Theory of NIR Analysis 216
Analysis of Botanical NIR Data 218
Application of NIR for Identification and Qualification of Botanicals 226
Dietary Supplement cGMPs 227
Single Laboratory Validated NIR Methods for Botanical Identification:
Definition of Target Materials 228
 Traceability 229
 Selectivity 229
 Ruggedness 229
 Instrument Performance 229
Screening for Economically Motivated Adulteration 232
 Polyphenols 234
NIR for Characterization of Natural Products by Determination of Key
Quality Parameters 234
Conclusion 234
References 235

INTRODUCTION

Current near-infrared (NIR) methods for the identification and qualification of
botanicals evolved through the hybridization of NIR developments in the agricultural industry with those in the pharmaceutical industry.

In 1968, at the US Department of Agriculture in Maryland, Shenk et al. [1] and, Ben-
Gera and Karl Norris [2,3] began to develop NIR methods to rapidly and simultaneously
determine protein and moisture in grains, and protein, moisture, and oil in soybeans. In
1976, Norris et al. [4–5] expanded their scope, working to develop rapid NIR methods
for a wide variety of quality parameters for forage, including correlating NIR spectra

to factors such as intake and digestibility. Today, NIR plays an integral global role in assessing grain quality and payment; characterizing nutritive value of animal feeds and forage; testing seed quality for breeding; and many other applications. Although testing of complex natural products is challenging due to differentiation of species, sourcing, and variation attributed to growing conditions and overall, the agricultural industry has successfully implemented NIR. Key elements to this success include compilation and management of large databases of materials; use of sophisticated chemometric methods to build and validate calibrations; proper focus on reproducible sample presentation; and attention to accuracy and precision of reference methods.

In the pharmaceutical industry, Rose et al. in 1982 [6] demonstrated that a large number of structurally similar penicillin-type drugs could be differentiated and identified using NIR. In 1984, Howard Mark and David Tunnell [7] developed Mahalanobis distance as an algorithm for discriminant analysis of raw materials. At that time, Emil Ciurczak [8] recognized the need for rapid identification (ID) methods as pharmaceutical companies were preparing to comply with the new European Pharmacopeia requirement of 100% identity testing of every incoming raw material container and studied the use of Mahalanobis distance algorithms to discriminate the NIR reflectance spectra of actives in multicomponent pharmaceutical dosage forms. Today, NIR plays an integral global role in the pharmaceutical industry for identification and qualification of incoming raw materials; monitoring blend homogeneity and moisture in process; and verification of final product assay in tablets and capsules [9–11]. The majority of incoming pharmaceutical raw materials are synthesized, high purity materials that exhibit minimal lot-to-lot variation. In addition, these materials are highly characterized so NIR method development is less challenging for pharmaceutical raw materials than for botanicals. The key developments in the pharmaceutical industry that facilitate the development of botanical identification methods is the extensive work invested in developing guidelines and protocols to validate NIR methods and to ensure compliance with regulatory guidelines. Harmonized protocols have been developed internationally [12–21].

BASIC PRINCIPLES AND THEORY OF NIR ANALYSIS

NIR analysis provides rapid, simultaneous results for multiple parameters with no sample preparation. Spectral patterns of samples in their *native* state are collected in seconds, providing information on both chemical and physical properties. Once methods are developed and validated, routine operators can perform the testing, facilitating the measurement of a far greater number of samples when compared to techniques that require time-consuming sample preparation. In addition, as samples are simply presented to the system without preparation, NIR analyzers can be placed close to the sample rather than having to bring samples to the laboratory. Analyzers for conforming identity and quality of incoming raw materials are often located near the receiving dock where incoming goods are delivered.

The physical origin of infrared and NIR spectra is based on molecular vibrations. In a simple diatomic model of atomic masses m_1 and m_2, the vibrating masses lead to changes in the internuclear distance. The lowest, or fundamental, frequencies of any two atoms connected by a chemical bond can be approximately calculated by assuming that band energies arise from the vibration of the diatomic harmonic oscillator model, and then applying Hooke's law,

$$v = \frac{1}{2}\Pi\sqrt{\frac{k}{\mu}}$$

where:
   v is the vibrational frequency
   k is the force constant
   μ is the reduced mass of the two atoms

Electron donating or withdrawing properties of functional groups in the substance influence the bond lengths, resulting in frequency differences specific to those functional groups, particularly for O—H, C—H, and N—H bonds. These frequency differences then contribute to the pattern, or spectrum, of the substance. Hydrogen bonding and other changes that occur in the substance result in changes to the spectral pattern so that one can monitor changes that occur in a sample, providing a wealth of information about a material, its matrix, and interactions between the material and its matrix.

When these molecular vibrators at ground state level $n = 0$ absorb light of a particular frequency, they are excited to a higher level $n = 1$. In the mid-infrared region, 4,000 to 200 cm^{-1}, the only allowed transitions are those that occur between consecutive energy levels $\Delta v = +1$ and cause a change in the dipole moment. By contrast, in the NIR region, 780–2,500 nm or 12,820–4,000 cm^{-1}, the selection rule allows transitions over more than one energy level, giving rise to multiple overtones for the corresponding fundamental band. The harmonic oscillator model cannot be used in the NIR region. Rather, the anharmonic oscillator model is used as energy levels are no longer equidistant. In addition to overtone bands, the NIR region also has combination bands that occur when two or more vibrational modes cause simultaneous energy changes; the frequencies of which are sums of multiples of each interacting frequency [10,22,23].

Practically speaking, overtones are multiples of the corresponding fundamental band with a slight shift in frequency due to the anharmonicity constant. The relative intensity of

---

TABLE 14.1
Relative Intensities of C—H Fundamental and Overtone Stretch Bands

Band	Wavelength region (cm^{-1})	Wavelength Region (nm)	Relative Intensity	Path Length	Optical Density
Fundamental (n)	2,849–2,959	3,380–3,510	100	0.01 mm	2
First overtone, ($2n$)	5,698–5,917	1,690–1,755	1	1.0 mm	2
Second overtone, ($3n$)	8,540–8,870	1,127–1,170	0.1	1.0 cm	2
Third overtone, ($4n$)	1,139–1,183	845–878	0.01	10.0 cm	2

Source: J. Workman, An introduction to near infrared spectroscopy, spectroscopynow.com, http://www
.spectroscopynow.com/details/education/sepspec1881education/An-Introductionto-Near-Infrared-
Spectroscopy.html?tzcheck=1, September 12, 2005.

each subsequent overtone band is orders of magnitude *less* intense than the fundamental band, depending upon the particular bond. Table 14.1 indicates the wavelength regions where the C—H fundamental and overtone stretch bands occur and the relative intensity of the bands. In this example, the first overtone band is two orders of magnitude *less* intense than the fundamental, while the sampling path length is two orders of magnitude greater than the corresponding fundamental measurement, and so on. The practical advantage of using NIR is the volume of sample being measured is far greater than that measured by mid-infrared. Greater measurement area affords a more representative view of the material being tested. This is particularly important for natural materials.

Another practical advantage of the NIR region is that, depending on the substances, the resulting combination bands can be more effective at resolving similar spectral features than fundamental and overtone bands. For example, the O—H of an alcohol, the O—H band of a carboxylic acid, and the O—H band of water tend to be more resolved in the combination band region when compared to the fundamental and overtone region [25]. The combination band region therefore is more useful for separation of similar botanicals or for quantification of active components in botanicals. In agricultural applications, high-precision NIR methods for protein routinely include the combination band region.

ANALYSIS OF BOTANICAL NIR DATA

NIR spectra of botanicals are information rich, containing information about the botanical itself, its gross physical properties and its matrix. Visual inspection of botanical NIR spectra may be an approach for troubleshooting *problem* samples, often highlighting gross anomalies such as the presence of unexpected residual solvent; unexpected processing aids or diluents; or the presence of unexpected synthetic pharmaceuticals. Figure 14.1 shows the second derivative, vector normalized spectra from approximately 5,900 to 5,500 cm⁻¹ of (1) neat maltodextrin; (2) a ChromaDex botanical reference material (BRM) for bilberry; (3) the bilberry BRM adulterated with 5% maltodextrin; (4) the bilberry BRM adulterated with 10% maltodextrin;

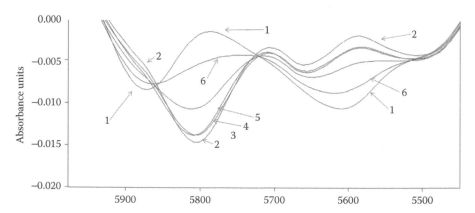

FIGURE 14.1   Dilution of a ChromaDex bilberry botanical reference material with varying amounts of maltodextrin is evident in the second derivative spectra between 5900 and 5500 cm⁻¹.

(5) the bilberry BRM adulterated with 25% maltodextrin; and (6) the bilberry BRM adulterated with 50% maltodextrin. The dilution of bilberry with maltodextrin is clearly seen in these spectra as distinct absorbance peaks are seen for pure malto-dextrin and for pure bilberry. The diluted samples trend in relation to their respective concentrations of maltodextrin and bilberry. Figure 14.2 shows the same spectra from approximately 4,900 to 3,900 cm⁻¹, and again, the dilution of bilberry with varying concentrations of maltodextrin is clearly visible.

Figure 14.3 shows the absorbance spectra of ChromaDex BRM black cohosh bark and ChromaDex BRM black cohosh root. Although one may visually discern some differences between the two spectra, the majority of the spectra are visually

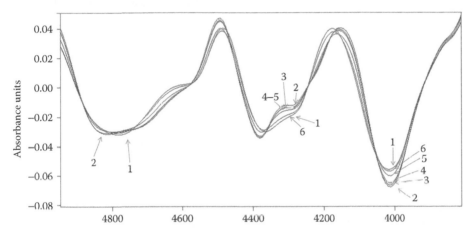

FIGURE 14.2 Dilution of a ChromaDex bilberry botanical reference material with varying amounts of maltodextrin is evident in the second derivative spectra between 5900 and 5500 cm⁻¹.

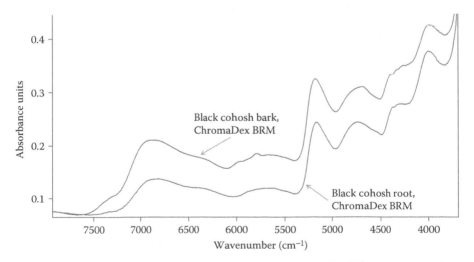

FIGURE 14.3 Absorbance spectra of different botanicals are not easily distinguished upon visual examination, with ChromaDex BRM black cohosh bark and root as an example.

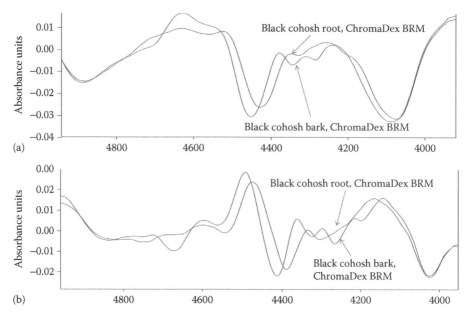

FIGURE 14.4 Data preprocessing such as first derivative with vector normalization (a) and second derivative with vector normalization (b) better show the spectral differences between botanicals with ChromaDex BRM black cohosh bark and root as an example.

quite similar. The spectral differences between black cohosh bark and black cohosh root become more noticeable with the application of data treatments as seen in Figure 14.4a the first derivative, vector normalized and Figure 14.4b the second derivative, vector normalized spectra.

For the first derivative spectra, the maximum peak absorbance is where the spectra intersect zero. With the second derivative spectra, the maximum peak absorbance is presented as the minima [26,27]. Data preprocessing is useful for method developers to visualize differences between materials. However, preprocessing is not always required to achieve separation of groups for prediction with chemometric methods. In fact, calculation of derivatives results in the loss of some spectral information, so depending on the level of separation required, method developers may chose to use or not to use data preprocessing in their chemometric method. The main reasons for using preprocessing is to reduce the effect of repacking error or particle size which results in a y-axis shift. Calculation of derivatives amplifies noise, so smoothing is typically used when calculating derivatives. Different instrument characteristics may also dictate the best type of preprocessing. Standard normal variate and vector normalization are other transformations used to reduce light-scattering effects of particle size variation.

If moisture variation is a concern, moisture region wavelengths can be de-selected when building the identity calibration model. The more lots used to represent the acceptable variation in the calibration set, the more robust the method will be and the less dependent on attributes of the spectra that are not related to identity such as moisture content.

Visual inspection has a place in the method development process; however, routine fit-for-purpose NIR methods for identification of botanicals must be based on chemometric methods. Visual assessment alone is not sufficient for routinely and reliably separating botanical materials. Even if one could reliably discern the differences between botanicals using visual means, the challenge is that with successive lots, it would not be practical for routine operators to discern *acceptable lot-to-lot variation* from *unacceptable lot-to-lot variation*, whereas chemometric methods enable establishment of numerical pass/fail criteria.

Chemometrics is defined as the application of mathematics or statistical methods to chemical data. For spectroscopic data-driven methods, this includes using multivariate statistics, applied mathematics, mathematical algorithms, and computer science [28].

For identification purposes, there are two main types of methods: unsupervised and supervised. The identity of the sample is unknown in unsupervised methods, and the test report typically presents a *hit* list of materials that are closest to the unknown. For pure samples, this is an option, but for natural products, it is an approach fraught with problems. In general, for complex natural products, hit lists are not very useful and can even be misleading for two main reasons. First, the list of what is *closest* is based on what is contained in the database, and if the database is not comprehensive, dissimilar materials might be the *closest*. Second, the major component of plant cell walls is cellulose, so it is not surprising that from a coarse point the NIR spectra of different botanicals look quite similar, potentially leading to dissimilar botanicals being cited as *closest* to each other. Furthermore, in an unsupervised method, if a material, X, included in the NIR identity method is in the wrong container labeled, Y, the analyst would get a report stating that identity of the material is X. The ID report would then pass the material as X, and would not flag that the material is mislabeled as Y. Perhaps the analyst would catch the mistake and fail the material, but that approach would place a large manual responsibility on routine operators to catch such errors.

Fit-for-purpose NIR methods for identification of botanicals are based on supervised methods validated with concrete pass/fail mathematical criteria. In addition, in a supervised method, the scenario above would result in a *fail*. With a supervised method, the analyst would be required to type in the material code or use a bar code reader to scan in the label information. Then, the analyst would scan the sample; the NIR identification method would analyze the spectral information, and then, the NIR-determined identity would be compared to the target. The sample X would fail the NIR identification because it would not fit the criteria for what is expected, Y.

Chemometrics identify small differences in the botanical spectra, resulting in the ability to uniquely identify that botanical from the others based on mathematical calculations. Fit-for-purpose NIR methods establish a very tight pass criteria by including in the calibration set BRM and/or the tested raw material lots as defined by the product specification, specific to supplier and verified by orthogonal methods.

Chemometric methods used for identity range from simple spectral comparison algorithms and correlation algorithms; distance calculations to mean spectra or to individual spectra in a group using Mahalanobis or Euclidean distance; and factor analysis using principal components. A vast amount of information is available in the literature about these different approaches [29–35].

Principal component analysis is defined by ASTM [36] as a mathematical procedure for resolving sets of data into orthogonal components whose linear combinations approximate the original data to any degree of accuracy. As successive components are calculated, each component accounts for the maximum possible amount of residual error in the set of data. Typically, the number of components is smaller than or equal to the number of variables or the number of spectra whichever is less. Fit-for-purpose method development seeks to find the optimum number of variables, as too many variables will result in overfitting the data and too few will not capture the acceptable lot-to-lot variation.

Although chemometrics is used to mathematically separate the different materials from each other, a prudent method developer will also evaluate the spectra to ensure that all spectra within a group are similar. Otherwise, there will be uncertainty as to whether the nature of an unusual spectrum in a group originates from a problem with sample presentation; to a mistyped name; to a problem with the instrument; to the material changing over time; and so on. Lois Weyer and S. C. Lo [37] and Jerome Workman, Jr. and Lois Weyer [38] studied functional group band assignment in the NIR region. Brian Dickens and Sabine Dickens [39] studied water dissolved in common solvents to better understand the shifting that can occur with water bands depending on their environment.

Appropriate use of NIR methods allows one to not only identify the chemical nature of the product but also obtain information about aggregate physical properties and overall quality. Unlike a Certificate of Analysis (COA) that reports values in relation to a defined specification, the NIR method looks at the entire material, often flagging unexpected adulterants, residual solvents, and other problems that a COA would not report. If a botanical has been diluted with maltodextrin, that botanical very well may pass a thin layer chromatography (TLC) test for identity, but it would fail an Fourier transform near infrared (FT-NIR) method for identity if the calibration set only included botanicals that were not prepared with maltodextrin. Unless the supplier disclosed that maltodextrin was present, and the calibration was developed accordingly, the FT-NIR would fail the material. Therefore, FT-NIR *identity* is actually *identity* + *quality*.

In general, generic spectral libraries are not useful because discrimination of botanicals requires discernment of very small differences in the spectra. Rather, *databases* with *traceable* samples are used to build an identity calibration method. The scope of the calibration set of traceable materials is dictated by the materials to be tested, but in general BRMs and/or actual raw material receipts that have been tested and approved are included. Processed botanicals and extracts require careful evaluation of methods of production and associated product specifications so an adequate description of the target can be defined. The goal is to include materials in one group that exhibit acceptable lot-to-lot variation.

For example, if a NIR identification method was developed and validated using aerial parts, an unknown samples of mixed parts or root samples would not be expected to pass both *identity* + *quality*. If the product specification did not specify a particular plant part, then samples representing all parts would be combined into one group. In another example, if a pure botanical is used to develop and validate the method, materials with maltodextrin or other *processing aids* would be expected to fail. Therefore, if the raw material supplier routinely includes processing aids in ground botanical samples, they would be required to disclose those ingredients, and NIR methods would be developed to cover the acceptable compositional range of those ingredients.

Establishing identity calibrations that are specific to a raw material and its quality specification does result in false negatives because the pass/fail criteria are quite tight. After the initial phases of qualifying a supplier and/or a material, the frequency of false negatives is dramatically reduced. Materials with different qualities or material from new suppliers must be qualified against the NIR method prior to routine testing. If material of a slightly different quality that fails the NIR method is later deemed to be acceptable after verification with traceable reference methods, the FT-NIR pass/fail criteria is dynamically updated to include that material. Version control is used to document the updated calibration method.

For each raw material, the NIR identity methods are designed to represent the scope of acceptable material. Figure 14.5 shows a 2D scores plot of different groups of botanicals including angelica root, butcher's broom, black cohosh root, chamomile flower, and California poppy herb. Each group is a separate cluster of points that represent the sample-to-sample variation in the samples used in the calibration set. Some groups are subject to larger lot-to-lot variation than others as seen by the differing sizes of each cluster on the plot.

The most successful FT-NIR identity methods for botanicals utilize a hierarchical approach [40–41]. All materials are first evaluated in a single calibration method.

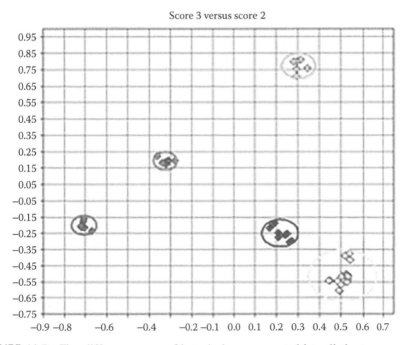

FIGURE 14.5 Five different groups of botanicals are separated into distinct groups using chemometrics and plotted in a 2D scores plot.

If there are materials within the group that are clustered together and are very similar, they may be promoted to a higher level of evaluation that allows for elimination/ selection of specific spectral regions, different preprocessing, and/or alternate chemometric algorithms. Ultimately, validation of the entire identity method includes evaluation of the main method and the sublevel methods all together to ensure specificity and to minimize false positive results in routine prediction mode.

Figure 14.6 shows the absorbance spectra of Angelica root, black tea extract, California poppy herb, green tea extract, lemon balm, peppermint leaf, and wild yam root. The green tea extract and the black tea extract contain high concentrations of polyphenols that cause their spectra to be visually rather different than the others. Only slight spectral differences can be seen between the other materials; however, the first derivative vector normalized spectra (Figure 14.7a) and the second derivative vector normalized spectra (Figure 14.7b) seen in Figure 14.7 allow improved visualization of the differences. In this example, one could envision the botanicals in one method to allow enhanced separation between them, and the tea extracts in a submethod.

Figure 14.8 shows a validation report and includes a numerical value for selectivity index, S for each group and its nearest neighboring group in a set of botanical raw materials. For the material, and its nearest neighboring group, S is the ratio of distance D between average spectra and the sum of the threshold values $T1$ and $T2$ (cluster radii). For $S < 1$, the two groups overlap and reliable identification cannot be performed using that method. If $1 < S < 2$, the groups are separated but at risk for overlap if lot-to-lot variation causes an increase in the size of the group. In the method development stage, materials with $S < 2$ are not reliably separated and are then moved to a sublibrary that allows elimination/selection of wavelength ranges, different preprocessing, and algorithm selection to specifically enhance spectral differences between the similar groups. In the example shown in Figure 14.8,

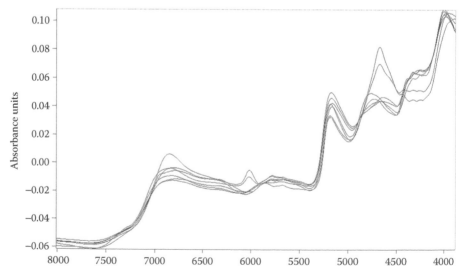

FIGURE 14.6   Absorbance spectra of seven different botanicals showing that the green tea extract and the black tea extract have distinct peaks related to high polyphenol content.

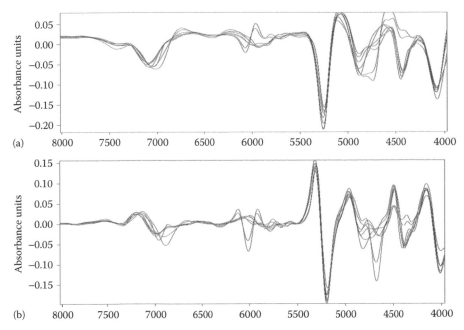

FIGURE 14.7 Preprocessed spectra (first and second derivative with vector normalization) show that the botanicals can be separated into two distinct submethods to enhance separation of similar materials.

	ID	Group1	Group2	IP-L	S	Threshold1	Threshold2	D
1	1	Angelica	Oregon Grape Root	IP1:	4.700767	0.019176	0.046626	0.309317
2	2	Black Tea Extract	Green Tea Extract	IP1:	2.108192	0.016989	0.102128	0.251121
3	3	California Poppy	Horehound Herb	IP1:	2.971932	0.011084	0.029679	0.121146
4	4	Green Tea Extract	Black Tea Extract	IP1:	2.108192	0.102128	0.016989	0.251121
5	5	Horehound Herb	Chamomile Flower	IP1:	1.742342	0.029679	0.025831	0.096717
6	6	Lemon Balm	Peppermint Leaf	IP1:	2.770407	0.020449	0.017857	0.106123
7	7	Oregon Grape Root	Wild Yam Root	IP1:	2.403041	0.046626	0.019375	0.158601
8	8	Peppermint Leaf	Spearmint Leaf	IP1:	2.522884	0.017857	0.011745	0.074683
9	9	Spearmint Leaf	Peppermint Leaf	IP1:	2.522884	0.011745	0.017857	0.074683
10	10	Wild Yam Root	Oregon Grape Root	IP1:	2.403041	0.019375	0.046626	0.158601
11	11	Chamomile Flower	Horehound Herb	IP1:	1.742342	0.025831	0.029679	0.096717

FIGURE 14.8 A selectivity report with numerical assessment of which groups are close and at risk of overlap potentially leading to false positive identification.

chamomile flower and horehound herb have an *S* value of 1.742342 and are at risk of false positive identification. When both groups are moved to a sublibrary as seen in Figure 14.9, a submethod is developed to focus on the small differences between the two, and the new method results in a selectivity index of 8.113062. All libraries and sublibraries are validated together before release for routine testing.

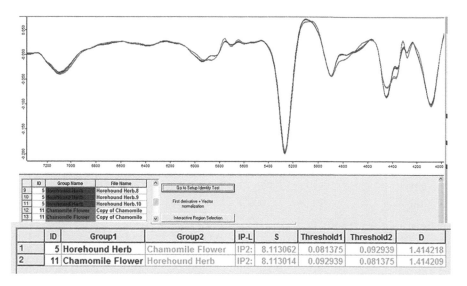

FIGURE 14.9 Hierarchial method development enables moving groups with low selectivity index to a sublibrary where specific parameters can be selected to improve the separation of similar materials.

APPLICATION OF NIR FOR IDENTIFICATION AND QUALIFICATION OF BOTANICALS

In 1990, Corti et al. [42] studied the application of NIR spectroscopy for the identification of drugs derived from plants. Several other publications appeared between 1996 and 1999 [43–46] that featured the use of NIR for rapid characterization of natural products and herbal medicines.

In the late 1990s, Dr. Jeff Golini of All American Pharmaceutical implemented NIR for identity testing [47–49]. With the ease of sampling and the speed of analysis, All American Pharmaceutical Company implemented a program to test every container of incoming raw material. Using a drum roller to enable a more representative sample for testing, several key problems were soon highlighted by NIR: (1) one supplier was shipping raw burdock root powder instead the material they paid for, burdock root extract; and (2) a supplier of *natural* Sida Cordifolia shipped some drums that passed NIR ID, but several other drums in the same shipment that were supposed to be of the same lot *failed NIR*. Further testing revealed that these drums contained synthetic ephedrine HCl, and (3) several lots of kola nut and guarana *failed NIR*, with further testing to reveal that they contained synthetic caffeine anhydrous when they were labeled as *all natural*. Jim Wagner [50] coined the phrase *quality firewall* to describe the use of NIR in the dietary supplement industry, and companies such as Pharmavite [51] described the use of NIR to test incoming raw materials as an integral part of their manufacturing operations.

Steur et al. [52] used principal component analysis to differentiate between the spectra of various citrus oils including grapefruit, orange, mandarin, lemon, and lime oils. Woo used NIR to classify the cultivation area of ginseng by NIR and inductively coupled plasma atomic emission spectroscopy (ICP-AES) [53]. Magali Laasonen et al. used NIR for identification of milled *Echinacea purpurea* roots, reporting demonstration of specificity against *Echinacea angustifolia* and *Echinacea pallida* [54]. Quansehng Chen et al. [55] used NIR to identify different tea (*Camellia sinensis* L.) varieties. Daniel Cozzolino's review article [56] highlights additional qualitative applications of NIR for identification and qualification. Devanand Luthria et al. [57] compared various spectral regions and preprocessing techniques for spectral fingerprinting of broccoli. Jesus A. Espinosa et al. [58] studied NIR to discriminate between different pine species and their hybrids.

DIETARY SUPPLEMENT cGMPs

On June 25, 2007, the FDA published the Current Good Manufacturing Practice (cGMP) in Manufacturing, Packaging, Labeling, or Holding Operations for Dietary Supplements: Final Rule. This rule stated that although dietary supplements fall under the category of food, specific cGMPs were required for dietary supplements to ensure that "process controls for dietary supplement manufacture include establishing and meeting specifications to ensure the finished dietary supplement contains the correct ingredient, purity, strength, and composition intended." Key excerpts from this regulation include the following [59]:

cGMP 111.70 (b) (1) requires you to establish component specifications for each component you use in the manufacture of a dietary supplement. You must establish an identity specification for each component that you use in the manufacture of a dietary supplement.... You must establish component specifications that are necessary to ensure that specifications for purity, strength, and composition of dietary supplements manufactured using the components are met.

cGMP 111.75 (a) (1) requires you, before you use a component that is a dietary ingredient, to conduct at least one appropriate test or examination to verify the identity of the dietary ingredient.

cGMP Section 111.75 (h) (1) requires you to ensure that the test and examinations you use to determine whether specifications are met are appropriate and scientifically valid methods.

cGMP section 111.75 (h) (2): requires that the tests and examinations that you use must include at least one of the following:

   (i) Gross organoleptic analysis;
  (ii) Macroscopic analysis;
 (iii) Microscopic analysis;
 (iv) Chemical analysis; or
  (v) Other scientifically valid methods.

cGMP 111.3 25. What Other Terms Did the Comments Want Defined?

    a. Identity. The "identity" of a dietary supplement refers to the dietary supplement's consistency with the master manufacturing record and/or that it is the same as described in the master manufacturing record.

b. *Purity. The "purity" of a dietary supplement refers to that portion or percentage of a dietary supplement that represents the intended product.*
c. *Strength. The strength of a dietary supplement relates to its concentration. By concentration, we mean the quantitative amount per serving (for example, weight/weight, weight/volume, or volume/volume). Therefore, for purposes of this final rule, strength does not refer simply to the quantity of an ingredient, rather it refers to the amount of a stated ingredient per a specified unit of measure.*
d. *Composition. A dietary supplement's "composition" refers to the specified mix of product and product-related substances in a dietary supplement.*

We decline to define "test," "scientifically valid analytical method," or "scientifically valid method" in this final rule. We also decline to define "validation" and "verification" because the final rule does not establish any requirements that use these terms.

In 2011, guidelines for botanical identification methods were finalized by an AOAC expert review panel (ERP) and approved by the Official Methods Board as an official AOAC standard, concluding the scope of work for the AOAC/US Food and Drug Administration/National Institutes of Health, Office of Dietary Supplements contract. There are three main parts to this AOAC Guideline including Part 1: AOAC Guidelines for Single-Laboratory Validation of Chemical Methods for Dietary Supplements and Botanicals; Part II: AOAC Guidelines for Validation of Botanical Identification Methods; and Part III: Probability of Identification: A Statistical Model for the Validation of a Qualitative Botanical Identification Methods [60,61].

The group agreed to define a botanical ID method as "one that established identity specifications for a botanical material and determines, within a specified statistical limit, whether the test material is a true example of the target botanical material—the botanical of interest. Inherent in the definition of a botanical ID method is that it achieves it goal by comparing the test method to a reference material, where the reference material can be any botanical material whose composition and/or characteristics are judged to accurately represent the target material" [60,61].

SINGLE LABORATORY VALIDATED NIR METHODS FOR BOTANICAL IDENTIFICATION: DEFINITION OF TARGET MATERIALS

Using the guidelines presented by the AOAC ERP, the validation of botanical identification methods begins with a clear definition of the material to be tested, the target material. For raw botanicals, this may include the specific part of the plant if the component specification identifies a specific part of the plant.

For industrial confirmatory testing, the raw material specification is used to define the target material. Definition of an extract is clearer when the quality of the extract is clearly defined regarding target analyte concentrations and limits for processing aids such as maltodextrin. In all cases, each supplier for a particular extract needs to be qualified against the calibration model as processes may vary from supplier to supplier.

For a *gravimetric* extract, where the final product is merely a weight-to-weight concentration, product quality varies batch to batch and definition of target material is challenging. For example, if 1,000 kg of a raw botanical is concentrated to 1 kg of final product extract, the yield may vary significantly from lot to lot. In these cases, the target material is set using several lots tested by orthogonal methods, and as new lots *fail* the NIR tests, they are tested and, if acceptable, added into the model to dynamically update the pass/fail criteria.

Often manufacturers are unaware of the levels of processing aid in the incoming raw material. The NIR method screens for unexpected dilution with processing aids. In addition, depending upon the level of concentration, NIR could detect the presence of residual solvents that are not disclosed to the manufacturer. Ethanol extractions are quite expensive for the supplier, and unscrupulous suppliers may attempt to substitute methanol or other solvents to increase their profits.

TRACEABILITY

All materials included in the calibration set, including BRMs, and tested lots must be traceable to reference methods to prove that the material used to set the pass/fail criteria is what it is supposed to be.

SELECTIVITY

Prior to approving the NIR method for routine use, it must be evaluated for specificity and selectivity. This is accomplished by verifying that similar materials in the identification method do not overlap, so that each material can be uniquely identified. Prior to release for routine testing, a negative test population must be run against the model to verify that similar materials do not result in false positives.

False negatives, on the other hand, are expected in routine NIR testing because the pass/fail criteria are based on very tight criteria which include the overall quality of a raw material. When a material *fails NIR*, an standard operating procedure (SOP) must be in place to pull that material so it can be tested by reference methods, and if acceptable, added to the group, thereby increasing the pass/fail limits.

RUGGEDNESS

The results of an NIR identity method should be repeatable when tested by different operators at different times.

INSTRUMENT PERFORMANCE

The instrument should be installed by qualified technical service personnel. Daily instrument suitability should be run and archived to document that the instrument was within performance specifications when predicting results. Guidelines in the US Pharmacopeia General Chapter <1119> define acceptable performance for NIR instrumentation [62].

Figure 14.10 highlights the typical process used to develop NIR methods for identification. Figure 14.11 outlines the key elements of a fit-for-purpose FT-NIR method for identification and qualification of incoming botanical raw materials. Several authors provide additional details on the development of cGMP compliant methods for identification of botanicals and dietary supplement ingredients [63–67].

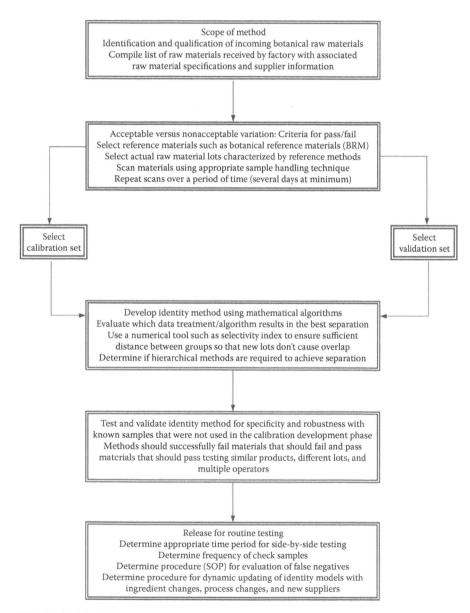

FIGURE 14.10 Typical process used to develop NIR methods for identification.

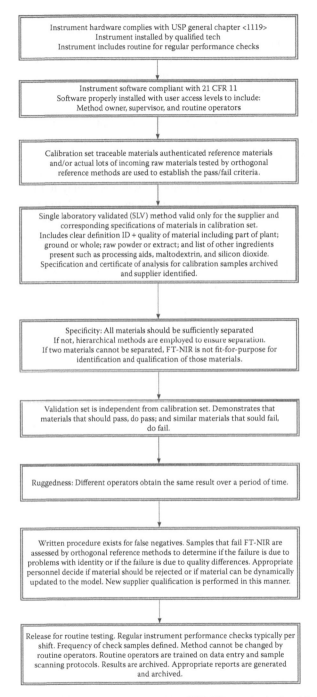

FIGURE 14.11   Key elements of fit-for-purpose FT-NIR methods for identification and qualification of incoming botanical raw materials.

SCREENING FOR ECONOMICALLY MOTIVATED ADULTERATION

Botanicals that are in short supply, expensive, or new to the market are targets for economically motivated adulteration (EMA). The most dangerous adulteration is spiking with undisclosed synthetic material, such as spiking *herbal sexual enhancement* supplements with prescription erectile dysfunction medications such as sildenafil, tadalafil, vardenafil, or synthetic derivatives. *Weight loss* supplements are sometimes spiked with sibutramine; and sports supplements are at risk of being spiked with steroids.

EMA also occurs with substitution of the material with similar materials that may or may not cause undesired side effects. Nevertheless, substitution of similar but less expensive materials results in a product falling short of label claim. Dilution with maltodextrin, silicon dioxide, and other *processing* aids may not be as harmful when ingested as other forms of EMA, but also result in a product falling short of label claim.

Conformity index algorithms [68] that compare the deviation of an NIR spectrum of a new material to established limits can be used to screen for EMA. The average spectrum of multiple scans/lots of a material known to be acceptable is calculated. Then, standard deviation of the absorbance values for that data set is calculated at each wavelength. The difference between the spectrum of an unknown sample and the reference is then compared at each wavelength. An absolute deviation is then weighted by the corresponding standard deviation on the respective wavelength and a conformity index value is calculated. Figure 14.12 shows the results of a conformity

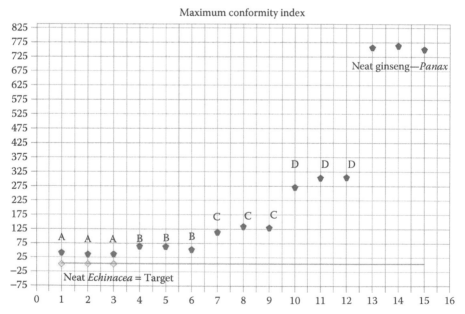

FIGURE 14.12 Target samples, neat *Echinacea*, is seen as diamonds on the conformity line. Test samples are indicated as pentagons. All test samples that fall above the pass line indicate that they fail and do not conform to pure *Echinacea*. The test samples are adulterated with varying levels of ginseng *Panax*. The last test sample is neat ginseng *Panax* indicating that it is the furthest away from Neat *Echinacea*.

index test using ChromaDex BRM for Echinacea as the target. Samples that do conform to the reference material will fall below the cutoff line, and those that do not conform will fall above the cutoff line. The three points labeled A represent spectra of three samples of the *Echinacea* BRM diluted 5% with Ginseng *Panax* ChromaDex BRM. B represents samples with a 10% dilution, C represents samples with a 25% dilution, and D represents samples with a 50% dilution of the *Echinacea* BRM with the Ginseng *Panax* BRM. The last three points on the plot represent three scans of the neat Ginseng *Panax* BRM. This conformity plots shows that adulteration of the *Echinacea* BRM with Ginseng *Panax* BRM was detected using conformity index. Figure 14.13 shows the specific portions of the NIR spectra of the diluted samples that did not conform to the reference materials.

Downey and Boussion used NIR to identify coffee bean variation [69], Bertran et al. [70] used NIR with pattern recognition as a screening method for authentication of virgin olive oils of close geographical origins, and Li et al. [71] developed an NIR method for rapid screening and authentication of Chinese Herbal Materia Medica.

Stephen Dentali highlighted the problem of bilberry adulteration [72]. Penman et al. [73] studied bilberry extracts adulterated with amaranth dye or Red Dye #2, noting that although the adulterated material did pass a general test for 25% anthocyanin standard, it did not produce a characteristic bilberry pattern of individual anthocyanins upon further examination.

Hector Juliani et al. [74] used NIR to screen for adulteration of African essential oils, differentiating between *ravensara aromatica* and *ravintsara* (*C. camphora*). Cynthia Kradjel discussed other examples of adulteration [75] and Frank Jaksch [76]

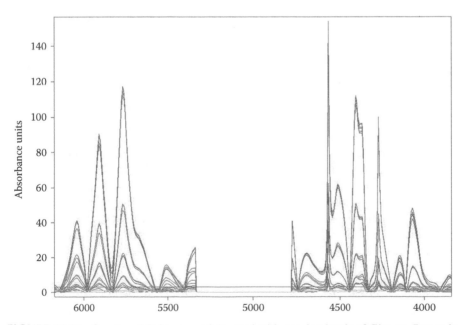

FIGURE 14.13 Spectra of *Echinacea* adulterated with varying levels of Ginseng *Panax* in the regions where the spectra fail the conformity index method.

presented technical aspects of botanical testing for identity and adulteration at the 2010 Nutrition and Health Forum.

EMA may also present in the form of attempting to pass off the raw botanical powder as an extract. NIR is used to qualify the lot-to-lot consistency of extracts.

POLYPHENOLS

Polyphenols are of interest due to their antioxidant properties. Total polyphenol content is measured in the laboratory using Folin Ciocalteau reagent, a reagent containing molybdenum +6 yellow. Polyphenols reduce Mo+6 to lower oxidation states, and the resulting color is measured in the visible range of the electromagnetic spectrum. This method is subject to potential interference due to background color and proper sample preparation is essential for good results. Schultz et al. [77] used NIR for simultaneous prediction of alkaloids and phenolic substances in green tea leaves, and Monica Harmanescu et al. [78] used NIR to measure total polyphenol content in medicinal plants from Romania. Walter Roy [79] studied procyanidins in powdered cinnamon from cinnamon tree and sticks, correlating with high performance liquid chromatography (HPLC) results for the monomer and oligomer fractions. Climaco Alvarex et al. [80] measured the (-)epicatechin content in unfermented and sun-dried Criollo cocoa beans as well as the fat, caffeine, and theobromine content.

NIR FOR CHARACTERIZATION OF NATURAL PRODUCTS BY DETERMINATION OF KEY QUALITY PARAMETERS

Magali Laasonen [81] suggested that NIR be used for quality control for the various steps in the manufacture of herbal medicinal products. Yuxiang Chen and Sorensen [82] studied marker constituents in radix glycyrrhizae and radix notoginseng by NIR. Ren and Chen [83] used NIR for simultaneous quantification of ginsenosides in American ginseng (Panax quinguefolium) root powder.

Gray et al. [84] used NIR to quantify root chicoric acid in purple coneflower; Schultz et al. [85] determined echinacoside content in *Echinacea* roots by attenuated total reflection-infrared (ATR-IR) and NIR spectroscopy; and Magali Laasonen et al. [86] simultaneously measured alkamides and caffeic acid derivatives to identify between *Echinacea purpurea*, *Echinacea angustifolia*, and *Echinacea pallida* roots.

Rager et al. [87] quantified constituents in St. John's Wort extracts by NIR spectroscopy; Sumapom Kasemsumran et al. [88] quantified total curcuminoids in Rhizomes of *Curcuma longa* by NIR with partial least squares (PLS) regression; and Sonal Tripathi et al. [89] determined curcuminoids in turmeric powder.

CONCLUSION

NIR spectroscopy is a technique that can be used for quality control to identify incoming raw materials when fit-for-purpose methods are properly developed including the use of traceable calibration samples and chemometrics for data analysis.

Methods are specifically developed for a material and its corresponding specifications using stringent pass/fail criteria. Materials of different quality that fail the NIR method may be evaluated using reference methods, and if suitable, dynamically updated in the NIR calibration.

If hierarchical methods cannot achieve separation of groups, and a subsequent review of the samples used in the calibration set proves that all calibration samples are of the proper identity and quality, then the identity of a material or set of materials cannot be reliably tested using NIR.

This technology is well suited to screening for EMA as sampling protocols can be optimized due to the ease and speed of measurement. Qualification of natural products can also be achieved through quantitative measurement of key parameters.

REFERENCES

1. J. S. Shenk, J. J. Workman, Jr., and M. O. Westerhaus, Application of NIR Spectroscopy to Agricultural Products, *Handbook of Near Infrared Analysis*, 3rd Edition, D. A. Burns and E. W. Ciurczak, Eds., CRC Press, Taylor & Francis Group, Boca Raton, FL, 2008: 347–386.
2. I. Ben-Gera and K. H. Norris, Determination of moisture content in soybeans by direct spectrophotometry, *Journal of Agricultural Research* 18, 1968: 135.
3. I. Ben-Gera and K. H. Norris, Direct spectrophotometric determination of fat and moisture in meat products, *Journal of Food Science* 33, 1968: 64.
4. K. H. Norris and R. F. Barnes, Infrared reflectance analysis of nutritive value of feedstuffs, in *Proceedings of the 1st International Symposium Feed Components*, Utah Agricultural Experimental Station, Utah State University, Logan, UT, 1976: 237.
5. K. H. Norris, R. F. Barnes, J. E. Moore, and J. S. Shenk, *Journal of Animal Science*, 43, 1976: 889.
6. J. J. Rose, T. Prusik, and J. Mardekian, Near infrared multicomponent analysis of parenteral products using the Infraalyzer 400, *Journal of Parenteral Science and Technology*, 36(2), 1982: 71.
7. H. L. Mark and D. Tunnell, Qualitative near infrared reflectance analysis using Mahalanobis distances, *Analytical Chemistry*, 57(7), 1985: 1449.
8. E. W. Ciurczak and T. A. Maldacker, Identification of actives in multi-component pharmaceutical dosage forms via NIR reflectance analysis, *Spectroscopy* 1(1), 1986: 36.
9. C. A. Anderson, J. K. Drennan, and E. W. Ciurczak, Pharmaceutical applications of near infrared spectroscopy, *Handbook of Near Infrared Analysis*, 3rd Edition, D. A. Burns and E. W. Ciurczak, Eds., Taylor & Francis Group, Boca Raton, FL, 2008: 585–612.
10. E. W. Ciurczak and J. K. Drennan, III. *Pharmaceutical and Medical Applications of Near Infrared Spectroscopy*. Practical Spectroscopy Series, Volume 31, Marcel Dekker, New York, 2002.
11. E. W. Ciurczak, Process analytical technologies (PAT) in the pharmaceutical industry, *Handbook of Near Infrared Analysis*, 3rd Edition, D. A. Burns and E. W. Ciurczak, Eds., Taylor & Francis Group, Boca Raton, FL, 2008: 581–584.
12. Food and Drug Administration, Guideline for industry: Validation of analytical procedures: Methodology, Rockville, MD, Food and Drug Administration. May 19, 1997, ICH-Q2H.
13. US Pharmacopeia, <1125> Validation of compendial methods. Vol. 24, Rockville, MD, US Pharmacopeia, 1999: 2149–2152.
14. US Pharmacopeia, <1119> Near infrared spectrophotometry, *Pharmacopeial Forum*, November–December, 28(6), 2002: 1119.

15. US Pharmacopeia, <1119> Near infrared spectrophotometry, (new) (2nd Supplement) *Pharmacopeial Forum*, September–October, 27(5), 2002: 1119.
16. European Pharmacopeia (EP), Physical and physicochemical methods, Near Infrared Spectrometry, Strasbourg, France, European Pharmacopeia, 1997.
17. The European Agency for the Evaluation of Medicinal Products (EMEA). Note the guidance on the use of near infrared spectroscopy by the pharmaceutical industry and the data to e forwarded in part ii of the dossier for a marketing authorization. EMEA, November 14, 2001: 1–13.
18. FDA and ICH Guidelines for Method Validation. Q2A "Text on Validation of Analytical Procedures" and Q2B "validation of Analytical Procedures: Methodology," 1996. http://www .fda.gov/Drugs/GuidanceComplianceRegulatoryInformation/Guidances/ucm265700.htm.
19. E. W. Ciurczak, Validation of spectroscopic methods in pharmaceutical analyses. *Pharmaceutical Technology*, 22(3), 1998: 70, 92–102.
20. H. Mark, G. E. Ritchie, R. W. Roller, E. W. Ciurczak, C. Tso, and S. A. MacDonald, Validation of a near infrared transmission spectroscopic technique, part A: Validation Protocols. *Journal of Pharmaceutical and Biomedical Analysis* 28(2), 2002: 251–260.
21. N. Broad, P. Graham, P. Hailey, A. Hardy, S. Holland, S. Hughes, D. Lee, K. Prebble, N. Slaton, and P. Warren. Guidelines for the development and validation of near-infrared spectroscopic methods in the pharmaceutical industry, *Handbook of Vibrational Spectroscopy*, J. M. Chalmers and P. R. Griffiths, Eds., Wiley, Chichester, 2002.
22. H. W. Siesler, Basic principles of near-infrared spectroscopy, *Handbook of Near Infrared Analysis,* 3rd Edition, D. A. Burns and E. W. Ciurczak, Eds., Taylor & Francis Group, Boca Raton, FL, 2008: 7–20.
23. M. Blanco and I. Villarroya. NIR Spectroscopy: A rapid-response analytical tool. *Trends in Analytical Chemistry*, 21(4), 2002: 240–250.
24. J. Workman, An introduction to near infrared spectroscopy, spectroscopynow.com, http:// www.spectroscopynow.com/details/education/sepspec1881education/An-Introduction-to-Near-Infrared-Spectroscopy.html?tzcheck=1, September 12, 2005.
25. C. Kradjel, An overview of Near Infrared spectroscopy: From an application's point of view. *Fresenius Journal of Analytical Chemistry*, 339, 1991: 65–67.
26. H. Mark and J. Workman, Jr., Chemometrics. Derivatives in spectroscopy. Part I—the behavior of the derivative. *Spectroscopy*, 18(4), 2003: 32–37.
27. H. Mark and J. Workman Jr., Chemometrics. Derivatives in Spectroscopy. Part II—The true derivative. *Spectroscopy*, 18(9), 2003: 25–28.
28. http://www.wordiq.com/definition/Chemometrics.
29. H. Mark, Chemometrics in near infrared spectroscopy. *Analytical Chimica Acta*, 223, 1989: 75–93.
30. H. Mark, *Principles and Practice of Spectroscopic Calibration*, 2nd Edition, Wiley, New York, 1991.
31. H. Mark, Qualitative NIR analysis, *Near Infrared Technology in the Agricultural and Food Industries*, AACC Publishers, Washington, DC, 2nd Edition, 2001.
32. G. E. Ritchie, H. Mark, and E. W. Ciurczak, Evaluation of the conformity index and the mahalanobis distance as a tool for process analysis: A technical note, *AAPS PharmSciTech* 4(2), Article 24, 2003:
33. J.-P. Conzen, *Multivariate Calibration*. Bruker Optics, Ettlingen, Germany, 2006.
34. H. Mark, Data analysis: Multilinear regression and principal component analysis, *Handbook of Near Infrared Analysis*, 3rd Edition, D. A. Burns and E. W. Ciurczak, Eds., Taylor & Francis Group, Boca Raton, FL, 2008: 152–188.
35. H. Mark, Qualitative discriminant analysis, *Handbook of Near Infrared Analysis*, 3rd Edition, D. A. Burns and E. W. Ciurczak, Eds., Taylor & Francis Group, Boca Raton, FL, 2008: 307–331.

36. H. Mark, *Near Infrared and Chemometrics*. Mark Electronics, Suffern, NY, 2012.
37. L. G. Weyer and S. C. Lo. *Spectra-Structure Correlations in the Near-Infrared*. Wiley, New York, 2002: 1817–1837.
38. J. Workman, Jr. and L. Weyer, *Practical Guide and Spectral Atlas for Interpretative Near-Infrared Spectroscopy*, 2nd Edition, Taylor & Francis Group, Boca Raton, FL, 2012.
39. B. Dickens and S. Dickens, Estimation of concentration and bonding environment of water dissolved in common solvents using near infrared absorptivity. *Journal of Research of the National Institute of Standards and Technology*, 104(2), 1990: 173.
40. A. Soderberg, C. Kradjel, and S. Lee, Development and validation of an FT-NIR method for identification of spectrally similar materials using a hierarchical approach. Poster presented at *Eastern Analytical Symposium*, Somerset, NJ, November, 2012.
41. J. Robustelli, A. Moffat, A. Soderberg, S. Lee, C. Kradjel, and E. Ciurczak, FT-NIR multi-level evaluation (ME) of ingredients, in-process blends, and final products of dry powder manufacturing processes of dietary supplements, Poster presented at *Eastern Analytical Symposium*, Somerset, NJ, November, 2012.
42. P. Corti, E. Dreassi, G. G. Franchi, G. Corbnini, A. Moggi, and S. Gravina, Application of near infrared spectroscopy to the identification of drugs derived from plants. *International Journal of Crude Drug Research*, 28, 1990: 185–192.
43. M. Kudo, A. C. Moffat, and R. A. Watt, The rapid characterization of natural products and herbal medicines by near infrared spectroscopy. *Journal of Pharmacy and Pharmacology*, 50(Suppl.), 1997: 258.
44. R. Watt, The characterization of herbal natural products—NIR spectroscopy back to its roots, *European Pharmaceutical Review*, 4(2), 1999: 15–19.
45. Y. A. Woo, H. J. Kim, and J. H. Cho, Identification of herbal medicines using pattern recognition techniques with near-infrared reflectance spectra, *Microchemical Journal*, 63(1), 1999: 61–70.
46. Y.-A. Woo, H.-J. Kim, J. H. Cho, and H. Chung, Discrimination of herbal medicines according to geographical origin with near infrared reflectance spectroscopy and pattern recognition techniques. *Journal of Pharmaceutical and Biomedical Analysis*, 21(2), 1999: 407–413.
47. C. Kradjel, D. Muller, A. Eilert, and J. Golini, Qualitative and quantitative NIR spectroscopic evaluation of the raw ingredients for health foods and nutritionals, *Pittcon*, Orlando, FL, March 1999.
48. C. Kradjel, NIR in the dietary supplement industry: Qualitative and quantitative analysis of ingredients, process blends and final products. *Handbook of Near Infrared Analysis*, 3rd Edition, D. A. Burns and E. W. Ciurczak, Eds., Taylor & Francis Group, Boca Raton, FL, 2008: 613–630.
49. J. Golini, Company tests herbs for drug content, *World Foods Magazine*, April 2000.
50. J. Wagner, The Quality Firewall. *Nutritional Outlook*, May 2000.
51. J. Wagner, Pharmavite, 2001 Manufacturer of the Year. *Nutritional Outlook*, November/December 2001.
52. B. Steuer, H. Schulz, and E. Laeger, Classification and analysis of citrus oils by NIR spectroscopy. *Food Chemistry*, 72(1), 2001: 113–117.
53. Y. A. Woo, Classification of cultivation area of ginseng by near infrared spectroscopy and ICP-AES. *Microchemical Journal*, 73(3), 2002: 299–306.
54. M. Laasonen, T. Harmia-Pulkkinen, C. L. Simard, E. Michiels, M. Rasanen, and H. Vuorela, Fast identification of *Echinacea purpurea* dried roots using near-infrared spectroscopy. *Analytical Chemistry*, 74(11), 2002: 2493–2499.
55. Q. Chen, J. Zhao, M. Liu, and J. Cai, Nondestructive identification of tea (*Camellia sinensis* L.) varieties using FT-NIR spectroscopy and pattern recognition. *Czech Journal of Food Sciences*, 26(5), 2008: 360–367.

56. D. Cozzilino, Near infrared spectroscopy in natural products analysis. *Planta Medica*, 75, 2009: 746–756.

57. D. L. Luthria, S. Mukhopadhyay, L.-Z. Lin, and J. M. Harnly, A comparison of analytical and data processing methods for spectral fingerprinting of broccoli. *Applied Spectroscopy*, 65(3), 2011: 250–259

58. J. A. Espinoza, G. R. Hodge, and W. S. Dvorak, The potential use of near infrared spectroscopy to discriminate between different pine species and their hybrids. *Journal of Near Infrared Spectroscopy*, 20(4), 2012: 437–447.

59. Federal Register, Current good manufacturing practice in manufacturing, packaging, labeling, or holding operations for dietary supplements; Final Rule. Federal Register, Volume 72, Number 121, June 25, 2007: 34752–34958.

60. J. Harnely, W. A. Quist, P. Brown, S. Caspar, D. Harbaugh-Reynaud, N. Hill, R. LaBudde et al., Guidelines for validation of botanical identification methods. *AOAC International*, 95, 2012: 268–272.

61. R. A. LaBudde and J. A Harnly, Probability of identification: A statistical method for the validation of qualitative botanical identification methods. *Journal of AOAC International*, 95, 2012: 273–285.

62. US Pharmacopeia, General Chapter <1119> Near infrared spectrophotometry, *Pharmacopeial Forum*, 30(6): 2137. http://www.pharmacopeia.cn/v29240/usp29nf-24s0_c1119.html.

63. C. Kradjel, Demystifying GMPs for nutritional supplements. PharmaManufacturing.com. 2008.

64. S. Dentali, An overview of GMP botanical identity testing, *AOAC Dietary Supplements Task Force and Community Meeting*, AOAC, Orlando, FL, September 26, 2010.

65. C. Kradjel, Development and validation of FTNIR methods for ID of incoming botanical raw materials for dietary supplement manufacturers. Presented at *124th AOAC International Meeting*, Orlando, FL, September, 2010.

66. C. Reynolds, F. Jaksch, and C. Kradjel, Establishing identity and quality testing programs for compliance with the dietary supplement cGMP's. Partner Series *Natural Products Insider*, Webinar, Virgo Publishing, December 14, 2010.

67. C. Kradjel and J. Richmond, FT-NIR tests for GMP compliance. *Natural Products Insider.* Virgo Publishing, October 2011.

68. A. Niemoeller, Conformity test for evaluation of near infrared (NIR) data, *Proceedings International Meeting on Pharmaceutics, Bopharmaceutics, and Pharmaceutical Technology*, Nuremberg, Germany, March 15–18, 2004.

69. G. Downey and J. Boussion, Authentication of coffee bean variation by near infrared reflectance spectroscopy of dried extract. *Journal of the Science of Food and Agriculture*, 71(1), 1996: 41–49.

70. D. Bertran, M. Blanco, J. Coello, H. Iturriaga, S. Maspoch, and I. Montoliu, Near Infrared spectrometry and pattern recognition as screening methods for the authentication of virgin olive oils of close geographical origins. *Journal of Near Infrared Spectroscopy*, 8(1), 2000: 45–42.

71. W. Li, Y. M. Tsang, F. S. C. Lee, H. W. Leung, and X. Wang, Development of near-infrared diffuse reflectance spectroscopy for rapid screening and authentication of Chinese materia madica. *Analytical Science*, 17(Suppl.), 2001: a439–a442.

72. S. Dentali, Adulteration: Spotlight on bilberry. *Nutraceuticals World.* July/August 2007: 72–75.

73. K. G. Penman, C. W. Halstead, and A. Matthias, Bilberry adulteration using the dye amaranth. *Journal of Agricultural and Food Chemistry*, 54(19), 2006: 7378–7382.

74. H. Juliani, J. Kapteyn, D. Jones, A. R. Koroch, M. Wang, D. Charles, and J. E. Simon, Application of near-infrared spectroscopy in quality control and determination of adulteration of African essential oils. *Phytochemical Analysis*, 17(2), 2006: 121–128.

75. C. Kradjel, Detecting adulterants and unwanted synthetics in ingredients. Food & Be verage Labs. Laboratory Equipment, 10, 2007. http://www.laboratoryequipment.com/ articles/2007/09/new-technology-offers-new-capabilities.

76. F. Jaksch, The technical aspects of botanical testing: Identity, adulteration and related assays. Drug, Chemical and Associated Technologies, *2010 Nutrition and Health Forum*. https://www.chromadex.com/Detailsp.aspx?Aid=454; http://dcat.org/Public/ Programs/Brochure/NHebro2010_3%209%2010_6.pdf.

77. H. Schultz, U. H. Engelhardt, A. Wengent, H. H. Drews, and S. Lapczynski, Application of near-infrared reflectance spectroscopy to the simultaneous prediction of alkaloids and phenolic substances in green tea leaves. *Journal of Agricultural and Food Chemistry*, 47(12), 1999: 5064–5067.

78. M. Harmanescu, A. Moisuc, F. Radu, S. Dragan, and I. Gergen, Total polyphenols con tent determination in complex matrix of medicinal plants from romania by NIR spec troscopy. *Bulletin, UASVM, Agriculture*, 65(1), 2008: 123–128.

79. W. Roy, Screening of cinnamon flavonoids by near infrared spectroscopy. Presentation at the *Eastern Analytical Symposium*, Somerset, NJ, November, 2012.

80. C. Alvarex, E. Perez, E. Cros, M. Lares, S. Assemat, R. Boulanger, and G. Davrieux, The use of near infrared spectroscopy to determine the fat, caffeine, theobromine, and (-) epicatechin contents in unfermented and sun-dried beans of Criollo cocoa. *Journal of Near Infrared Spectroscopy*, 20(2), 2012: 307–315.

81. M. Laasonen, Near Infrared Spectroscopy: A quality control tool for the different steps in the manufacture of herbal medicinal products. Academic Dissertation, University of Helsinki, Helsinki, Finland, 2003.

82. Y. Chen and L. K. Sorensen, Determination of marker constituents in radix Glycyrrhizae and radix Notoginseng by near infrared spectroscopy. *Fresenius Journal of Analytical Chemistry*, 367, 2000: 491–496.

83. G. Ren and F. Chen, Simultaneous quantification of ginsenosides in american ginseng (*Panax quinguefolium*) root powder by visible/near infrared reflectance spectroscopy. *Journal of Agricultural and Food Chemistry*, 47(7), 1999: 2771–2775.

84. D. E. Gray, C. A. Roberts, G. E. Rottinghaus, H. E. Garrett, and S. G. Pallardy. Quantification of root chicoric acid in purple coneflower by near infrared spectroscopy. *Crop Science*, 41, 2001: 1159–1161.

85. H. Schultz, S. Pfeffer, R. Quilitzsch, B. Steuer, and K. Reit, Rapid and non-destructive determination of the echinacoside content in Echinacea roots by ATR-IR and NIR spec troscopy. *Planta Medica*, 68(10), 2002: 926–929.

86. M. Laasonen, T. Wennberg, T. Harmia-Pulkkinen, and H. Vuorela, Simultaneous analy sis of alkamides and caffeic acid derivatives for the identification of Echinacea purpurea, *Echinacea angustifolia, Echinacea pallida* and *Parthenium integrifolium* roots. *Planta Medica*, 68, 2002: 568–572.

87. I. Rager, G. Roos, P. C. Schmidt, and K. A. Kovar, Rapid quantification of constitu ents in St. John's wort extracts by NIR spectroscopy. *Journal of Pharmaceutical and Biomedical Analysis*, 28 (3/4), 2002: 439–446.

88. S. Kasemsumran, V. Keeratinijakal, W. Thanapase, and Y. Ozaki, Near infrared quantita tive analysis of total curcuminoids in rhizomes of curcuma longa by moving window partial least squares regression. *Journal of Near Infrared Spectroscopy*, 18(4), 2010: 263–269.

89. S. Tripathi, K. G. Patel, and A. M. Balna, Nondestructive determination of curcuminoids from turmeric powder using FT-NIR. *Journal of Food Science and Technology*, 47(6), 2010: 678–681.

15 High-Performance Thin-Layer Chromatography for the Identification of Botanical Materials and the Detection of Adulteration

Eike Reich, Débora A. Frommenwiler,
and Valeria Maire-Widmer

CONTENTS

Introduction to HPTLC 242
  High-Performance Thin-Layer Chromatography versus Conventional
  Thin Layer Chromatography? 242
  Theoretical Background 242
    Apparatus 243
    Standardization of Methodology 244
HPTLC for Identification of Botanicals 245
  Why HPTLC for the Identification of Botanicals? 245
  Methods and Their Validation 245
    Transfer Validation of an External TLC Method 247
    Transfer Validation of a Nonstandardized HPTLC Method 248
    Transfer Validation of a Certified and Standardized HPTLC Method 248
Practical Aspects and Examples 248
  HPTLC Fingerprints 248
  Identification of Botanicals by HPTLC 250
  Setting Specifications for a Fingerprint (Example: Thyme Leaf) 251
Annex: SOP for HPTLC 255
References 257

INTRODUCTION TO HPTLC

HIGH-PERFORMANCE THIN-LAYER CHROMATOGRAPHY VERSUS CONVENTIONAL THIN LAYER CHROMATOGRAPHY?

The terms *thin-layer chromatography* (TLC), *high-performance thin-layer chromatography* (HPTLC), *instrumental TLC, modern TLC*, and *planar chromatography* are commonly used interchangeably without much discrimination. Planar chromatography is the most general of the above terms. It refers to liquid chromatography with a *flat* stationary phase, which can be a sheet of paper (paper chromatography) or a thin layer of fine particles coated onto a solid support (TLC/HPTLC).

TLC can be performed manually with a very simple setup, but there are instruments at various levels of sophistication available. In *instrumental TLC*, suitable instruments are employed to achieve reliable qualitative and quantitative results. In an attempt to miniaturize and optimize TLC for solvent consumption and analysis time, HPTLC plates were introduced in 1975 by Merck. They feature finer particles of smaller size distribution in a more homogenous layer. With the launch of the *Journal of Planar Chromatography—Modern TLC* [1] in 1988, the term *planar chromatography* became also synonymous with *modern TLC*. Early papers reported almost exclusively work performed with instruments on HPTLC plates but recently the term *HPTLC* is also used to describe work on TLC plates (instrumental TLC).

In distinguishing TLC and HPTLC, most textbooks focus on the differences due to the plates [2]. The same approach is taken by leading pharmacopoeias [3,4] which introduced the option of using HPTLC plates in 2005. We will follow the definition of the International Association for the Advancement of HPTLC (the HPTLC Association) [5]:

> HPTLC is the High-Performance version of Thin Layer Chromatography and a state-of-the-art technique for plant analysis. It features significantly shorter developing times, lower solvent consumptions and improved resolution. Highly reproducible results and traceable records are achieved through a standardized methodology and the use of suitable instruments (typically controlled by software) for all steps of the analysis. A system suitability test (SST) is used to qualify results. The stationary phase is a HPTLC glass plate or aluminum sheet coated with a uniform thin layer (thickness typically 200 μm) of porous particles (size 2–10 μm) with an average particle size of 5 μm. The layer typically consists of silica gel with a pore size of 60 Angstroms, a polymeric binder and a so-called fluorescence indicator (F_{254}). The standard format of the plate is 20 × 10 cm.

In contrast, we will use the term *TLC* only when referring to very simple chromatography on TLC plates performed without instrumentation.

THEORETICAL BACKGROUND

As opposed to column chromatography, which analyzes samples sequentially online, the planar chromatographic process is an off-line combination of individual steps [6], each defined by a set of parameters:

1. The sample is applied onto the stationary phase and then dried. When the plate is introduced into the chromatographic chamber and the mobile phase migrates through the plate, the chromatographic system is established. Chromatography begins when the mobile phase reaches the applied sample and is able to dissolve it. The size of the applied zone affects the achievable separation. Improper application techniques can cause chromatographic separation by the sample's solvent during the application process.
2. The plate is an open system, which is exposed to environmental factors, particularly to the relative humidity in the laboratory. As a result, silica gel as stationary phase can have a varying amount of water adsorbed on its surface. This can strongly affect the chromatographic result.
3. The mobile phase is moved by capillary action and its velocity is not constant. With increasing developing distance, the velocity decreases. This has significant effects on the separation efficiency.
4. Depending on the application position relative to the level of liquid in the chamber, chromatographic results change due to different mobile phase velocity when chromatography begins.
5. The existence of a gas phase in the chromatographic chamber has major effects on the separation [7]. The degree of chamber saturation and the preconditioning of the stationary phase can change observed R_F values of zones influencing resolution and sequence. Chamber geometry and handling also play important roles. After development, the plate is dried causing sample components to move from deeper area of the layer to the surface. When equal amounts of samples are analyzed on multiple tracks of the plate, irregular drying can cause intensity differences.

All these factors contribute to the flexibility of the method. At the same time, good reproducibility is at stake unless the process is tightly controlled and kept constant.

APPARATUS

The apparatus for HPTLC consist of

- A device suitable for application of samples as bands providing control of dimension and position of the application as well as applied sample volume.
- A suitable chromatographic chamber (typically a twin trough chamber) providing control of saturation and developing distance.
- A device suitable for controlling the activity of the stationary phase via relative humidity.
- A device suitable for reproducible drying of the developed plate.
- Suitable devices for defined reagent transfer and heating as part of the derivatization procedure.

- A device suitable for reproducible electronic documentation of chromatograms under UV 254 nm, UV 366 nm, and white light
- A densitometer or image evaluation software for quantitative determinations

Any progress in TLC and HPTLC is directly linked to the advances in available instrumentation and software [8,9]. This includes also added convenience. However, as far as reliable analytical work is concerned, fancy equipment is not an absolute necessity! Suitable equipment for development can, for example, be a fully automatic developing chamber with control of chamber saturation, developing distance, activity of the stationary phase, and drying. The simple (but maybe tedious and laborious) alternative would be a regular twin trough chamber, a timer for saturation control, a desiccator with a saturated salt solution for controlling the activity of the stationary phase, pencil and ruler and timer for monitoring the developing distance, a rack and a hair dryer on cold setting mounted at a fixed distance in front of the plate as well as a timer for drying.

STANDARDIZATION OF METHODOLOGY

Although state-of-the-art instrumentation can help to gain independence from human and environmental factors, it is also important to have a standardized methodology in place when predictable and reproducible HPTLC results are expected. It is not only the control of parameters discussed in the section Theoretical Background that matters but also the fact that for each step of the HPTLC process, a number of parameters must be defined. Most of those factors can be freely selected over a wide range. For example, to apply 5 µL of a sample, the analyst must decide where on the plate it should be and how this is accomplished, e.g., as spot or band of which dimension. Many labs have standard operating procedures (SOPs) in place which do not specify all important parameters. For example, a SOP may require the use of a saturated chamber (e.g., 1 h saturation time) but may fail to describe how much solvent is used, whether filter paper (which size) is used and which way the plate is positioned for development. When three labs compare results obtained with a published HPTLC method, each lab will have to guess at any missing parameters and may make a different choice. Even though each lab will probably be able to identify sample A against reference B on the same plate, a comparison of all results will show significant differences.

Standardization of methodology in a global sense is primarily a matter of agreement [10]. The selected parameters should be in line with general chapters of the pharmacopoeias and useful to optimize the overall result. The HPTLC Association has published on their Web site a detailed SOP that fulfills these requirements [11] (see Annex 1). All work in a laboratory should follow a very similar SOP.

TLC can also be standardized and optimized and the same instruments used for HPTLC can also work with TLC plates. Results will be similar because it is not the plate that makes the principal difference, but the underlying concepts. In such a comparison, HPTLC will make points with shorter developing time, lower solvent consumption, and increased number of samples analyzed on the same plate.

HPTLC FOR IDENTIFICATION OF BOTANICALS

WHY HPTLC FOR THE IDENTIFICATION OF BOTANICALS? [12,13–15]

HPTLC combines the advantages of the planar off-line principle (simplicity, low cost per sample, flexibility, multiple samples analyzed in parallel, static two-dimensional result suitable for multiple detection) with high reproducibility and complete traceability of data for current good manufacturing practice (cGMP) compliance. Digital images of chromatograms allow the comparison of multiple samples against reference images for similarity and differences. (HP)TLC is an established analytical technique recognized by regulatory authorities. Almost all plant monographs of pharmacopoeias include (HP)TLC methods for identification. Existing TLC methods can easily be converted to HPTLC. Method development and validation is straightforward and method transfer from lab to lab as well as transfer validation is smooth. Validated methods are easily accessible from the HPTLC Association.

Identification always requires definition and specification of acceptance criteria. Botanicals are naturally variable and often chemically not well defined. Therefore, there is certain fuzziness to the analytical target of an identification test. HPTLC can deal with this situation very well because the human eye is a great detector. After little training, it is possible to distinguish small differences between multiple samples and decide about their relevance. Modern approaches to pattern recognition combined with powerful fingerprinting techniques might be superior, particularly if statistically sufficient sample populations are investigated. But from a practical and pragmatic point of view, it is much easier to establish acceptance criteria based on visual HPTLC data of several reference materials.

METHODS AND THEIR VALIDATION

For the identification of a certain plant material by HPTLC, a suitable and valid method is needed. Sources for methods include official compendia monographs (European Pharmacopoeia, United States Pharmacopoeia (USP) Dietary Supplements Compendium, Pharmacopoeia of the People's Republic of China, etc.) and other compilations (American Herbal Pharmacopeia, HPTLC Association, etc.) as listed in Table 15.1.

Most of these sources provide HPTLC methods, HPTLC and TLC methods, or exclusively TLC methods, and it has to be decided whether what is called an HPTLC method really is based on a standardized methodology. The relevant information can be taken from the respective general chapters and the quality of the provided data. If several methods are found, it must be decided which of those is most suitable: under cGMP the selected method must be fit for the purpose of identification each time a given material changes custody. TLC methods found in the pharmacopoeias are treated as validated, even though they might never have undergone a formal validation. The fundamental difference of such compendia methods is their context, a suite of definitions and multiple tests, which describe the quality of a material. If a TLC method is taken out of the context, it may need to be used with caution.

TABLE 15.1

Sources for TLC and HPTLC Methods for Identification of Medicinal Plants

Source	Number of TLC/HPTLC Methods for Identification of Plant Material
European Pharmacopoeia (PhEur) [16]	196
Swiss Pharmacopoeia [17]	23 (in addition to European Pharmacopoeia)
British Pharmacopoeia [18]	11 (in addition to European Pharmacopoeia)
USP Dietary Supplements Compendium [19]	46 (illustrated: 35)
Pharmacopoeia of the People's Republic of China [20]	450 (part I: Chinese Materia Medica and Crude Drugs)
The Japanese Pharmacopeia [21]	79
The Korean Pharmacopoeia [22]	72
TLC Atlas of Chinese Crude Drugs in Pharmacopoeia of the People's Republic of China [23]	222
Quality Standards of Indian Medicinal Plants [24]	344
Malaysian Herbal Monograph [25]	20
Thai Herbal Pharmacopoeia [26]	39
American Herbal Pharmacopoeia (AHP) Monographs [27]	31
The International Association for the Advancement of High Performance Thin Layer Chromatography [28]	176
Herbal Medicines Compendium [29]	27 (17 in publication)
	Total: 1666

Several concepts of validation have been proposed by different organizations including the International Conference on Harmonisation of Technical Requirements for Registration of Pharmaceuticals for Human Use (ICH) [30], AOAC [31], and the International Union of pure and applied chemistry (IUPAC) [32]. They evaluate the performance characteristic of a method. For identification of botanicals, however, there are two additional elements complicating the procedure significantly: natural variability and the possibility that a material (e.g., powders, extracts) can actually be a mixture of more than one botanical. Aside of mere identity, a suitable method may need to be able to address purity as well. This problem is excellently addressed by AOAC: under the National Institute of Health initiative on dietary supplements, the organization developed guidelines for dietary supplements and botanicals [33], including guidelines for validation of botanical identification methods using the statistical model probability of identification (POI). Like all statistical models, POI requires numerous samples to be included. Outside of academia and for anything but the major botanical materials, it might be very difficult to justify such thorough validation in practice. In 2008, a practical and rather pragmatic approach was published [34]: when the analytical goal (e.g., identification of botanical X and discrimination of the adulterant botanical Y) is defined, suitable methods (e.g., from pharmacopoeias or the primary literature) are reviewed and candidates are evaluated based on botanical reference materials (BRMs) using standardized HPTLC methodology. If no suitable method exists, one has to be developed from scratch [7]. During validation of the selected method, the stability of the analyte and the chromatographic result are assessed together with specificity, precision (repeatability, intermediate precision, reproducibility), and robustness. This finding can be used to derive the SST to qualify a chromatographic result. On each plate, at least two general chemical markers, which are representative for the principal compound class targeted by the chromatographic system at hand, are used. The SST is also very helpful when a method is transferred to another lab.

In a cGMP environment, all HPTLC is performed according to a standardized methodology, which is documented in an SOP. Ideally, all HPTLC methods for identification are based on the same general template and differ only in samples and references, mobile phase, and detection mode. All HPTLC methods are also part of the quality management system.

In order to include a new method into such system from an outside source, a transfer validation is required. Three cases can be considered.

Transfer Validation of an External TLC Method

Even if the external method is not optimized, it can be converted into a standardized HPTLC method. This is done in a preliminary experiment utilizing reference materials as defined by the external method together with a reliable BRM. Assuming that the TLC method is optimized and based on a standard procedure, it is possible to tell which deviations (if any) from the standard HPTLC methodology may need to be appropriate at a later point in time (e.g., unsaturated chamber, preconditioning, multiple development). As a rule of thumb, the volumes of samples applied onto the HPTLC plate should be reduced to about one-fifth of those in the TLC method. The outcome of the experiment depends on the quality/validity of the TLC method and its type of result description:

- *The HPTLC results match the description:* The external TLC method was optimized. The first part of transfer validation is completed successfully → an SST needs to be defined and repeatability should be assessed.
- *The HPTLC results differ somewhat from the description:* The TLC method was either not optimized or based on incompatible methodology. In this case, it must be decided whether correlation between the two sets of data is possible. If yes, the first part of transfer validation is completed successfully → an SST needs to be defined and repeatability should be assessed. If not, a deviation from the standard procedure as specified by the original method may be considered.
- *No agreement at all is seen:* The TLC method does not provide sufficient details. A second experiment may be considered. The results of this second experiment will most likely be quite similar to those of the first but may still not be in line with what the external method states. One common reason for this outcome is that either the BRM is wrong or the method description is based on a different material. Transfer validation has failed, but the HPTLC method may be still be fit for the purpose if it is subsequently validated. The suitability of the selected BRM has to be verified.

Transfer Validation of a Nonstandardized HPTLC Method

According to our definition, HPTLC methods should include an SST. If this is the case, transfer validation is quite simple. The first experiment of the transfer validation will apply standard conditions and use all specified reference material and if already specified, a reliable BRM. The possible outcomes are as follows:

- SST is passed and results match → transfer validation is accomplished.
- SST is passed and results do not match → the SST is not suitable or/and the BRM does not match → problems need to be resolved and repeatability should be assessed.
- STT fails and results do not match → differences in methodology need to be resolved.

Transfer Validation of a Certified and Standardized HPTLC Method

This type of methods includes not only result descriptions but also reference images (obtained in the analysis of BRMs) that can be used for comparison [35]. Transfer validation may be accomplished without a BRM. Any sample of the correct species can be used together with the SST. There are only two outcomes:

- SST and results match within accepted natural variability documented by the method → transfer validation is accomplished successfully.
- SST fails but results are still similar to those documented by the method → transfer validation has failed.

In this situation, the experiment should be repeated after careful evaluation of the entire process to locate any deviation from the standardized methodology. If the results now match those described in the external method, the experiment is repeated twice to verify these findings.

If the results of the first and second experiment are in agreement but different from those described in the external method, analysis should now be repeated at least three more times to assess repeatability. In this case, it is advisable to also use a verified BRM. Based on the repeatability study, the SST and the result description for the now transferred method can be adjusted.

PRACTICAL ASPECTS AND EXAMPLES

HPTLC FINGERPRINTS

As a result of the HPTLC analysis, each sample on the plate is separated into a sequence of zones, a chromatogram beginning at the application position ($R_F = 0.0$) and ending at the solvent front ($R_F = 1.0$). The chromatogram, which can be evaluated visually or documented as an electronic image or/and a densitometric curve (Figure 15.1), is called HPTLC fingerprint.

Electronic images are the most convenient way of conveying information contained in the fingerprint, that is, position, color, and intensity of zones. Generating

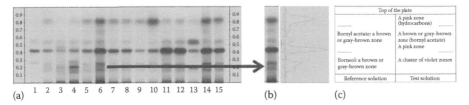

FIGURE 15.1 HPTLC of various pine oils (a) and fingerprint of the sample on Track 6 (b), Description of the fingerprint of dwarf pine oil from PhEur 8.0 (c). (Table from *European Pharmacopoeia*, 8th edition. Directorate for the Quality of Medicines and Healthcare of the Council of Europe: Strasbourg, France, 2013, p. 1231.)

such electronic images is part of the HPTLC process (defined by an SOP) and therefore also standardized, any *photoshopping* is strictly prohibited! Evaluation or comparison of a fingerprint should be performed on the computer screen only. Visual impressions, for example, from viewing the plate under UV or in daylight, may be quite different and printers can have major effects on the result.

Other possibilities for describing fingerprints are texts about what the analyst can see, or like in the European Pharmacopoeia a schematic table containing information about zones positioned in the three-thirds of a chromatogram (Figure 15.1c). If a fingerprint has been qualified by the SST on the same plate, its electronic image technically becomes independent from the original plate. This means the (electronic) HPTLC fingerprint of a BRM, or better a set of fingerprints of BRMs representing the natural variability of a given botanical, could be used for direct comparison as part of a validated identification process.

The central element of any identification is the decision about whether acceptance criteria are being met. For HPTLC fingerprints that means to judge whether two fingerprints are similar or different. All elements of the fingerprint, that is, position, color, and intensity of all zones, should be considered. In view of natural variability, this seems to be a difficult task and not very objective. So, computer-aided approaches such as principal component analysis or other pattern recognition tools in combination with statistics are widely discussed and at least in theory, likely embraced. But the capabilities of the human eyes and brain should not be underestimated. On the left side of Figure 15.2, six fingerprints of different samples of black cohosh (*Actaea racemosa*) are shown, and in the center and right of the figure are fingerprints of Chinese *Actaea* species which are considered to be adulterants. Although it is impossible to describe precisely with words what is seen in the picture, most people will probably agree on a number of statements:

- The fingerprints of all species are variable but also quite *similar* within each.
- The fingerprint of black cohosh is clearly different from that of other *Actaea* species.
- *Actaea dahurica* and *Actaea heracleifolia* cannot be discriminated

If this example is accepted as an *empirical proof* for the principal suitability of HPTLC as a tool for identification, other possibilities for HPTLC could be explored. All samples in Figure 15.2 represent roots/rhizomes of individual plants. In daily

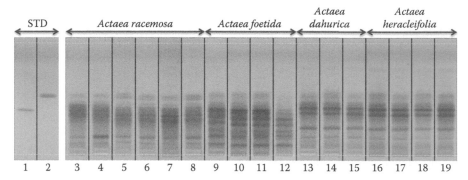

FIGURE 15.2 Similarity and differences of samples: comparison of fingerprints representing samples of different *Actaea* species; HPTLC was performed according to the monograph on Black cohosh (*Actaea racemosa*) (PhEur 8.2). For details see text. This figure is printed in grayscale and has been inverted for clarity. (Data from *European Pharmacopoeia*, 8.2th edition. Directorate for the Quality of Medicines and Healthcare of the Council of Europe: Strasbourg, France, 2013; 3702 ff.)

routine, batches of botanical raw materials are analyzed based on representative samples obtained according to a sampling plan (e.g., USP) [38]. That means parts from different plants will end up in the same (pooled) sample. The same situation is encountered when analyzing powdered plant materials. To correctly identify such samples, it is important to know how well the selected method is able to detect the presence of mixtures with other (adulterant) material. Pooled samples of target material and known adulterant are mixed in defined proportions. Based on the detectability of zones present in the adulterant and absent in the target material, a limit test can be performed. In the case of black cohosh, adulteration with 5% of any of the Chinese *Actaea* species or 10% of *Actaea americana* can be detected. This test was validated [39] and then adopted into the PhEur monograph on black cohosh.

IDENTIFICATION OF BOTANICALS BY HPTLC

Figure 15.3 illustrates some important aspects of *suitable* methods of identification. Figure 15.3a shows the result obtained with a TLC method that was taken out of its context, a pharmacopeial monograph [40]. A chemical marker, a BRM, and the sample(s) to be identified are applied onto some TLC plate. The sample would be passed or failed based on whether reference and test sample have the same *fingerprint* and a zone in the samples matches that of a marker. Although in this example all samples are quite similar, it is difficult to conclude (only by making some adjustment to what is actually really seen) that a zone matching that of the marker is present in all of them. Figure 15.3b shows HPTLC, giving much greater detail based on the same monograph. Different ginseng species can clearly be distinguished: *Panax ginseng* (Tracks 6, 7), *Panax quinquefolius* (Tracks 8, 9), *Panax pseudoginseng* (Tracks 10, 11), and *Panax japonicus* (Tracks 12, 13). The HPTLC method therefore is likely to give suitable results when applied in quality control.

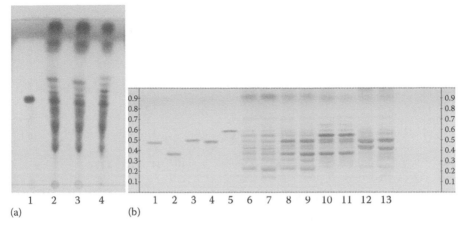

FIGURE 15.3 TLC analysis of *Panax ginseng* (a) and HPTLC analysis of 4 different ginseng species (b); for details see text.

The routine use of HPTLC methods depends on the preference, needs, and available instrumentation of a laboratory, but it is recommended to

- Establish a separate method for each botanical.
- Define an SST to qualify data on each plate.
- Perform comparison against reference images including multiple BRMs (covering natural variability) and known adulterants.

Some labs prefer the use of one or more BRMs on each plate and rely on direct comparison to avoid rigorous standardization and accept that reproducibility is compromised to a certain extent. In the end, this is a valid but expensive approach because all references (several BRMs of the target and adulterant species) must be on each plate. BRM have limited shelf life and also natural batch-to-batch variability. Consequently, the basis for comparison can change over time and has to be addressed in the quality documentation of the method.

Companies having to deal with limited populations of many different botanicals at the same time may find it costly to have individual methods for each. If adulteration is not an issue (e.g., samples may come always from the same controlled environment), it is possible to harmonize methods for common targets as flavonoids (Figure 15.4), essential oils, or fatty oils. Before adding a new botanical to such group, method specificity has to be established. Although this is an extra effort, parallel analysis of actually unrelated sample could save costs.

SETTING SPECIFICATIONS FOR A FINGERPRINT (EXAMPLE: THYME LEAF)

Traditionally, many species of the *Lamiaceae* family are used as medicinal plants, aromatic plants, and/or spice. These plants often contain essential oils in relatively large amount and it is possible to identify many of them by their typical smell. Examples include lavender, marjoram, peppermint, thyme, and lemon balm. Many of the essential oils are monographed, for example, in the European Pharmacopoeia,

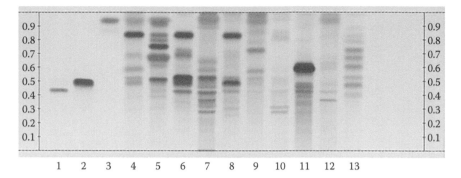

Track	Sample
1	Rutin
2	Chlorogenic acid
3	Quercitrin
4	Roman chamomile, herb
5	Feverfew, herb
6	Mate, leaf
7	Ginkgo, leaf
8	Yarrow, herb
9	Houttuyniae, herb
10	Green tea, leaf
11	Forsythia, fruit
12	Motherwort, herb
13	Lime flower, herb

FIGURE 15.4 Identification of multiple species on the same HPTLC plate with a common method: Mobile phase: Ethyl acetate, formic acid, glacial acetic acid, water 100:11:11:27, derivatization natural product reagent/polyethylene glycol (NP/PEG). This figure is printed in grayscale and has been inverted for clarity.

and it seems logical that monographs of the corresponding plants include tests for identification as well as assays based on the essential oils. Gas chromatography is the analytical method of choice when essential oils are to be investigated, but also HPTLC can provide good data. As in Figure 15.5, it is not difficult to distinguish essential oils from different species. But specifications must be drafted with caution, because essential oils are generally technical products and their composition is not exclusively determined by the corresponding plant. That means, essential oils obtained by small-scale steam distillation of a plant as part of sample preparation for identification or assay may be different from essential oils extracted by solvents (e.g., toluene or dichloromethane) from the same plant and again different from the food, cosmetic, or pharmaceutical grade technical products that are commercially available and bear the same name (e.g., peppermint oil). It may thus be questioned whether identification of an essential oil containing plant is properly achieved when specifications of a technical product are taken as reference point. Also from the point

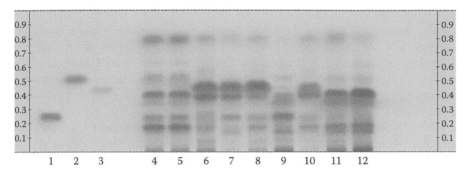

FIGURE 15.5 HPTLC of essential oils; mobile phase: toluene: ethyl acetate (95:5 v/v), derivatization with anisaldehyde reagent. Track 1: linalool; track 2: linalyl acetate; track 3: thymol; tracks 4–5: sage oil; tracks 6–8: thyme oil; track 9: marjoram oil; track 10: oregano oil; tracks 11–12: rosemary oil.

of practicality, a steam distillation as part of identification does not make the work flow easy.

Another common feature of many Lamiaceae drugs is the considerable content of rosmarinic acid. Old monographs therefore often included a TLC—identification based on the detection of this marker. In combination with other tests, this may have been sufficient for identification of a material, but a common feature is hardly a suitable target for differentiation of species.

In an approach to harmonize monographs of closely related plants, we have proposed to the European Pharmacopeia replacing the TLC on essential oils by a specific HPTLC fingerprint based on flavonoids. The mobile phase ethyl acetate: formic acid: water (15:1:1 v/v/v) on HPTLC silica gel 60 F_{254} allows distinguishing many plant drugs, including thyme, sage, oregano, melissa, spearmint, rosemary, and more. Figure 15.6 gives an example.

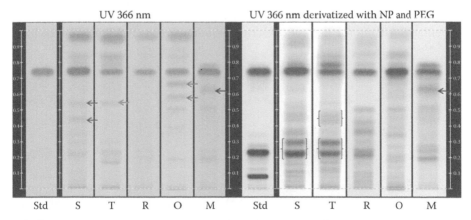

FIGURE 15.6 HPTLC fingerprint (flavonoids) of leaves: sage (S), thyme (T), rosemary (R), oregano (O), and melissa extract (M), rutin, hyperoside, and rosmarinic acid—bottom to top (Std). For details see text. This figure is printed in grayscale and has been inverted for clarity.

Even though the five species give fingerprints with many common features, there are also striking differences, if multiple detection modes are taken into account.

Under UV 366 nm (Figure 15.6a), the Oregano sample shows two unique blue fluorescent zones (arrows). In the same detection mode Melissa features one blue fluorescent zone positioned between those zones of Oregano (arrow). Sage features two other blue fluorescent zones (arrows), while thyme shows only the upper one of those (arrow). After derivatization with natural product reagent, under UV 366 nm (Figure 15.6b), additional differences appear: thyme shows two double yellow zones (brackets) which are absent in oregano, while in sage only the lower of those appears (brackets). The blue fluorescent zone (arrow) in Melissa is missing in all other species.

In Figure 15.6, only one sample each of the five species is considered, which (possibly by chance) gives different fingerprints. In order to establish specifications for proper identification, natural variability must also be taken into account. Therefore, we have looked at multiple samples. Triplicate accessions of 26 individual thyme plants and additional 37 single accessions of *Thymus vulgaris* (total 115 samples) grown by ITEIPMAI (Chemillé, France) were analyzed. To assess reproducibility of the analysis, the triplicates were chromatographed on different plates and then compared side by side using an image comparison tool that is part of the utilized HPTLC software. Evaluation was performed after derivatization with natural products reagent under UV 366 nm. For Figure 15.7, six sets of triplicates have been selected to illustrate two things: (1) The triple accessions within each set have practically identical fingerprints proving that the method is highly reproducible, and (2) there are significant differences between the sets in the fingerprint region just below the rosmarinic acid.

All of the 115 samples can be grouped into six chemotypes. If white light is included in the comparison as an additional detection mode, two of those chemotypes (4 and 5) split into subgroups (Figure 15.8) differentiated by the absence (a) or presence (b) of a pink zone (arrows). The significance of these findings is not clear particularly because we did not find any correlation between the flavonoid fingerprints and the chemotypes of *T. vulgaris* established by gas chromatographic analysis of the essential oils (e.g., thymol, carvacrol, linalool, geraniol, alpha-terpineol, and thujanol).

From the view point of the European Pharmacopoeia, all investigated samples are acceptable. Therefore, the result description for the HPTLC identification excludes

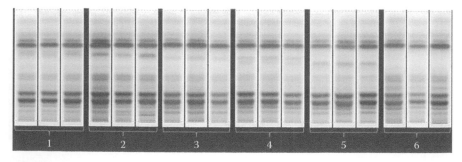

FIGURE 15.7 Comparison of multiple thyme samples; triplicate accessions (each analyzed on a different plate) of six different thyme plants, representing different chemotypes. For details see text. This figure is printed in grayscale and has been inverted for clarity.

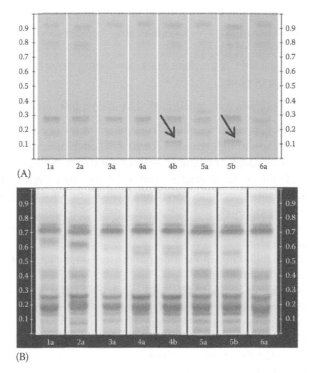

FIGURE 15.8 The chemotypes of thyme; derivatization with NP/PEG reagent, white light (A) and UV 366 nm (B). This figure is printed in grayscale and has been inverted for clarity.

the faint blue zones in question so all chemotypes are included. Other *Thymus* species will not comply.

ANNEX: SOP FOR HPTLC

Source: http://www.hptlc-association.org

Purpose: This SOP provides general guidance for analysis by HPTLC.

Definitions: HPTLC is performed on 20 × 10 cm HPTLC glass plates coated with silica gel 60 F254. Suitable (preferably software-controlled) instruments must be employed for sample application, chromatogram development, derivatization, and documentation. If no automatic developing chamber is available, a 20 × 10 cm twin trough chamber may be used.

Note: Record temperature and relative humidity in the laboratory.

1. *Preparation of plates*
    a. Obtain HPTLC plate silica gel 60 F254 (20 × 10 cm). Record the batch number.
    b. Inspect plate under UV 254 nm for any damage of the layer. If damage is detected discard plate.

   c. With a soft pencil label the plate in the upper right corner with: your initials—date (dd/mm/yy)—consecutive number for the day. For example, ER-23/02/10-01.

   d. On the right side of the plate mark developing distance at 70 mm from lower edge of plate.

Note: Left-handed persons may label/mark the plate on left side.

2. *Preparation of chamber (manual development only)*
   a. Obtain a twin trough chamber for 20 × 10 cm plates.
   b. Fit the rear trough of chamber with a filter paper of corresponding size.
   c. Pour 20 ml of developing solvent over the filter paper into the rear trough ensuring complete wetting. Pour 10 ml of developing solvent into the front trough.
   d. Close the lid of the chamber and allow 20 min for saturation.

3. *Sample application*
   a. Select the following application parameters on the application device.
      i. Band length 8 mm.
      ii. Number of tracks 15 (20 × 10 cm plate)/7 (10 × 10 cm plate).
      iii. First application position X: 20 mm.
      iv. Application position Y: 8 mm.
      v. Distance between tracks: automatic (minimum 11 mm).
      vi. Sample solvent type: methanol.
   b. Disable any unused tracks.
   c. Apply the application volumes as according to the standardized procedure for selected herbal drugs.

4. *Plate conditioning (manual development only)*
After sample application, place the plate for 45 min in a suitable desiccator containing a saturated solution of $MgCl_2$.

5. *Manual and automatic development*
   a. Manual development.
      i. Slowly open the lid of the saturated chamber and insert the conditioned plate into the front trough so that the back of the plate rests against the front wall of the chamber and the layer faces the inside of the chamber. Close the lid.
      ii. Let the mobile phase ascend until it reaches the mark.
      iii. Open the lid and remove the plate. Place it upright in a rack under a fume hood.
      iv. Dry plate with cold air from a hair dryer for 5 min.
   b. Automatic development. Use the following settings of the automatic chamber:
      i. Enable pre-drying.
      ii. Saturation with filter paper 20 min.
      iii. Humidity control 10 min with $MgCl_2$.
      iv. Migration distance 70 mm.
      v. Drying time 5 min.

     vi.   10 ml of developing solvent.
    vii.   25 ml of saturation solvent.

Note: If no humidity control is available follow step 4.

6. *Derivatization.* Follow the description in the individual monograph. Examples are as follows:
   a. Flavonoids—Derivatization by dipping.
     i.   Heat the dry plate for 5 min at 100°C.
    ii.   While hot dip plate for 1 sec into a solution of 0.5% NP reagent in ethyl acetate. Then, after 2 min of waiting, dip the plate for 1 sec into a solution of 5% macrogol 400 in dichloromethane.
   b. Flavonoids—Derivatization by automatic spraying.
     i.   Heat the dry plate for 5 min at 100°C.
        A. While hot spray the plate with 3.5 ml of a solution of 1% NP reagent in methanol then with a solution of 5% macrogol 400 in methanol, or
        B. While hot spray the plate with 3.5 ml of a solution of 0.5% NP reagent in ethyl acetate then with 3.5 ml solution of 5% macrogol 400 in dichloromethane.
7. *Documentation.* 30 min after the second derivatization step, take an image of the derivatized plate under UV 366 nm.
8. *Reporting.* Create a copy of software-based report or use own reporting documents.

REFERENCES

1. *Journal of Planar Chromatography—Modern TLC.* (accessed September 08, 2014) http://www.akademiai.com/content/120518.
2. F. Geiss. *Fundamentals of Thin Layer Chromatography: Planar Chromatography.* Huethig: Heidelberg, Germany, 1987.
3. *United States Pharmacopeia, 31st edition—The National Formulary, 26th edition,* General Chapter <621> Chromatography; The Unites States Pharmacopeial Convention: Rockville, MD, 2012.
4. *European Pharmacopoeia.* 7th edition, Method of Analysis 2.2.27 Thin-layer chromatography. Directorate for the Quality of Medicines and Healthcare of the Council of Europe: Strasbourg, France, 2012.
5. International Association for the Advancement of High Performance Thin Layer Chromatography (HPTLC Association). (accessed September 08, 2014) http://www.hptlc-association.org/about/hptlc_association.cfm.
6. E. Reich and V. Widmer. Thin Layer Chromatography. *Ullmann's Encyclopedia of Industrial Chemistry,* Wiley-VCH Verlag GmbH & Co. KgaA, Weinheim, Germany, 2012.
7. E. Reich and A. Schibli. *High-Performance Thin-Layer Chromatography for the Analysis of Medicinal Plants.* Thieme Medical Publishers: New York, 2006; pp. 36–40; 129; 175–192.
8. J. Sherma. A field guide to instrumentation. Thin-layer chromatography densitometers: An update. *Journal of AOAC International* 88, 2005: 85A–90A.

9. CAMAG. Steps of the TLC/HPTLC procedure. (accessed September 08, 2014) http://www.camag.com/en/tlc_hptlc/products/steps_of_the_tlchptlc_procedure.cfm.

10. E. Reich and A. Schibli. A Standardized approach to modern high-performance thin-layer chromatography (HPTLC). *Journal of Planar Chromatography* 17, 2004: 438–443.

11. International Association for the Advancement of High Performance Thin Layer Chromatography (HPTLC Association). Standard operating procedure for HPTLC. (accessed September 08, 2014) http://www.hptlc-association.org/media/7V94HYRV/SOP_for_HPTLC_ZHAW_new.pdf.

12. E. Reich and A. Blatter. Modern TLC: A key technique for identification and quality control of botanicals and dietary supplements. *Inside Laboratory Management* 3, 2004: 14–18.

13. E. Reich and V. Widmer. Plant analysis 2008—Planar chromatography. *Planta Medica* 75, 2009: 711–718.

14. E. Reich and J. Sherma. Thin layer chromatography in botanical analysis. *Journal of AOAC International* 93, 2010: 1347–1348.

15. A. Ankli, V. Widmer, and E. Reich. *Thin-Layer Chromatography in Phytochemistry*, M. Waksmundzka-Hajnos, J. Sherma, and T. Kowalska, Eds. CRC Press: Boca Raton, FL, 2008; Vol. 99, pp. 37–57.

16. *European Pharmacopoeia*, 7th edition, 7th supplement. Directorate for the Quality of Medicines and Healthcare of the Council of Europe: Strasbourg, France, 2013.

17. *Pharmacopoeia Helvetica*, 11th edition, 1st supplement. Swissmedic: Bern, Switzerland, 2012.

18. *British Pharmacopoeia*, Volume III. The Stationary Office, Medicine and Healthcare products Regulatory Agency: London, 2009.

19. *United States Pharmacopeia, 31st edition—The National Formulary, 26th edition.* The United States Pharmacopeial Convention: Rockville, MD, 2008.

20. *Pharmacopoeia of the People's Republic of China*, Vol. I. The State Pharmacopoeia Commission of P.R. China, People's Medical Publishing House: Beijing, People's Republic of China, 2010.

21. *The Japanese Pharmacopeia*, 16th edition. The Ministry of Health, Labour and Welfare, Tokyo, Japan, 2011.

22. *The Korean Pharmacopoeia*, 9th edition. Korea Food and Drug Administration: Seoul, Korea, 2007.

23. TLC Atlas of Traditional Chinese Crude Drugs in Pharmacopoeia of the People's Republic of China, Vol. I. The State Pharmacopoeia Commission of P. R. China, People's Medical Publishing House: Beijing, People's Republic of China, 2009.

24. *Quality Standards of Indian Medicinal Plants*, Vol. 1–10. Indian Council of Medical Research, Ansari Negar: New Dehli, India, 2012.

25. Malaysian Herbal Monograph, Vol. 2. Forest Research Institute Malaysia: Kuala Lumpur, Malaysia, 2009.

26. *Thai Herbal Pharmacopoeia*, Vol. 1–3. Department of Medical Sciences: Bangkok, Thailand, 2009.

27. *American Herbal Pharmacopoeia*, Monographs. (accessed September 08, 2014) http://www.herbal-ahp.org/order_online.htm.

28. International Association for the Advancement of High Performance Thin Layer Chromatography (HPTLC Association). (accessed September 08, 2014) www.hptlc-association.org.

29. Herbal Medicines Compendium. The United States Pharmacopeial Convention: Rockville, MD, 2013. (accessed September 08, 2014) https://hmc.usp.org/.

30. International Conference on Harmonisation of Technical Requirements for Registration of Pharmaceuticals for Human Use (ICH). Quality Guidelines. Analytical Validation Q2. (accessed September 08, 2014) http://www.ich.org/products/guidelines/quality/article/quality-guidelines.html.

31. AOAC International. AOAC guidelines for single laboratory validation of chemical methods for dietary supplements and botanicals. (accessed September 08, 2014) http://www.aoac.org/imis15_prod/AOAC_Docs/StandardsDevelopment/SLV_Guidelines_Dietary_Supplements.pdf.

32. M. Thompson, S.L.R. Ellison, and R. Wood. Harmonized guidelines for single-laboratory validation of methods of analysis (IUPAC Technical Report). *Pure and Applied Chemistry* 74, 2002: 835–855.

33. AOAC International: *Appendix K: Guidelines for Dietary Supplements and Botanicals* (2013). (accessed September 08, 2014) http://www.eoma.aoac.org/app_k.pdf.

34. E. Reich and A. Schibli. Validation of high-performance thin-layer chromatographic methods for the identification of botanicals in a cGMP environment. *Journal of AOAC International* 91, 2008: 13–20.

35. CAMAG method library, (accessed September 08, 2014) http://www.camag.com/en/lp/visioncats.cfm.

36. European Pharmacopocia, 8th edition. Directorate for the Quality of Medicines and Healthcare of the Council of Europe: Strasbourg, France, 2013, p. 1231.

37. European Pharmacopoeia, 8.2th edition. Directorate for the Quality of Medicines and Healthcare of the Council of Europe: Strasbourg, France, 2013; 3702 ff.

38. Unites States Pharmacopeia USP 36-NF 31, General Chapter <561>Articles of Botanical Origin. Sampling. The United States Pharmacopeial Convention: Rockville, MD, 2012.

39. A. Ankli, E. Reid, and M. Steiner. Rapid high-performance thin-layer chromatographic method for detection of 5% adulteration of black Cohosh with *Cimicifuga foetida, C. heracleifolia, C. dahurica,* or *C. americana. Journal of AOAC International* 91, 2008: 1257–1264.

40. *The Japanese Pharmacopeia*, 16th edition. Ginseng. The Ministry of Health, Labour and Welfare: Tokyo, Japan, 2011, p. 1646.

16 Sensory and Chemical Fingerprinting Aids Quality and Authentication of Ingredients and Raw Materials of Vegetal Origin

Marion Bonnefille

CONTENTS

Introduction 262
Instruments for Chemical and Sensory Fingerprinting 262
    Electronic Nose 262
    Electronic Tongue 264
Chemical and Sensory Fingerprinting of Vegetal Products 265
    Quality Control of Vegetal Oils 265
        Objectives 265
        Experimental Conditions and Samples 266
        Differentiation of Oil Aroma Qualities 266
        Quality Control Model 267
        Conclusion 267
    Follow-Up of Oils Shelf Life 268
        Objectives 268
        Equipment and Method 268
        Samples and Experimental Conditions 268
        Oils Aroma Profiling and Characterization 268
        Conclusion 269
    Sensory Profiling of Olive Oils 270
        Objectives 270
        Samples and Experimental Conditions 270
        Virgin Olive Oil Aroma Profiling 270
        Conclusions 271
    Olive Oils Ranking Based on Bitterness Intensity 272
        Objectives 272
        Samples and Sensory Panel Evaluation 272
        Experimental Method 273

Taste Profiles Comparison..273
Quantification of the Bitterness Level...274
Conclusion...275
Sensory Profiling and Selection of Spices Based on Origin 275
Objectives..275
Samples and Analytical Method..275
Aroma Profile Comparison and Recognition 276
Quality Control of Asafoetida Powder Aroma 277
Conclusion...277
General Conclusion...278

INTRODUCTION

Over the last years, the use of *electronic sensing* or *e-sensing* instruments has spread out in particular in the food and beverage area. The expression *electronic sensing* refers to the capability of reproducing human senses for assessing the organoleptic characteristics of products. By providing fast and reliable measurements of the chemical and sensory features of the whole products, hence the term *fingerprint*, these instruments are particularly suitable for evaluating quality and authenticity of raw materials or manufactured products. Based on gas sensor technology or gas chromatography (GC), electronic noses analyze the volatile compounds in products headspace and give a fingerprint of their odor. In the same way, the electronic tongue detects all dissolved compounds responsible for taste in liquids and can thus assess the overall taste profile of products.

After presenting the electronic nose and electronic tongue instruments, this chapter will focus on their applications for the analysis of vegetal oils.

INSTRUMENTS FOR CHEMICAL AND SENSORY FINGERPRINTING

ELECTRONIC NOSE

The specificity of electronic nose analyzers is that their working principle mimics human olfaction by capturing the global odor profile of a product. Basically, an electronic nose consists of a sample delivery system, a detection system, and a computing system for pattern recognition. When headspace injection mode is used, the sample delivery system enables the generation of the headspace (volatile compounds) by heating the sample at a constant temperature under agitation. The system then injects this headspace into the detection system of the electronic nose. The sample delivery system is essential to guarantee constant operating conditions. The detection system is the *reactive* part of the instrument. Gas sensor arrays technology or GC can be employed to detect volatile compounds. With gas sensors electronic noses (Figure 16.1), a change of electrical properties is recorded at the surface of sensors when in contact with volatile compounds, and these data constitute the input for further data processing. Each sensor is sensitive to all volatile molecules but each in their specific way. Therefore, the array of several different sensors allows to characterize the overall volatile and odor profile of the samples in a unique way. With GC electronic noses (Figure 16.2), a fraction of

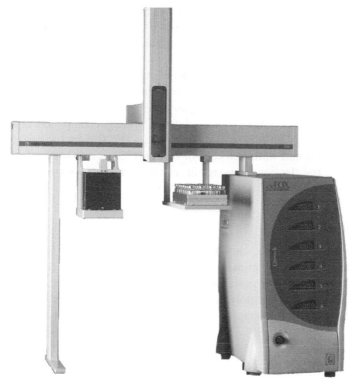

FIGURE 16.1 FOX electronic nose (Alpha MOS, France) using gas sensor technology.

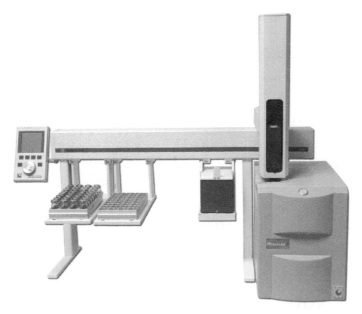

FIGURE 16.2 HERACLES electronic nose (Alpha MOS, France) using fast GC technology.

the product headspace or a liquid fraction can be analyzed. The raw data consist of retention times and peak areas. The advantage of GC electronic noses is that in addition to the odor fingerprint analysis, they give information about the chemical compounds involved.

The software computing system works to combine the multiple raw data, which represents the input for the multivariate statistical processing. This part of the instrument performs global fingerprint analysis and provides results and representations that can be easily interpreted such as radar plot, principal component analysis (PCA) to compare olfactory profiles, partial least square (PLS) model to quantify concentrations or scores, and statistical control cards (SQC) to monitor conformity checking. Moreover, the electronic nose results can be correlated with those obtained from other techniques (sensory panel, GC, GC/MS).

ELECTRONIC TONGUE

In its working principle, the ASTREE electronic tongue from Alpha MOS (Figure 16.3) mimics the human sense of taste. In humans, chemical compounds responsible for taste are perceived by human taste receptors, and in the electronic tongue, the organic polymer sensors detect the dissolved compounds. Like human receptors, each E-tongue sensor has a spectrum of reactions different from

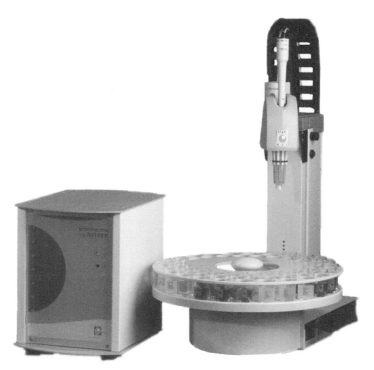

FIGURE 16.3 ASTREE electronic tongue (Alpha MOS, France) using ChemFET sensors technology.

the other. The working principle of the E-tongue sensors is based on noncovalent and reversible chemical bonds—as ionic or van der Waals ones—with dissolved molecules that in turn generate potentiometric variations recorded as electrical signals. The information given by each sensor is complementary, and the combination of all sensors generates a unique fingerprint. Detection thresholds for the vast majority of the compounds are similar or better than those of human receptors.

In humans, taste perception and recognition is based on pattern recognition of activated sensory nerve patterns—called taste broad tuning—by the brain and on the taste fingerprint of the product.

In the E-tongue, sensor data are interpreted into taste patterns by statistical software.

CHEMICAL AND SENSORY FINGERPRINTING OF VEGETAL PRODUCTS

The chemical and sensory quality of oil products employed as raw materials in the food industry is crucial because it could negatively impact the quality of final products. To guarantee oils of acceptable and consistent quality, many issues are to be addressed by oil producers and food manufacturers.

First, oils quality is very dependent upon raw materials origin and quality as well as on the extraction and refining process. For example, fish oils need to be reliably evaluated to avoid that undesirable off-odors remain after the deodorizing step. Besides, oils are particularly sensitive to oxidation and degradation induced by light or heat, which generally alters their sensory features over time and can be responsible for rancidity development. Therefore, it is very important to monitor the shelf life of oils, the packaging or the storage conditions that will best preserve their initial features. Also, some oils are submitted to numerous regulations on both physicochemical parameters and sensory characteristics, these regulations being aimed at guaranteeing oils quality and origin. Producers and industrials thus need the proper analytical tools to be able to comply with the regulation requirements.

In the following sections, several case studies conducted with electronic nose or electronic tongue instruments will be described: sensory quality control of oils, shelf life follow-up of various oils under different storage conditions, sensory profiling of South American olive oils aimed at overseas markets, ranking of olive oils based on bitterness intensity, sensory profiling, and selection of spices based on origin.

Quality Control of Vegetal Oils

Objectives
Vegetal oils are common ingredients entering the composition of many food products. In order to guarantee an appropriate flavor and a conform quality, it is crucial for both manufacturers and users to test the organoleptic features of these ingredients from vegetal origin. The objective of this study was to compare and differentiate the aroma of various sunflower oils samples using a metal oxide electronic nose,

then to identify the quality of blind samples. The final industrial goal was to set up a quality control model for a rapid assessment of production batches in routine operation.

Experimental Conditions and Samples

Twelve samples of sunflower oils of different grades were evaluated (Table 16.1). Six of them were also evaluated by a sensory panel (three good, three bad) and six other were *blind* samples of unknown quality. The *bad* quality corresponds to oils having a rancid odor as opposed to *good* oils that are not oxidized. All oils were analyzed with a FOX sensor based electronic nose under specific parameters listed in Table 16.2. The six oils also graded by the sensory panel allowed to set up a qualitative model for characterizing the quality of blind samples. All samples were analyzed three times each (three replicates) with the electronic nose. The FOX instrument integrates an array of eighteen metal oxide sensors (MOS) and analyzes the headspace of the oils, that is, the volatile compounds emitted by the oils.

Differentiation of Oil Aroma Qualities

The electronic nose measurements show a high repeatability (residual standard deviation <5%).

In order to rapidly differentiate the oil samples based on their odor profile, a PCA is performed by taking into account all gas sensors measurements (Figure 16.4).

The two qualities (good/bad) are clearly discriminated on this graph. Blind samples are also plotted (in black) on this qualitative model, which allows to see similarities and differences with the samples representative of good/bad grades.

TABLE 16.1
Samples of Vegetal Edible Oils Analyzed

Sample Label	Quality Assessed by the Sensory Panel
G1, G2, G3	Good
B1, B2, B3	Bad
U1 to U6	Unknown (blind samples)

TABLE 16.2
Analytical Conditions Applied with FOX Electronic Nose for the Analysis of Vegetal Edible Oils

Parameter	Value
Quantity of sample	3 g in a10 mL vial
Headspace generation	15 min at 80°C
Carrier gas	Synthetic dry air
Injected volume	5 mL
Syringe temperature	90°C
Acquisition time	120 s

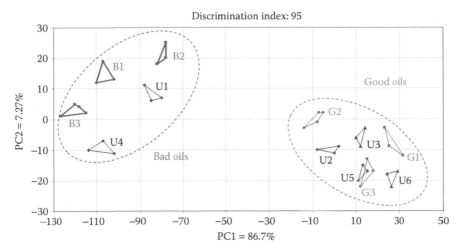

FIGURE 16.4   Odor map of vegetal edible oils based on PCA.

Quality Control Model

For routine analysis purposes, a quality control card can be set up based on E-nose measurement (Figure 16.5). To establish the range of acceptable grade (gray band), good samples were taken as the reference quality. Bad oils were plotted on this model: they fell outside this gray band of conform quality. Therefore, they are correctly recognized as out of specifications. Based on this quality control card, it was also possible to determine that blind samples U2, U3, U5, and U6 have a good quality, whereas U1 and U4 have a bad quality.

Conclusion

Fingerprint analysis using a gas sensor electronic nose is a fast means to determine the quality of oil products in correlation with sensory panel assessment, and it assures the consistency of testing methods.

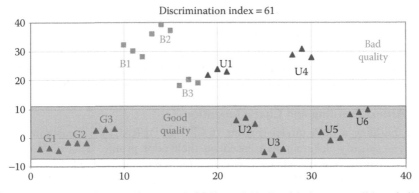

FIGURE 16.5   Statistical quality control (SQC) model built with the two qualities of oils for conformity control purposes.

FOLLOW-UP OF OILS SHELF LIFE

Objectives

Over ageing and oxidation, the sensory properties of vegetal oils may suffer changes and alteration, in particular with the development of rancidity. To make sure that their organoleptic features will not negatively impact the sensory quality of the final product over ageing, it can be useful to study the shelf life of several types of oils before using them, in order to select the most stable. This study describes the comparative analysis of the sensory profile of three types of oils over time, using HERACLES electronic nose.

Equipment and Method

The oils samples were analyzed with HERACLES electronic nose (Alpha MOS, France) which is based on fast chromatography. It features two metal columns of different polarities (nonpolar RXT-5 and slightly polar RXT-1705, length = 10 m, diameter = 180 μm, Restek) mounted in parallel and coupled to two flame ionization detectors (FID) producing two chromatograms simultaneously. It allows headspace or liquid injection modes. An integrated solid adsorbent trap thermoregulated by Peltier cooler (0°C–260°C) pre-concentrates light volatiles before injection. The HERACLES e-nose was additionally equipped with AroChemBase (Alpha MOS, France). AroChemBase consists of a software module aimed at characterizing the compounds by GC. The module contains a library of chemical compounds with their chemical data and sensory descriptors for odor-related molecules. The characterization of chemical compounds separated by GC is based on Kovats index matching. Kovats retention index is a concept used in GC to convert retention times into system-independent constants. The retention index of a certain chemical compound is its retention time normalized to the retention times of adjacently eluting n-alkanes. Although retention times vary with the individual chromatographic system (e.g., with regard to column length, film thickness, diameter, carrier gas velocity and pressure, and void time), the derived retention indices are quite independent of these parameters and allow comparing values measured by different analytical laboratories under varying conditions. Tables of retention indices can help identify components by comparing experimentally found retention indices with known values.

Samples and Experimental Conditions

Three vegetal oils (palm, soybean, colza) aged at ambient temperature and under daylight conditions were used in this study (Table 16.3). Samples were analyzed at day 0, then at days 5, 10, and 15 with HERACLES electronic nose (Table 16.4).

Oils Aroma Profiling and Characterization

Upon building a PCA model (Figure 16.6) from HERACLES measurements, the oils were clearly differentiated based on their odor profile. This odor map also showed that the sensory characteristics of the three oils evolved differently over time. This indicates that they have different behaviors toward photooxidation. Using the AroChemBase and Kovats indices on MXT-5 and MXT-1701 columns, it could be observed that major peaks corresponded to alkanes, which is typical of a high

TABLE 16.3
Samples of Palm, Soybean and Colza Oil Analyzed During Shelf Life

Sample Label	Description
P0	Palm oil at day 0
P5, P10, and P15	Palm oil at days 5, 10, and 15
S0	Soybean oil at day 0
S5, S10, and S15	Soybean oil at days 5, 10, and 15
C0	Colza oil at day 0
C5, C10, and C15	Colza oil at days 5, 10, and 15

TABLE 16.4
Analytical Conditions Applied with HERACLES Electronic Nose for the Analysis of Vegetal Oils

Parameter	Value
Sample mass	2 ± 0.02 g in a 20 mL vial
Sample incubation	20 min at 60°C
Injected volume	5 mL
Injection speed	250 μL/s
Injector temperature	200°C
Injector carrier pressure	25 kPa
Trap initial temperature	20°C
Trap pressure	80 kPa
Trap final temperature	240°C
Columns temperature program	40°C for 2 s
	3°C/S → 250°C (0 s)
	5°C/S → 280°C (0 s)
Acquisition duration	90 s
Detectors temperature	280°C

level of oxidation. The low level of aldehydes explains why oils have a weak odor. Each oil had a unique composition in volatile compounds that is characteristic of photooxidation kinetics related to fatty acids composition. In colza oil, an important amount of 2-methylbutanal could be found and also hexanal and pentanal in lower quantities.

Conclusion

The fingerprint analysis brought the ability to rapidly compare the sensory quality of oils. By building quality control cards, it would be also possible to check batch-to-batch conformity of ingredients or final products and assure the consistency of testing method.

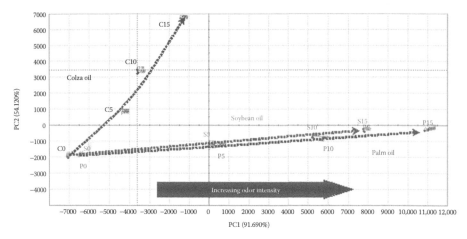

FIGURE 16.6   General odor map (PCA) of fresh and aged oils obtained with HERACLES e-nose measurements.

SENSORY PROFILING OF OLIVE OILS

Objectives

Being considered as a healthy ingredient, extra virgin olive oil is widely consumed, especially in European Mediterranean countries. It is also one of the most
regulated food products in Europe, and it is classified into eight grades, based
on numerous physicochemical and sensory parameters. Therefore, for importation
purposes, high-grade olive oils produced outside Europe must comply with strict
European regulations. Among the quality parameters tested, sensory features are
particularly critical. The objective of this study was to compare and characterize
the aroma of virgin olive oils produced in America and bound to the European
market.

Samples and Experimental Conditions

The HERACLES fast GC electronic nose (Alpha MOS, France) was used to analyze
the complex headspace of 10 virgin olive oils produced in South America (OO1 to
OO10). Each olive oil sample was analyzed in four repeats based on the analytical
conditions described in Table 16.5.

Virgin Olive Oil Aroma Profiling

The superimposed chromatogram profiles of olive oils highlight differences of
chemical composition between the various samples (Figure 16.7). Nevertheless, it
is difficult to compare the overall aroma profiles of the different oils from the chromatograms. For this reason, qualitative data processing using multivariate statistic
models such as PCA is applied to facilitate aroma profile comparisons. By computing HERACLES measurement in a PCA, the various olive oils were clearly differentiated. In the upper right part of this PCA (Figure 16.8), samples OO1, OO5, and
OO7 are very close which indicates they have a very similar aroma.

TABLE 16.5

Analytical Conditions Applied with HERACLES Electronic Nose for the Analysis of Extra Virgin Olive Oils

Parameter	Value
Carrier gas	Hydrogen
Quantity of sample	2 g in 20 mL vial
Headspace generation	20 min at 60°C
Injected volume	5 mL
Syringe temperature	70°C
Trap sampling temperature	40°C
Trap desorption temperature	250°C
Injector temperature	200°C
FID temperature	280°C
Columns temperature program	40°C (10 s) to 270°C (4 s) by 5°C/s

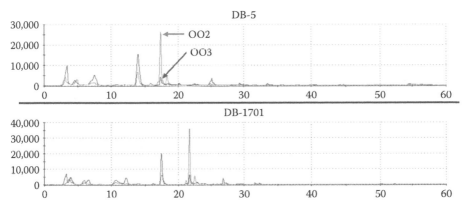

FIGURE 16.7 Superimposition of two chromatograms of different oil samples on the two columns of HERACLES.

Rapid human evaluation allowed to identify the main sensory attributes in these oils: green, fruity, and ripened.

- Samples OO8 and OO2 were found to be very *green* and *fruity*.
- Samples OO1, OO5, OO6, and OO7 were determined as *fruity* and *ripened*.
- Samples OO3 and OO4 were considered as having no particular sensory defect.

Consequently, the intensity of these attributes on the different oils could be related to the PCA mapping, by determining a trend axis for each attribute.

Conclusions

In this case, sensory fingerprinting proved to be useful for ensuring that the imported vegetal oils comply with the sensory standards applying in a specific market.

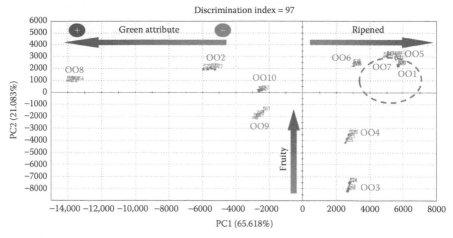

FIGURE 16.8   PCA of olive oils showing aroma differences.

Olive Oils Ranking Based on Bitterness Intensity

Objectives

Bitterness, along with pungency and fruitiness are all sensory characteristics present and well-balanced in the best olive oils and are a token of their authenticity. In olive oil, the flavonoid polyphenols contribute to the bitter taste and resistance to oxidation. These polyphenols are strong antioxidants and have been shown to provide beneficial health effects. Therefore, bitterness is one of the best indications of an extra virgin olive oil's antioxidant and anti-inflammatory value. In this study, it is proposed to assess the bitterness level of olive oil, usually evaluated by a sensory panel, by means of an electronic tongue.

Samples and Sensory Panel Evaluation

Eight olive oils (Table 16.6) have been analyzed with ASTREE electronic tongue. Six of them, characterized by a sensory panel score for the bitterness level, were used to set up a calibration model. This model will then allow to determine the bitterness score of two blind samples, using the same scoring scale as the sensory panel given as follows:

- *Scores <2:* Low bitterness
- *Scores between 2 and 3:* Middle bitterness
- *Scores between 3 and 4:* Strong bitterness

The sensory panel consisted of nine assessors. They evaluated the oils according to CE 2568/91. This European standard describes the chemical and organoleptic characteristics of olive oils and the different qualities (e.g., virgin or refined) as accepted on the European market. It also defines the method to evaluate these chemical and organoleptic characteristics. For the taste evaluation of olive oil, the panelists absorb a sip of about 3 mL of oil (at 28°C ± 2°C), spread it in the whole mouth, and then swallow it. Mouth is rinsed between two evaluations. The scores attributed to the oils correspond to the average value of the panelists' scorings.

TABLE 16.6

Olive Oils Samples Description and Sensory Panel Scores

Sample Label	Bitterness Score (Sensory Panel)	Additional Description
S1	1.5	Low bitterness, green
S2	1.5	but sweet
S3	2.5	Middle bitterness, ripe
S4	2.5	
S5	3.5	Strong bitterness,
S6	3.5	strongly green
B1	Blind samples	
B2		

Experimental Method

Because the electronic tongue sensors are altered by an oily matrix, the compounds responsible for the taste were collected from the oils by liquid–liquid extraction, prior to the analysis.

The liquid–liquid extraction was conducted as follows:

- Addition of 30 mL of a H_2O/EtOH (80/20) solution to 30 mL of olive oil
- Stirring (1 min) followed by 30 min of ultrasonic
- Settling and collection of the aqueous phase
- Filtration with polytetrafluoroethylene (PTFE) filters (first 0.45 μm then 0.2 μm)

The solution finally analyzed with the e-tongue constituted of 20 mL of the upper extracted solution and 3 mL of ethanol solution. Between two analyses, the ASTREE sensors were rinsed with a H_2O/EtOH (70/30) solution. Samples were analyzed three times each, to take into account an average measurement, at ambient temperature, with an acquisition time of 120 s.

Taste Profiles Comparison

With ASTREE electronic tongue, the samples analyzed can be ranked on each of the seven sensor axis, on a 0–12 intensity scale. Three of the seven sensors are directly linked with a taste attribute given as follows:

- Sensor SRS corresponds to a sourness intensity axis
- Sensor STS corresponds to a saltiness intensity axis
- Sensor UMS corresponds to a umami intensity axis

The electronic tongue measurements were then computed using PCA (Figure 16.9) to compare the global taste profile of all olive oils. The taste map shows a clear distinction of the oils based on an axis related with their level of bitterness: from low

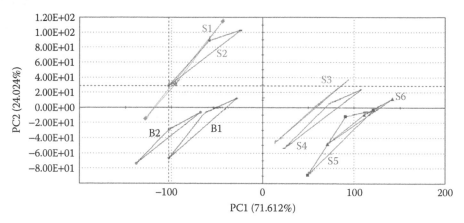

FIGURE 16.9 Taste map (principal components analysis) of the eight olive oils obtained with ASTREE measurements.

bitterness (samples S1 and S2) on the left to strong bitterness on the right (samples S5 and S6). Oils having a similar bitterness level are very close one to another on the map, which indicates they have a very similar taste profile.

Quantification of the Bitterness Level

In order to determine the bitterness intensity of the blind samples, a calibration model based on PLS regression (Figure 16.10) was set up by correlating the sensory panel results with the electronic tongue measurements. For building the calibration curve, the electronic tongue data obtained on STS sensor were not taken into account because it appeared not to be relevant for bitterness ranking. A high level of correlation is obtained between the electronic tongue measurements and the sensory panel evaluation (correlation coefficient = 0.9585). Upon projecting the blind samples on this quantification model, it could be determined that B1 had a bitterness intensity

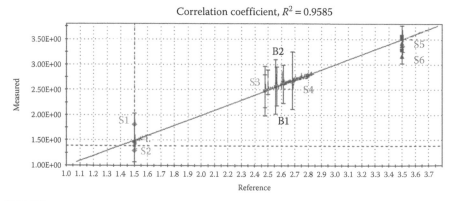

FIGURE 16.10 Calibration curve (partial least square regression) for the determination of the bitterness level of olive oils with ASTREE electronic tongue.

of 2.55 and B2 of 2.57. The sensory panel confirmed these results because panelists attributed a score of 2.5 for both B1 and B2 oils.

Conclusion

A suitable method of liquid–liquid extraction of taste compounds coupled with ASTREE electronic tongue analysis allows to rank olive oils based on their bitterness, in the same way as sensory panel evaluation. The correct determination of the bitterness intensity of unknown olive oils suggests that this technique can be successfully applied for routine analysis.

SENSORY PROFILING AND SELECTION OF SPICES BASED ON ORIGIN

Objectives

Asafoetida, or asafetida, is a gum oleoresin exuded from the living underground rhizome or tap root of several species of Ferula. The species is native to the mountains of Afghanistan and is mainly cultivated in nearby India or near countries. Asafoetida has a pungent, strong smell when raw. Cooked in dishes, it delivers a smooth flavor and is widely used as a spice in Indian cuisine. The flavor and odor of asafoetida can be very different based on variety and origin. The objective of this study was to compare the aroma profile of asafoetida powders produced from plants of three different geographical areas and to set up a qualitative model for selecting similar products based on defined quality standards.

Samples and Analytical Method

Fourteen samples of asafoetida powder were assessed (Table 16.7) among which seven were of known origin and were used for sensory profiling and qualitative model building. The seven others were blind samples whose quality was later determined after the qualitative model previously set up. The analysis was conducted with HERACLES fast GC electronic nose with headspace analysis (Table 16.8).

TABLE 16.7
Asafoetida Powder Samples Description

Sample Label	Type	Country of Origin	Additional Details
SAM	Siyabandi	Afghanistan	Fresh
AFG-D	Kabuli	Afghanistan	Reference quality
TJKI-KB	Kabuli	Tajikistan	
UZB-KB	Kabuli	Uzbekistan	
AFG-PN	Pirnaksir	Afghanistan	Fresh
KZ-PN	Pirnaksir	Kazakhstan	Fresh
OLD PN	Pirnaksir	Afghanistan	Aged
1 to 7	Blind samples	Unknown	

TABLE 16.8
Analytical Conditions Applied with HERACLES Electronic Nose for the Analysis of Asafoetida Powders

Parameter	Value
Carrier gas	Hydrogen
Quantity of sample	10 ± 0.1 g in a 100 mL vial
Headspace generation	20 min at 30°C
Injected volume	1 mL
Syringe temperature	50°C
Trap sampling temperature	40°C
Trap desorption temperature	250°C
Injector temperature	170°C
FID temperature	280°C
Columns temperature program	40°C (2 s) to 270°C (2 s) by 5°C/s

Aroma Profile Comparison and Recognition

Upon comparing the overall odor profiles of asafoetida powders on an odor map (Figure 16.11) produced from HERACLES measurements, clear differences can be observed between the three types of products: Kabuli, Siyabandi, and Pirnaksir. To be able to identify the type of asafoetida among unknown products, a recognition model based on discriminant factorial analysis (DFA) was set up. By projecting the measurements of the blind samples, this model then allows to predict to which type they belong (Table 16.9). The type of samples 1 and 2 could not be identified because these products seem to be mixture of different powders. The five others were successfully identified.

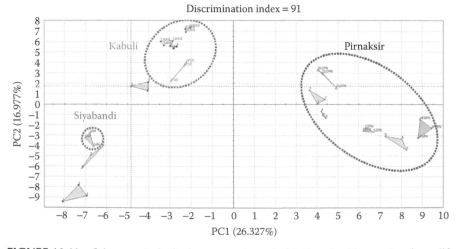

FIGURE 16.11   Odor map (principal components analysis) of asafoetida powders from different origins.

TABLE 16.9

Identification of Unknown Samples Type Using DFA Recognition Model for Asafoetida Powders

Sample Label	Recognition	Identified Type	Percentage of Recognition (%)
1	No	/	/
2	No	/	/
3	Yes	Kabuli	98
4	Yes	Pinarksir	92
5	Yes	Pinarksir	99
6	Yes	Siyabandi	99
7	Yes	Pinarksir	93

Quality Control of Asafoetida Powder Aroma

By considering Kabuli asafoetida as the reference quality, the electronic nose measurements allowed to define a quality control card based on statistical quality control (SQC) data processing (Figure 16.12). This model determines the area of conform quality within a range of acceptable variability (light gray band on Figure 16.12). All future samples mapped in this area are considered as conform to the desired quality, whereas samples plotted outside this area are non-conform. According to this model, only sample 3 can be considered as conform to Kabuli standard quality, all other blind sample and known samples do not comply with this quality.

Conclusion

Aroma profiling based on fast GC electronic nose analysis is a fast means of assessing the type or origin of incoming vegetal raw materials and thus to guarantee

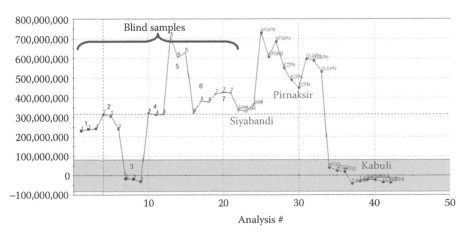

FIGURE 16.12 SQC model for assessing asafoetida powder quality against a standard grade.

the authenticity and quality of the products from new suppliers. The rapidity and reliability of this method makes it particularly suitable for routine analysis and suppliers quality monitoring.

GENERAL CONCLUSION

Traditionally, chemical methods and sensory panels are individually employed to analyze, respectively, the chemical composition and sensory attributes of food products. To be detailed and complete, the analysis process can thus be time-consuming and expensive. The case studies detailed above have shown that new testing methods using instruments such as electronic nose and electronic tongue can be promising for evaluating simultaneously the chemical and sensory features of food ingredients and food products.

17 The Hidden Face of Botanical Identity
An Industrial Perspective on Challenges from Natural Variability and Commercial Processes on Botanical Authenticity

Leila D. Falcao, Camille Durand, Alexis Lavaud, Marc Roller, and Antoine C. Bily

CONTENTS

Introduction 279
Chemical Complexity of the Botanical Raw Materials 281
Impact of Industrial Processes on the Chemical Profile of Botanicals 282
    Post-Harvest Treatments 282
    Extraction Process 284
    Purification and Enrichement 286
Fingerprint Analyses for Botanical Identity 287
Conclusion 288
Perspectives 288
Acknowledgments 288
References 289

INTRODUCTION

Humans have always made use of the plant kingdom for their nutrition, health, and personal care needs. Since the discovery of organic synthesis and its industrialization in the nineteenth century, synthetic compounds have played an important role in all three markets. Synthetic ingredients were used to replace botanicals in many applications. Today, consumers are increasingly and actively demanding a return to natural solutions for health, food, and personal care uses.

In ancient times, extraction and processing of natural products were used for food, herbal remedies, and cosmetic applications. Their use is widely emphasized in the archaeological literature, with major evidence of wine and olive derivate products [1]. The identification of tartaric acid in pots is a potential indicator for a grape-based alcoholic beverage. Resinated wine was being produced on a fairly large scale in the Neolithic period (5400–5000 BC) at the site of Hajji Firuz Tepe, Iran, and Terebinth tree resin (*Pistacia atlantica* Desf.) was intentionally added for its antimicrobial properties [2]. These archeological findings show evidence of an intentional extraction process used to create an additive for protecting the wine from turning to vinegar, dating to the sixth millennium BC.

In the Late Bronze Age, a variety of herbs, spices, and flowers were already available for medicinal, cosmetic, and flavoring purposes [3]. Dioscorides, a Greek physician, pharmacologist, and botanist, writes about the medicinal uses of sage herb, a botanical of great importance for medicinal uses in this period: "decoction of the leaves and stems has the power to incite urination and menstruation, to induce abortion, and to heal the wounds caused by sting-rays" [1].

Today, in addition to the well-established extraction processes for sugar (from beet or cane), and decaffeinated tea and coffee in the food industry, many other formulations using botanical extracts have been developed. Water is the solvent of choice for the extraction of polar compounds and water soluble applications. For other applications, organic and inorganic solvents are required and, depending on the solubility characteristics of the extractable material and the regulatory framework, different solvents are selected. Recent trends in *green extraction techniques* have largely focused on finding solutions that carefully select and minimize the use of solvents, enabling process intensification for the production of high quality extracts [4,5].

To guarantee the safety of botanical preparations, the identity of the raw material needs to be verified and the processed material (botanical preparation or ingredient) also has to be tested for identity. Adulteration of plant material and plant extracts, whether accidental or intentional, is a frequent source of media attention in the dietary supplement and food markets.

Botanicals and processed botanical preparations are generally harmless at commonly used dosages. A regulatory framework also exists to ensure that botanical use does not represent a risk to consumers. Current regulations in the United States [6] and Europe [7] oblige manufacturers to identify the botanical raw material ingredient to avoid fraud and ensure the absence of consumer health risks. Fingerprint analysis for quality evaluation and control of botanicals is being recognized by different official organizations as a tool for obtaining reliable quality data for unknown constituents and botanical plant extracts. Fingerprinting can be accomplished using different methods of analysis, mainly HPTLC, HPLC, GC, MS, NMR. The American Herbal Products Association recommends the use of botanical reference materials to validate chromatographic methods when used for botanical identification [8]. Ideally, the testing methods should include the use of the botanical reference with an associated herbarium voucher [9]. Members of the industry are constantly faced with the challenge of identifying a botanical extract or preparation on the basis of its chemical content. The complexity of this task is largely due to the following factors:

1. The chemical complexity of the botanical raw materials: Available botanical reference materials do not always cover the diversity of chemical compositions.
2. The industrial processes applied to botanicals (extraction solvents, extraction process, purification steps, carriers): These processes have an impact on the chemical profile of botanicals.

In this chapter, we will cover these two root causes of complexity and discuss solutions to overcome them.

CHEMICAL COMPLEXITY OF THE BOTANICAL RAW MATERIALS

The quality of the plant material depends on a variety of factors, including the natural, intraspecific phytochemical variation, with the occurrence of chemotypes; ontogenetic variation; the plant part being examined; and environmental influences during growth [10]. It has also become clear that there is a complex network of interactions underlying phenotypic diversity.

American ginseng root (*Panax quinquefolium*) is a good illustration of this type of variation. Used in popular medicine to improve cognitive function, both ginsenosides Rg1 and Rb1 have been shown to increase choline acetyltransferase levels in the rat brain [11]. Schlag and McIntosh [12] studied the composition of saponin ginsenosides from 10 populations grown in Maryland. The results presented in this study supported earlier evidence that the relative ginsenoside Rg1 to ginsenoside Re content varies by region and these differences are likely influenced more by genotype than environmental factors [12].

Basil is used in traditional medicine, as a culinary herb, and a well-known source of flavoring. Intraspecific variation in the quantity and quality of the essential oils and quantitative and qualitative characteristics in different accessions of basil were studied by Zheljazkov et al. [13]. Authors reported a significant variation among genotypes with respect to oil content and composition. Oil content of the accessions varied from 0.07% to 1.92% in dry herbage. The range of chemotypes showed that basil oil may contain compounds such as (–)-linalool, eugenol, methyl chavicol, methyl cinnamate, or methyl eugenol.

In general, to validate raw materials references (i.e., for basil or American ginseng), the use of several reference chemotypes or an averaged combination of them is indicated to include variation and cover the chemical diversity of genetic resources for botanical identification.

Tribulus terrestris L. herb extracts are used as ingredients in food supplements in the United States, Europe, and Asia, with claims of general stimulating action (physical performance). The steroidal saponins (protodioscin, prototribestin, pseudoprotodioscin, dioscin, tribestin, and tribulosin) of *Tribulus terrestris* and their biological activity depend on the concentration and the distribution of individual saponins, which in turn is influenced by the geographical origin of plant material [14]. Dinchev et al. [14] showed that the chemical profiles of Vietnamese and Indian herb samples vary significantly from those collected in Europe (Bulgaria, Turkey, Greece, Serbia, Macedonia, Georgia, and Iran), suggesting the existence of one chemotype common to East South Europe and West Asia and the presence of other chemotypes in India and Vietnam. The use of a botanical of each origin as reference or an averaged combination of different origins could be a solution to include chemotype variation in botanical identification.

Maca root (*Lepidium peruvianum* Chacon, previously known as *L. meyenii* Walpers) has traditionally been used by Peruvian natives, even prior to the Inca civilization, for both nutritional and putative medicinal purposes, mainly fertility. Zheng et al. [15] first showed scientific evidence *in vivo* of the *aphrodisiac* activity of a purified *Maca root* extract (MacaPure®) by oral administration in normal mice and in rats with erectile dysfunction. These effects were later confirmed in women [16] and men [17]. Different ecotypes exist for maca root and vary in the color of their hypocotyls, which range from white to black [18]. Gonzales et al. [18] indicated that different maca color types have different properties. More recently, these differences have been associated with variations in the concentration of distinct bioactive metabolites, such as macaene, macamides, sterols, and glucosinolates [19]. Melnikovova et al. [20] compared the macaene and macamides content in field and greenhouse grown maca tubers. The greenhouse grown tubers did not contain macaene and macamide. This recent finding shows the importance of growing conditions on the chemical fingerprint. The use of specific compounds as markers (such as macaenes and macamides) allows for botanical identification and the detection of phenotypical variations.

IMPACT OF INDUSTRIAL PROCESSES ON THE CHEMICAL PROFILE OF BOTANICALS

Our goal is not to list all the processes used, but rather to provide several examples on how these treatments impact the phytochemical profiles and their interpretation for identity determination.

POST-HARVEST TREATMENTS

Post-harvest treatments (chopping, drying, steaming, cooking, fermentation, storage, etc.) are commonly used on botanicals and are of primary importance in traditional herbal remedies because they allow the material to be stored for further use. Once the material has been harvested and post-harvest treated/milled, the industrial manufacturing process generally begins with an extraction phase (Figure 17.1).

Panax ginseng root is a good example of how a post-harvest treatment can affect the chemical content of a raw material. Asian ginseng (*P. ginseng* C. A. Meyer) can be classified into two main categories based on drying conditions: natural or oven drying (White ginseng) or steaming the root at high temperature prior to drying (Red ginseng). Red ginseng exhibits a red color due to the presence of caramelized sugars. Examining the distribution of ginsenosides using HPLC is a powerful tool for differentiating and identifying different species (*P. ginseng* C. A. Meyer and *P. quinquefolium*), plant parts (root, aerial part), and post-harvest treatment (red or white ginseng) [21].

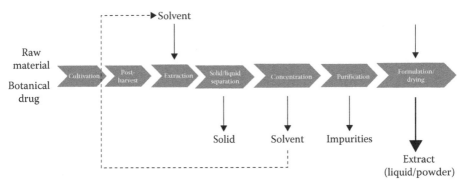

FIGURE 17.1 Illustration of the main industrial manufacturing steps to obtain a powder/liquid extract from a raw material/botanical drug.

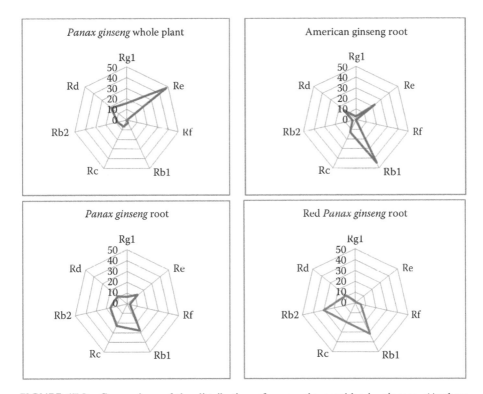

FIGURE 17.2 Comparison of the distribution of seven ginsenosides in ginseng. (Authors data, unpublished.)

To illustrate these differences, and using only ginsenosides described by the European Pharmacopeia/United States Pharmacopeia (ginsenoside Rg3 not evaluated), we analyzed the differences in ginsenoside distribution within the species and the transformation pattern of ginsenosides in steam processing. Distributions for the seven main ginsenosides (Rg1, Re, Rf, Rb1, Rc, Rb2, and Rd) were examined (Figure 17.2).

There are noticeable compositional differences between the types of ginseng. The amounts of Rb1, Re, and Rd in American ginseng are generally higher compared to Asian ginseng, while the amounts of Rg1, Rb2, and Rc are greater in Asian ginseng [21,22]. Moreover, there is a variation in ginsenoside content between leaves and roots of Asian ginseng. Re, Rd, and Rg1 levels were higher in the leaves than other ginsenosides, whereas Rb1 tends to accumulate in the roots [23].

Ginsenosides varied in kind and content between raw and processed ginseng. The results show that during the high temperature steam treatment, polar ginsenosides (Rg1, Re) can convert to low polar form by hydrolyzing sugar chains [21].

Carotenoids are also affected by post-harvesting conditions. Carotenoids are naturally occurring colored compounds abundant in many plants. Dutta et al. [24] indicate that the biosynthesis of carotenoids in fruits and vegetables increases considerably during the ripening process. For example, the β-carotene content of lettuce triples between young and mature leaves. Carotegenesis continues even after the harvest, but processing and storage can affect the bioavailability of carotenoids in fruits. Carotenoids, as unsaturated compounds, can easily isomerize and oxidize, resulting in loss of color and biological activity. Cutting, chopping, and pulping increase exposure to oxygen and trigger the release of enzymes that catalyze their degradation [25].

Isatis tinctoria L. (woad, Brassicaceae family) is an ancient dye and medicinal herb with anti-inflammatory properties [26] that was widely cultivated in Europe from the twelfth to the seventeenth century as a source of indigo [27]. Oberthür et al. [27] studied the influence of different post-harvest treatments on the amounts of isatans A and B, and indican in *I. tinctoria* L. leaf samples. Although the concentrations of isatans A and B in freeze-dried samples measured up to 10.5% and 0.27%, respectively, these two indigo precursors were not detectable in the samples dried at ambient temperature and at 40°C. Moreover, the authors also revealed that the isatan A/isatan B ratio was dependent on time of harvest of the *I. tinctoria* samples (isatan A/isatan B ratio variation from 1–3/1 to 5–10/1 for harvests in June and September, respectively).

EXTRACTION PROCESS

Extraction is a key operation in the manufacturing process of a botanical ingredient. The aim is generally to concentrate an active fraction from the plant. It is technologically defined as the transfer of one or more components of a botanical matrix from its source material to a solvent phase, to bring out specific molecules [5,28]. The transfer of these active molecules to the surroundings occurs through a diffusion that is mainly the result of a concentration gradient of solute between the inner solutions near the solid phase (more concentrated) and the liquid phase [29].

Because phytochemical compounds vary in their nature, the first step of the extraction process involves choosing an appropriate solvent and extraction technique. The solvent should be capable of dissolving the desired solute [5]. The polarity of the solvent used may modify the phytochemical composition of the extract as compared to another extract with another solvent (of differing polarity) obtained from the same raw material.

Dent et al. [30] present the effect of extraction solvents on the chemical composition of sage (*Salvia officinalis* L.). Sage contains several biologically active compounds such as phenolic acids, flavonoids, and diterpenoid compounds. Phenolic components are responsible for the antioxidant properties of sage along with their proven antimicrobial [30] action and antimutagenic activity [30]. The polyphenols found in sage have different polarities, and different extraction solvents were tested, such as water, ethanol, and acetone. The study of Dent et al. [30] demonstrated that the recovery of phenolic compounds was dependent on solvents used, extraction time, and temperature. Their results indicated that hydroalcoholic extraction (30% ethanol) at 60°C was the most efficient condition for the extraction of polyphenols from dry sage leaves.

Tomsone et al. [31] compared different extraction technologies (conventional and Soxhlet) and different extraction solvents (hexane, HE; ethyl acetate, EA; diethyl-ether, DI; propanol, PR; acetone, AC; ethanol 95%, ET; ethanol/water/acetic acid, EWA; and ethanol/water, EW mixtures). Three different genotypes of horseradish were used, and each time, the total polyphenol content was analyzed. As results, this study showed that the recovery of polyphenols from raw material was influenced by their affinity with the chosen solvent; the chemical composition of the different extracts varied according to extraction process and solvent.

Naturex Labs have tested several solvents and different process for rosemary leaf (*Rosmarinus officinalis* L.) extraction. Among them, hydrodistillation process was applied on rosemary leaves resulting in oil extract. Others process as percolation (using different solvents: water, ethanol, and acetone) and a liquid/liquid purification (using acetone and water) were also applied, and the HPTLC profile of all obtained extracts is given in Figure 17.3. HPTLC analysis of the different extracts

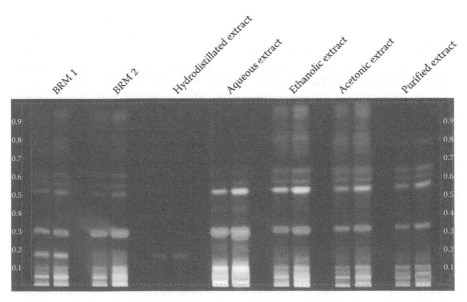

FIGURE 17.3 Impact of different extraction solvent on HPTLC profile. Study carried out in Naturex Labs. BRM1 = Botanical raw material 1 (batch: 192/03/M6); BRM2 = Botanical raw material 2 (batch: B225/012/A11).

was analyzed using a polar method that allows the detection of phenolic acids as rosmarinic as described by Wagner et al. [32].

Using HPLC and GC techniques, Scalia et al. [33] compared the phytochemical profiles of chamomile flower head (*Matricaria recutita* L.) extracts obtained from supercritical fluid extraction, Soxhlet extraction, steam distillation, and maceration. The recovery of the flavonoid apigenin obtained by supercritical CO_2 was 71.4% compared to Soxhlet extraction and 124.6% compared to maceration. However, the highly polar flavonoid apigenin-7-glucoside could not be completely extracted using only CO_2. Different extraction technologies result in both qualitative and quantitative compositional differences.

Xie et al. [34] compared *P. ginseng* C. A. Meyer root extracts prepared in the laboratory with industrial quality extracts using HPTLC and HPLC. These authors described higher levels of the primary ginsenosides (e.g., Rb1, Re, and Rg1) in the laboratory-scale extract compared to the industrial extract, while minor ginsenosides were considerably higher in the industrial extract. This demonstrates the fact that chemical changes occur during the extraction process. In the case of ginseng, these changes are quite similar to the changes previously described in the steamed ginseng [21].

Zhang et al. [35] developed a quality control procedure for these medicinal processed plants. To develop the chromatographic fingerprinting for *Radix Scrophulariae*, crude and processed plants were compared in order to define the main characteristic peaks on the liquid chromatography (LC) traces. This comparison accounts for the concentration of specific secondary metabolites following the extraction process. The LC profile of the botanical extract and raw material should have similarities. For exemple, 5-hydroxymethylfurfural (5-HMF) was not detected in the raw material and appeared as one of the main markers in *Radix Scrophulariae* extracts. Hydroxymethylfurfural is an organic compound derived from dehydration of certain sugars. It is commonly described in various processed foods (coffee and bread) [36]. Saccharides are hydrolyzed under hydrothermal conditions, producing mono- and oligosaccharides. The resulting monosaccharides are further decomposed to secondary products, such as 5-HMF. Knowledge of these mechanisms is a key to understanding the fingerprint profiles of extracts and correctly interpreting identity testing results [37].

Zhang et al. [35] highlighted in their study the differences in the main marker compounds between raw and processed botanical. This observation is a key point for botanical identification because differences on botanical profiles could be from sample scale botanical, as example, when we compare the botanical reference profile from a lab scale (which has a shorter time of processing) with a processed botanical from industrial scale (which has generally much more extraction time than lab-scale preparation) [38].

Purification and Enrichement

After primary extraction, several purification processes can be applied to botanical extracts in order to remove undesirable compounds and/or increase selected compounds. The purification processes commonly used in industry include chromatographic purification (adsorption, ion-exchange, and affinity), extraction (liquid–liquid or solid–liquid), membrane filtration, precipitation, crystallization, and so on. All of these processes involve the selection of a particular fraction of the botanical chemical content and removal of other fractions. When specific markers are removed during the extraction process, the determination of botanical identity is a true challenge. Egb761 is a standardized *Ginkgo biloba* leaf extract [39] and is a good example of an industrial process that uses several purification steps; it implicates more than 50 operations, which include the removal of polyphenols by filtration, and liquid–liquid extraction for removal of undesirable ginkgolic acids [39].

The examples cited above illustrate how the profiles of purified extracts may differ from those of crude extracts. The excipient and encapsulant used for formulation and drying also add a subsequent layer of complexity because they limit the recovery of phytochemicals during sample preparation prior to profile analysis. In the case of purified extracts, the fingerprinting technology does not apply and has to be replaced by other technologies that allow naturality determination such as isotopic ratio methods used for identify caffeine as of natural or synthetic origin [40].

FINGERPRINT ANALYSES FOR BOTANICAL IDENTITY

Fingerprinting for botanicals is a scientific process in which the chemical profile of a botanical raw material is compared to a processed botanical (extract) to verify their authenticity (botanical identification). Because processed botanicals could show a slight different chemical profile of the crude raw material, to avoid false negative results in the botanical authentication process, the integration of different analytical approaches is a necessary root strategy. Hydrophilic and lipophilic botanical extracts, for example, cannot be successfully analyzed using the same sample preparation prior to analysis.

Botanical products have batch-to-batch quality variability due to raw material variation and current manufacturing processes. The rational evaluation and control of product quality and consistency are essential to ensure the efficacy and safety [34].

Chromatographic fingerprinting has been extensively studied for years and still remains the main high-throughput screening method of analysis for the industry of botanical plant extract. Multivariate statistical analysis has showed its efficacy and applicability in the quality evaluation of many kinds of industrial products. Indeed, in order to develop a robust and reliable chromatographic fingerprinting method of analysis, it remains necessary to couple the analytical method with a statistical/chemometric assessment of the processed plant from different origins and from various batches of manufacturing. The aim of this approach is to define the main marker compounds that allow for botanical extract fingerprinting.

Reliable liquid chromatographic methods of fingerprinting have been established for Caowu [41] and Radix Scrophulariae (*Scrophularia ningpoensis* Hemsl.). The combined use of multivariate statistical analysis and chromatographic fingerprinting is also useful to evaluate batch-to-batch quality of botanical drug products. For example, HPLC fingerprinting was applied to monitor the consistency of a peony extract standardized in total glycosides [34]. Thanks to a pattern recognition software interface, raw material and standardized plant extracts can be compared in order to ensure the consistency of raw material quality and manufacturing practices [34].

CONCLUSION

In summary, the identification of botanical raw materials and finished products is multifaceted and many influencing factors need to be taken into account when ensuring conclusive positive authentication. A good strategy for botanical identity testing embraces a combination of methods, from the integration of natural variation knowledge to the chemical changes that can occur during harvesting, drying, and processing. The development of a large range of analytical assays is required, but is not always sufficient when ensuring a strategy for identification. A successful strategy for botanical identification encompasses many elements from the information, knowledge, and understanding of the different steps of botanical processing to the manufacturing processes. Having all the information from the supplier of the starting raw material including the growing conditions, location, drying process, postharvest treatment, and sterilization treatment, and information from the industrial manufacturer of the processed extracts including solvent used, extraction procedure, and formulation agents, are all critical information required to ensure proper identification, as it plays a major role and need to be considered when interpreting a botanical fingerprint. Considering all the aspects previously stated, and also keeping in the mind the overall invariable complexity and diversity of the plant kingdom, many issues can arise and lead to misinterpretation. Transparency between botanical industry members is mandatory, in order to use this set of information, to develop and adjust protocols that can ensure reliable interpretation of analysis and help solve the many challenges botanical identity represents in the industry today.

PERSPECTIVES

Novel approaches are in progress to understand the relationship between chemical fingerprint and biological activity of botanical species. This novel strategy was particularly developed to assess the efficacy of Traditional Chinese Medicine based on the correlation analysis of the chemical fingerprint and biological effect [42]. The same plants, from which herbal medicines are derived, are often harvested in different geographical regions, and their compositions and concentrations vary with the production areas, as previously described. It has been found that their therapeutic effects may often be influenced by their origin and manufacturing process. A direct relationship between quality marker components and biological activity can thus be developed to ensure reliability in quality evaluation. This approach not only uses marker compounds between raw materials and processed plants, but also defines efficacy-related marker components [43].

ACKNOWLEDGMENTS

The authors are warmly grateful to Eve Landen and Jacquelyn Thompson for the final linguistic revision of the English text.

REFERENCES

1. M. J. Cuyler. Rose, Sage, Cyperus and e-ti: The adornment of olive oil at the palace of Nestor. Kosmos Jewellery, Adornment and Textiles in the Bronze Age Aegean: Proceedings of the 13th International Aegean Conference /13e Rencontre égéenne internationale, University of Copenhagen, Danish National Research Foundation's Centre for Textile Research, April 21–26, 2010. Nosch M.-L., and Laffineur R. (eds.), Peeters Publishers, Leuven, Belgium, 2012: 655–662.

2. P. E. McGovern, D. L. Glusker, L. J. Exner, and M. M. Voigt. Neolithic resinated wine. *Nature* 381, 1996: 480–481.

3. R. Arnott. Healing and medicine in the Aegean Bronze Age. *Journal of the Royal Society of Medicine* 89, 1996: 265–270.

4. F. Chemat. *Eco-Extraction Du Végétal—Procédés Innovants Et Solvants Alternatifs.* Dunod, Paris, 2011: 322p.

5. N. Lebovka, E. Vorobiev, and F. Chemat. *Enhancing Extraction Processes in the Food Industry.* CRC Press, 2012.

6. 21CFR111.70. Code of Federal Regulations (CFR). http://www.accessdata.fda.gov/ scripts/cdrh/cfdocs/cfcfr/CFRSearch.cfm?fr = 111.70 T.1, 2013.

7. European regulation (EC) no 178/2002 of the European parliament and of the council of 28 January 2002 laying down the general principles and requirements of food law, establishing the European Food Safety Authority and laying down procedures in matters of food safety. *Official Journal of the European Communities* 31, 2002: L 31/1–L 31/24.

8. L. G. Saldanha, J. M. Betz, and P. M. Coates. Development of the analytical methods and reference materials program for dietary supplements at the National Institutes of Health. *Journal of AOAC International* 87, 2004: 162–165.

9. W. L. Applequist and J. S. Miller. Selection and authentication of botanical materials for the development of analytical methods. *Analytical and Bioanalytical Chemistry* 405 (13), 2013: 4419–4425.

10. C. Franz, R. Chizzola, J. Novak, and S. Sponza. Botanical species being used for manufacturing plant food supplements (PFS) and related products in the EU member states and selected third countries. *Food & Function* 2, 2011: 720–730.

11. K. N. Salim, B. S. McEwen, and H. M. Chao. Ginsenoside Rb1 regulates ChAT, NGF and trkA mRNA expression in the rat brain. *Molecular Brain Research* 47, 1997: 177–182.

12. E. M. Schlag and M. S. McIntosh. Ginsenoside content and variation among and within American ginseng (*Panax quinquefolius* L.) populations. *Phytochemistry* 67, 2006: 1510–1519.

13. V. D. Zheljazkov, A. Callahan, and C. L. Cantrell. Yield and oil composition of 38 Basil (*Ocimum basilicum* L.) Accessions grown in Mississippi. *Journal of Agricultural and Food Chemistry* 56, 2007: 241–245.

14. D. Dinchev, B. Janda, L. Evstatieva, W. Oleszek, M. R. Aslani, and I. Kostova. Distribution of steroidal saponins in *Tribulus terrestris* from different geographical regions. *Phytochemistry* 69, 2008: 176–186.

15. B. L. Zheng, K. He, C. H. Kim, L. Rogers, Y. Shao, Z. Y. Huang, Y. Lu, S. J. Yan, L. C. Qien, and Q. Y. Zheng. Effect of a lipidic extract from *Lepidium meyenii* on sexual behavior in mice and rats. *Urology* 55, 2000: 598–602.

16. N. A. Brooks, G. Wilcox, K. Z. Walker, J. F. Ashton, M. B. Cox, and L. Stojanovska. Beneficial effects of *Lepidium meyenii* (Maca) on psychological symptoms and measures of sexual dysfunction in postmenopausal women are not related to estrogen or androgen content. *Menopause* 15, 2008: 1157–1162.

17. G. F. Gonzales, A. Cordova, K. Vega, A. Chung, A. Villena, C. Goñez, and S. Castillo. Effect of Lepidium meyenii (MACA) on sexual desire and its absent relationship with serum testosterone levels in adult healthy men. *Andrologia* 34, 2002: 367–372.

18. G. F. Gonzales, S. Miranda, J. Nieto, G. Fernandez, S. Yucra, J. Rubio, P. Yi, and M. Gasco. Red maca (*Lepidium meyenii*) reduced prostate size in rats. *Reproductive Biology and Endocrinology* 3, 2005: 2–16.

19. C. Clement, D. Diaz, I. Manrique, B. Avula, I. A. Khan, D. D. P. Aguirre, C. Kunz, A. C. Mayer, and M. Kreuzer. Secondary metabolites in maca as affected by hypocotyl color, cultivation history, and site. *Agronomy Journal* 102, 2010: 431–439.

20. I. Melnikovova, J. Havlik, E. F. Cusimamani, and L. Milella. Macamides and fatty acids content comparison in maca cultivated plant under field conditions and greenhouse. *Boletin Latinoamericano y del Caribe de Plantas Medicinales y Aromáticas* 11, 2012: 420–427.

21. Y. C. Zhang, Z. F. Pi, C. M. Liu, F. R. Song, Z. Q. Liu, and S. Y. Liu. Analysis of Low-Polar Ginsenosides in Steamed Panax Ginseng at High-Temperature by HPLC-ESI-MS/MS. *Chemical Research in Chinese Universities* 28, 2012: 31–36.

22. D. G. Popovich, C. R. Yeo, and W. Zhang. Ginsenosides derived from Asian (*Panax ginseng*), American ginseng (*Panax quinquefolius*) and potential cytoactivity. *International Journal of Biomedical and Pharmaceutical Sciences* 6, 2012: 56–62.

23. X. M. Li and L. Brown. Efficacy and mechanisms of action of traditional Chinese medicines for treating asthma and allergy. *Journal of Allergy and Clinical Immunology* 123, 2009: 297–306.

24. D. Dutta, U. R. Chaudhuri, and R. Chakraborty. Structure, health benefits, antioxidant property and processing and storage of carotenoids. *African Journal of Food, Agriculture, Nutrition and Development* 4, 2004: 1510–1520.

25. D. B. Rodriguez-Amaya and M. Kimura. *Harvest Plus Handbook for Carotenoid Analysis* 2, 2004: 1–58.

26. A. Brattström, A. Schapowal, M. A. Kamal, I. Maillet, B. Ryffel, and R. Moser. The plant extract Isatis tinctoria L. extract (ITE) inhibits allergen-induced airway inflammation and hyperreactivity in mice. *Phytomedicine* 17, 2010: 551–556.

27. C. Oberthür, H. Graf, and M. Hamburger. The content of indigo precursors in *Isatis tinctoria* leaves: comparative study of selected accessions and post-harvest treatments. *Phytochemistry* 65, 2004: 3261–3268.

28. P. Mafart and E. Beliard. *Génie Industriel Alimentaire. Tome II, Techniques Séparatives.* Tec & Doc, Lavoisier, Paris, 1992: 273p.

29. K. Allaf, C. Besombes, B. Berka-Zougali, M. Kristiawan, V. Sobolik, and T. Allaf. Instant controlled pressure drop technology in plant extraction processes. *Enhancing Extraction Processes in the Food Industry*, 2011: 255–303.

30. M. Dent, V. Dragovi-ç-Uzelac, M. Peni-ç, M. Brn-ìi-ç, T. Bosiljkov, and B. Levaj. The effect of extraction solvents, temperature and time on the composition and mass fraction of polyphenols in dalmatian wild sage (*Salvia officinalis* L.) extracts. *Food Technology and Biotechnology* 51, 2013: 84–91.

31. L. Tomsone, Z. Kruma, and R. Galoburda. Comparison of different solvents and extraction methods for isolation of phenolic compounds from Horseradish roots (*Armoracia rusticana*). *Proceedings of the World Academy of Science, Engineering and Technology* 31, 2012: 903–908.

32. H. Wagner, S. Bladt, E. M. Zgainski. *Plant Drug Analysis: A Thin Layer Chromatography Atlas.* Springer, London, (Translated by Th. A. Scott) 1984: 307p.

33. S. Scalia, L. Giuffreda, and P. Pallado. Analytical and preparative supercritical fluid extraction of chamomile flowers and its comparison with conventional methods. *Journal of Pharmaceutical and Biomedical Analysis* 21, 1999: 549–558.

34. P. Xie, S. Chen, Y. Z. Liang, X. Wang, R. Tian, and R. Upton. Chromatographic fingerprint analysis—A rational approach for quality assessment of traditional Chinese herbal medicine. *Journal of Chromatography A* 1112, 2006: 171–180.

35. Y. Zhang, G. Cao, J. Ji, X. Cong, S. Wang, and B. Cai. Simultaneous chemical finger-printing and quantitative analysis of crude and processed Radix Scrophulariae from different locations in China by HPLC. *Journal of Separation Science* 34, 2011: 1429–1436.

36. J. A. Rufian-Henares and S. P. De la Cueva. Assessment of hydroxymethylfurfural intake in the Spanish diet. *Food Additives and Contaminants* 25, 2008: 1306–1312.

37. W. T. Chang, Y. H. Choi, R. Van der Heijden, M. S. Lee, M. K. Lin, H. Kong, H. K. Kim, R. Verpoorte, T. Hankemeier, and J. Van der Greef. Traditional processing strongly affects metabolite composition by hydrolysis in *Rehmannia glutinosa* roots. *Chemical and Pharmaceutical Bulletin* 59, 2011: 546–552.

38. C. Durand, L. Falcao, V. Leroux, and A. Bily. The hidden face of botanical identity: Adulteration and process effect on raw material authentication. *127th AOAC Annual Meeting & Exposition*, Chicago, September, 2013: P-W-004.

39. T. A. Van Beek. Chemical analysis of *Ginkgo biloba* leaves and extracts. *Journal of Chromatography A* 967, 2002: 21–55.

40. O. J. Weinkauff, R. W. Radue, R. E. Keller, and H. R. Crane. Caffeine evaluation, identification of caffeine as natural or synthetic. *Journal of Agricultural and Food Chemistry* 9, 1961: 397–401.

41. Y. L. Qiao, Y. H. Zhang, W. Zhang, and J. L. Zhang. A rapid resolution liquid chromatographic method for fingerprint analysis of raw and processed caowu (*Aconitum kusnezoffu*). *Journal of AOAC International* 92, 2009: 653–662.

42. D. L. Zhu, H. Zhang, W. J. Tian, X. F. Chen, R. An, Y. F. Chai, and X. H. Wang. Study on quality control of Liu Wei Di Huang Wan based on components efficacy relationship. *Scientia Sinica Chimica* 40, 2010: 786–793.

43. J. Wang, H. Kong, Z. Yuan, P. Gao, W. Dai, C. Hu, X. Lu, and G. Xu. A novel strategy to evaluate the quality of traditional chinese medicine based on the correlation analysis of chemical fingerprint and biological effect. *Journal of Pharmaceutical and Biomedical Analysis* 83, 2013: 57–64.

18 Aspects of Quality Issues Faced by Botanicals Used as Cosmetic Ingredients

Jean-Marc Seigneuret

CONTENTS

Introduction 293
The Use of Plants in Cosmetics 294
Quality and Safety Guaranties 294
　　Wild Plants 295
　　Cultivated Plants 296
　　　　Good Agricultural Practices 296
Obtaining Plants with an Optimal Content in Active Principles 299
Eco-Responsible Production 302
　　Conventional Farming (Sustainable Farming) 302
　　Organic Farming 303
Conclusion 303

INTRODUCTION

Every plant and every plant cell is a microfactory that synthesizes thousands of different molecules. These substances represent a broad variety of chemical families from carbohydrates, lipids, and proteins to more complex compounds such as flavonoids, terpenes, and alkaloids.

Not all these molecules are necessarily of interest to cosmetics manufacturers. In fact, sometimes a single molecule, and more often a small family of molecules or a combination of different molecules having a highly targeted function, will offer the desired properties. In other words, only a very limited share of the molecules produced by any given plant is of true cosmetic value.

This being the case, and given today's ongoing focus on eco-responsibility and sustainability, manufacturers of plant-based ingredients for the cosmetics industry must gear their operations to three vital imperatives: first, using plants that offer all quality and safety requirements; second, using plants that synthesize the greatest possible quantity of beneficial molecules; and third, producing the plant biomass as close to processing facilities as possible.

To achieve this, methods for producing these plants for cosmetic use have to respect a number of principles which we will discuss subsequently.

THE USE OF PLANTS IN COSMETICS

Greek mythology tells of a goddess, Panacea, who used plants to heal all ills. Her name became the modern word for *universal remedy*. For a very long time, people have used plants to treat disease as well as to care for and beautify the skin. There were countless formulations for ointments, balms, and lotions that incorporated specific plants, reflecting empirical knowledge about their effects and properties. However, with the advent of chemical synthesis in the laboratory, traditional plant-based remedies were abandoned.

In recent decades, interest in medicinal plants has revived. Species are once again studied and sought after. This trend has been supported by a combination of factors. In the nineties, the cosmetics sector abandoned the use of ingredients derived from animals in the wake of bovine spongiform encephalopathy. Furthermore, campaigns warning against excessive consumption of medications were conducted in developed countries, convincing customers to look for *softer* products. From a scientific point of view, studies provided evidence to confirm the plant properties described empirically by the Ancients. In addition, a collective realization of the richness of nature has taken place as well as a global awareness of sustainable development giving plants the importance they deserve in contrast to nonrenewable resources.

However, the rush to develop new cosmetic applications for plants must be accompanied by utmost scientific rigor and complete environmental responsibility. The actors of this development must never lose sight of which plants should be picked in the wild, and which should be cultivated, of where, when, and how should plants be gathered and of how can plants be processed into a form that preserves their properties and enables them to exert the desired effects.

QUALITY AND SAFETY GUARANTIES

Plants can be collected from two sources: in the wild or from cultivated crops. Each source has its advantages but above all, each one calls for certain essential precautions. Whether the plant is wild or cultivated, it must be identified using a number of requisite criteria to eliminate any risk of confusion.

First, plant nomenclature uses a binary system consisting of a genus name and a species name. Valid names are defined by an international committee of botanists. Each Latin name is accompanied by the name of the botanist (in abbreviated form) who first described or defined the plant. For example, the scientific name of German chamomile is *Matricaria recutita* L. where L. stands for Carl Linnaeus, the eighteenth century Swedish botanist credited with introducing this binary classification.

Second, the specific plant part actually used must be defined. Whole plants are rarely used and the phytochemical composition can differ significantly between the various parts of a given plant. Such variations can result in different parts of a plant having different effects or in a risk of toxicity from some parts of the plant that contain toxic substances. For example, the root of comfrey (*Symphytum officinale* L.) is widely used in cosmetics, although the plant's flower contains toxic pyrrolizidine alkaloids.

An important thing regarding the part of the plants used is that people utilizing them must be able to recognize them through visual observation. Identification is based on the description of the macroscopic and microscopic characteristics of the plant part (e.g., shape of flowers and leaves for the macroscopic description and types of cells and cell wall characteristics for the microscopic description). The identification process also allows any unwanted foreign matter to be detected.

This identification must be achieved using various chromatography techniques: thin-layer chromatography, high-performance liquid chromatography, and gas chromatography. It ensures that plants chemical composition conforms to their standards. Chemical identification is a useful counterpart to botanical identification: a nonconforming aspect may indicate another species, adulteration, or mixture with another plant. How relevant the identification is depends on the representativeness of the sample. Indeed, within a given batch of plants, any contamination may be extremely localized. It is therefore advisable to follow a sampling plan specified for the purpose by the pharmacopeia.

Since very recently, we have a new possibility at our disposal of identifying plants, that is, genomic identification. Today, it is imperative to authenticate the species, variety, and geographic origin of plants. Modern techniques centered on the plant genome are being developed and starting to find real-world applications to authenticate the species, variety, and geographic origin. It will soon be possible to identify the variety and species of plants using PCR (polymerase chain reaction)-amplified and sequenced short fragments of deoxyribonucleic acid (DNA), especially mitochondrial DNA. The same method will also allow detection of the presence of other plants. The same imprint can be used to track the plant all along the processing chain and through to the commercial product; it can serve as the *signature* of the product. Genomic identification will ensure traceability, detect adulteration, strengthen the brand image, and help support claims of ethical production and preservation of biodiversity. It will become fully operational once a reliable and extensive database becomes available.

WILD PLANTS

Wild plants have been extensively used, although not always in accordance with biodiversity conservation principles. There is no doubt that some plants have become extinct or are endangered due to reckless exploitation by humans.

Rosewood (*Aniba rosaeodora* Ducke) is an eloquent example of this, one that triggered awareness of the dangers of irresponsible exploitation. In the early nineteenth century, French Guiana was one of the world's largest producers of rosewood essential oil. Production was carried out by traveling distillers who would cut down all the trees within an acceptable haulage radius of their processing facilities. Once they had depleted an area, they would move to an adjacent tract and repeat the process disregarding the need to regenerate the species by planting new saplings. Without the introduction of protected status for the plant, accompanied by a replanting program, rosewood would have disappeared from Guiana's equatorial forest.

Today, the collective awareness of the need to preserve biodiversity led to the Convention on International Trade in Endangered Species of Wild Fauna and Flora (CITES). The text of the agreement was first adopted in 1973 in Washington. It has now been signed by 175 countries.

CITES classifies endangered plants in three appendices according to the severity of the threat of extinction. Appendices cover plants which it is prohibited to pick, those subject to restrictions as to the part picked, and authorizations covering cultivated plants only. However, the existence of a treaty does not mean that simple common sense should not prevail when picking wild plants that do not feature on the CITES list.

Picking practices must ensure the long-term survival of plant populations in the wild, along with the related habitats. The target species' population density must be sufficient at the collection site.

Picking practices must be nondestructive for both the plant and the environment. For example, when collecting tree roots, the main roots must not be sliced or unearthed. When a species is used for its bark, the latter must be removed only on a single side of the tree, in longitudinal strips.

Those who gather the plants must possess sufficient knowledge of the target species. They must be able to distinguish that particular species from related species and/or species of similar morphological characteristics. They must also be instructed on all aspects of environmental protection and the conservation of plant species. They must understand the benefits for society of ensuring sustainable harvesting of wild plants.

CULTIVATED PLANTS

Efforts to ensure regulatory compliance, safety, and quality coupled with concerns about eco-responsibility and the need to conserve biodiversity all make the use of cultivated plants a wise choice.

Cultivating plants within the framework of contracts between growers and manufacturers is a way to be sure of target species identification and to manage the risk of unwanted contamination. Moreover, cultivation permits varietal selection to breed plants that exhibit favorable agronomic characteristics (productivity, acclimation to the environment, resistance to disease) in addition to optimized concentrations of active ingredients.

With the current focus on reducing carbon footprints, an advantage not to be underestimated is that cultivated crops can be established as close as possible to the industrial processing site. The idea of transporting plants over thousands of kilometers is now as intolerable as it is unsustainable, given that only a very small portion of the plants will actually end up in a cosmetic product.

Good Agricultural Practices

Adhering to good agricultural practices (GAPs) is a form of quality assurance for plant materials destined to become raw materials for cosmetics manufacture. GAPs can improve the quality, innocuousness, and effectiveness of plant-based finished products. They also aim to encourage and support sustainable cultivation and harvesting of high-quality medicinal plants using methods that promote the conservation of plants and the environment in general. Certain fundamental principles must be followed at every step in the growing of the plant, from seed to seedling and from crop management to harvest, not to mention the role of personnel.

Seeds and seedlings must be of appropriate quality and, to the greatest possible extent, free of contamination and diseases. Quality at this stage promotes healthy growth of the plant.

As far as *cultivation* is concerned, good agricultural management principles must be applied, including suitable crop rotation to meet the plants' environmental requirements. The emphasis should be on *conservation agriculture* techniques. These aim to conserve, enhance, and utilize natural resources more efficiently through the stewardship of available soil, water, and biological resources, combined with external inputs.

This approach helps to preserve the environment and ensure strong, sustainable agricultural yield. The choice of the crop location must consider the risk of contamination from ambient pollution of the soil, air, or water. If necessary, soils must be analyzed to determine toxic metals concentrations because some plants fix them selectively. Analyzing these risks makes it possible to limit the concentrations of toxic (heavy) metals in the cultivated plants. In addition, the ecological impact of cultivation activities must be assessed and monitored because introducing crops of nonindigenous medicinal plant species can jeopardize the biological and ecological balance of the area.

Crop management practices will be guided by the plant's growth characteristics and by the part of the plant actually being used. In cases of absolute necessity, authorized agrochemicals may be used to protect medicinal plant crops. They must be applied at the minimum effective dose in accordance with regulations in effect. Applications must respect a specified minimum interval between treatment and harvest. Treatments must be recorded in a crop record. Full compliance with all these conditions will ensure that the levels of pesticide residues in the plants will not exceed authorized limits.

When harvesting, to ensure the best possible quality of plant material, medicinal plants must be picked at the optimal time. That time depends on what part of the plant is being used. Moreover, the concentration of biologically active ingredients varies with the plant's stage of development. The timing of picking is determined by the quality and quantity of biologically active ingredients, rather than by the total volume of the plant part to be picked. The optimal period can be pinpointed by monitoring the concentration of actives over the course of the plant's life.

When collecting the plants, care must be taken to ensure that no foreign matter, weeds, or toxic plants are mixed in with the crop of medicinal plant material.

Medicinal plants must be picked under optimal conditions. The material collected must be transported to a drying room without delay to prevent microbial fermentation and mould growth. To limit the risks of deterioration of the plant, it is important to consider the distance between the plot and the drying and storage locations. The acceptable distance varies with the part of the plant collected. For example, seeds are usually collected dry, so the transport distances can be relatively long. Likewise for underground parts which, moreover, are often gathered in autumn or winter. In contrast, aerial parts picked in late spring or summer require much more limited transport distances—less than 30 min.

Growers must have sufficient knowledge about each medicinal plant, namely, their botanical identification, cultivation characteristics, and environmental requirements, and the techniques for picking and storage. They must be instructed on all aspects of environmental protection, conservation of medicinal plant species, and appropriate agricultural practices.

Initial post-harvest processing is less known but also a critical step. The plants harvested are said to be in a *fresh* state and contain a variable percentage of water, ranging from 15% for a seed, to about 80% for an aerial part (leaf, flower), to over 90% for the fruit. In order to be preserved without being altered between harvest and extraction, the parts of the plant used must imperatively be frozen or dried.

After being harvested, the plant is very quickly subject to general natural biochemical or microbiological deterioration. Storing in cold or dry conditions slows these phenomena.

Extraction from a fresh plant requires almost immediate use, as well as specific precautions in order to avoid deteriorated molecules, which are not initially present in the plant and which could be toxic. For example, this technique can be used in the case of extraction of essential oil where the hydrodistillation machine is used at the harvest site. Furthermore, the use of steam solves the enzyme and microbiological problems.

Cold, specifically freezing, prevents enzyme reactions and microbiological fermentations. The enzymes and the microorganisms are therefore preserved and are reactivated when defrosted, or during extraction, as long as a destruction phase is not carried out. This freezing technique can be expensive if the storage time is long and thus makes the carbon footprint bigger.

The drying technique is the most commonly used to preserve plants. However, this one must also be perfectly handled. If it is carried out badly or in the wrong conditions, it can be dangerous to the quality and safety. Drying allows the quick elimination of water in order to avoid enzyme reactions which lead to deterioration and microbial growth. The percentage of water generally permitted and recognized by experience is 12%. Below this level, the quantity of free water is insufficient for the biochemical and microbiological mechanisms. The risk of the occurrence of undesirable, and even toxic, molecules can therefore be handled perfectly as long as the drying is carried out quickly and efficiently at an adequate temperature which must be a compromise between speed and the heat sensitivity of the plants' active ingredients.

An interesting example is the case of *erysimum*, a plant from the Brassicacea family, which contains a glucosinolate type molecule (sinigrin) that has numerous uses. This molecule is stored in the cell's vacuole where it remains stable. During the harvest, the plant tissues may be lacerated destroying the cellular compartmentalization and therefore putting the sinigrin in contact with a degrading enzyme (myrosinase). The sinigrin therefore breaks down into a series of molecules, for certain types into volatile and strong-smelling toxins, among others. It must be noted that this process is part of the plant's natural defense strategy in the event of attack by parasites (insects) or microorganisms, as the degraded molecules play a repulsive or antimicrobial role. However, for the cosmetic use of erysimum, it is vital to preserve the sinigrin integrity, which is molecularly very stable in the absence of myrosinase, in order to maintain its activity and safety of use.

From this deterioration problem point of view, the erysimum drying temperature and conditions are optimal at 75°C, the necessary temperature to completely prevent myrosinase activity (Figure 18.1).

This shows the need to carry out the drying in controlled conditions with the appropriate tools. Even though natural sun drying is very environmentally advantageous, it

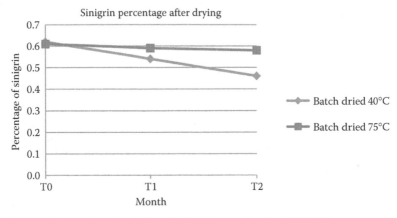

FIGURE 18.1   Internal study of Alban Muller International and PMA28.

does not generally produce good-quality plants as the oxidation phenomenon accelerates in direct light and climate variations and external variants cannot be closely controlled. However, the use of drying equipment powered by solar power, season, and climate permitting may be an option.

In addition, the artificial drying material must be adapted to the drying quantities and especially must not generate PAH (polycyclic aromatic hydrocarbon) contaminants; highly toxic molecules produced by burning fossil fuels.

The quality of storage is vital for the preservation of the plant's quality. Even if the plant is perfectly dry at the start, it will deteriorate if the necessary hygroscopicity and temperature conditions are not met. The increase in humidity in the plant can provoke certain plants to develop microorganisms, generally fungus, producing mycotoxins such as aflatoxins and ochratoxins. These are highly toxic molecules, as LD_{50} in animals is very often lower than μg/kg, and this is the reason why it is important to have the correct storage conditions. For certain plants that come from countries with hot, humid climates, notably rich in lipophilic components, systematic analysis is strongly recommended.

It is also important not to neglect attacks from harmful creatures such as rodents and insects. Trapping systems must be implemented in order to avoid deterioration and contamination of stored plants.

OBTAINING PLANTS WITH AN OPTIMAL CONTENT IN ACTIVE PRINCIPLES

The growing of plants for medicinal purposes enables the use of certain varieties of a species that are the most resilient to environmental, agricultural, and industrial pressures. This plant breeding is mainly carried out in accordance with traditional agricultural methods following two protocols.

On the one hand, *mass selection*, which is a simple and cheap method that farmers have used intuitively throughout history. It consists of selecting seeds from the plants that respond best to the relative requirements and to replant them the following year. This method has enabled the domestication and improvement of plant species generally consumed by man. However, very often the initial desired plant characteristics are not preserved over successive harvests.

On the other hand, *cross breeding*, which avoids this drawback. This is the reason why more modern techniques are now used and are based on plants' genetic information, even though this is more expensive and takes longer. The aim is to create a new variety with the maximum number of desirable properties (rapid growth, increased yield, disease resistance, high concentration of active principles, etc.). This is achieved by cross-breeding plants which, generally, only have one of the desired criteria. The techniques used will depend on the plants' atomic and genetic nature and either sexual or vegetative reproduction will be used.

Today, research into molecular markers helps the selector to have a better understanding of the important genetic characteristics in order to maximize the efficiency of the selection programs. It enables rapid testing of the varieties and to keep those that have the desired characteristics. Selection aided by markers is indeed used to identify the agronomically desirable characteristics, such as yield or resistance to disease.

Plant breeding is an important tool for the future of plant use in cosmetics, an industry which must be exemplary in terms of sustainable development and protection of the environment. This selection must be guided by three major concerns. First, level of yield and frequency of harvest, which remain the continuous areas of research. The higher the production of biomass the lower the production cost and therefore the more competitive. Second, the plants' ability to be autonomous is equally fundamental. For example, the selection of varieties that are disease resistant enables the reduction or the elimination of fungicides. This selection, of resistant characteristics, enables a reduction in material input; that is to say products will only be used if necessary and at the lowest level possible, therefore reducing the risk of plant contamination. Finally, the technological quality of plants must be at the heart of the selection programs in order to meet the processors' different needs. The concentration in active principles must be as high as possible in order to reduce the quantity of plants to dry, transport, and process, therefore significantly reducing the environmental impact. The carbon footprint of plant extracts used in cosmetics is significantly affected by the post-harvest processes such as drying, grinding, transportation, and the process of manufacture of the plant extract. Therefore, a higher active ingredient concentration enables the use of less of the plant and reduces the carbon footprint (Figure 18.2).

This example shows that it is possible to obtain good-quality extracts equivalent to the concentration of active ingredient by using three times less plant mass. Also the risk of undesirable contaminants is mathematically reduced by 3.

Furthermore, the carbon footprint of the obtained extract for the variety selected is 65% lower than for the extract produced with the usual variety. In fact, the best concentration of an active plant ingredient has a direct effect on the carbon footprint all along the production chain of the extract, from farming to waste reprocessing (Figure 18.3).

Composition	German Chamomile Flower (*Matricaria recutita*)		Native Extract (Hydroalcoholic Extraction)		Standardized Liquid Extract	
	Usual Variety (%)	Selected Variety (%)	Usual Variety (%)	Selected Variety (%)	Usual Variety (%)	Selected Variety (%)
Carbohydrates (polysaccharides)	60.00	60.00	63.00	56.00	1.60	1.60
Proteins	20.00	20.00	5.00	4.00	<0.10	<0.10
Lipids	3.00	3.00	<0.10	<0.10	<0.01	<0.01
Active ingredients: flavonodis	1.00	3.00	5.00	15.00	0.10	0.30
Mineral products	10.00	10.00	17.00	15.00	<0.10	<0.10
Solvent: glycerine					98.00	98.00
Quantity of plant per kg of extract					100 g	100 g

FIGURE 18.2 Cosmetic extracts obtained from the flower of a selected variety of German Chamomile (*Matricaria recutita* L.). (Data from internal study of Alban Muller International and PMA28.)

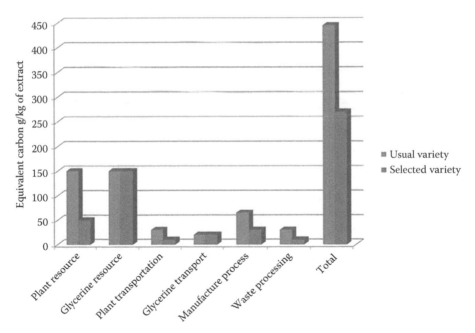

FIGURE 18.3 Carbon footprint of German chamomile cosmetic extracts depending on the variety.

ECO-RESPONSIBLE PRODUCTION

By whatever means the plants are farmed, the main concern is protecting the environment. In this context, the ideal would be organic farming. Today, however, economic, competition, and, at times, environmental pressures mean that the organic process cannot always be used without sustainable risks. The unpredictable aspect of yield, and therefore price, the uncertainty of quality vis-à-vis the concentration in active principles, the risk arising from uncontrollable factors leading to the loss of harvest make organic plants a luxury raw material. Only the most committed cosmetic brands that take on the high additional costs of organic ingredients and use it to their advantage, by using the significant added value in their marketing, can permit themselves to use organic plants.

Unfortunately, the use of organic plants in cosmetics is very often overstated and they are used in cosmetic products simply to promote the product and not as a veritable active ingredient. The organic cosmetic labels require the use of a certain level of organic plant, which leads the less scrupulous companies to use the least expensive organic plant extracts that have little real effect.

Organic farming, in the domain of cosmetic and medicinal plants, therefore does not yet offer perfect supply security which is necessary and indispensable to this industrial domain. However, we can hope to envisage, following the extensive research currently being carried out on crop management, in the relatively close future, the joining of the two techniques of sustainable and organic farming.

CONVENTIONAL FARMING (SUSTAINABLE FARMING)

The majority of plants used in cosmetics come from conventional farming. Depending on the country, there is a very big difference in crop quality. Only a few countries, with strictly applied and monitored agricultural regulations, have a sufficiently high security level for risk-free use in cosmetics. For plants from a number of origins, it will be necessary to carry out rigorous quality checks based on several criteria already quoted: identification, falsification, contamination by other plants, presence of contaminants (pesticides, toxic metals, mycotoxins, PAHs), and other depending on the geographical origin.

Plant processors increasingly demand sustainable farming from their partners whose constraints mean that only some countries are able to meet such demands in practice. This method of farming argues for a farming method which has the aim of maximizing the producers' economic output while also controlling the input quantity (especially chemical substances such as fertilizers or other pest control products) in order to limit their negative environmental impact as much as possible.

Today, it is a sustainable production method that enables us to produce good-quality cosmetic plants that have a safe level of contaminants and are economically viable, if the GAPs are adhered to.

ORGANIC FARMING

Organic farming is based on four major principles:

- It must support and improve soil, plant, animal, human, and environmental health which is one and indivisible.
- It should be based on living ecological cycles and systems, work with them, emulate them, and help to sustain them.
- It should build on relationships that ensure fairness with regard to the common environment and life opportunities.
- It should be managed in a precautionary and responsible manner to protect the health and well-being of current and future generations as well as the environment.

This method of production is notably based on the avoidance of synthetic chemical and genetically modified (GM) products, the recycling of organic matter, crop rotation, and the biological fight.

The aims are undeniably tempting, but currently remain largely utopian unfortunately. They cannot always meet the pressures of a modern and innovative cosmetics industry. Furthermore, in the current conditions of use, they do not necessarily guarantee the safe usage of plants vis-à-vis environmental pollutants. Monitoring on this level remains necessary.

CONCLUSION

Today, the cosmetic industry can have good-quality plants at its disposal, whether they are picked or from sustainable or organic farming. However, this quality can only be achieved if every link of this chain is environmentally and biodiversity friendly, which is good for the planet and human well-being.

Scientific research on physiology, plants' genetic, but also on ecological cycles and systems, now enables us to acquire plants of an optimal quality, which are safe to use and are economically viable which also adhere to the ideal principles of organic farming: the principle of health, the principle of ecology, the principle of fairness, and the precautionary principle.

Index

Note: Locators followed by '*f*' and '*t*' refer to figures and tables, respectively.

A

Absorbance spectra, 224, 224*f*
Accelerator mass spectrometry (AMS), 135–137
Acetylated polysaccharides, 164
Acquity Sample Manager module, 95
Actaea racemosa, See Black cohosh
Actaea species, 70–74, 72*t*
    marker compounds, 74, 75*t*
    marker in Asian, 76*f*
Actein, 166*f*
Adulterants, 8, 9, 34–38, 43, 46–47
    nontargeted detection of
        chemometrics, 93–94
        classical statistical analysis, 94–95
        materials and methods, 95–97
        sample complexity, 92
Adulterated food products, 5, 8–9,
        See also Adulterants; Adulteration
    nonprotein nitrogen-containing
        compounds, 8
Adulteration
    American ginseng with Asian ginseng, 92
    by closely related species, 45*t*
    by distantly related species, 46*t*
    identification, 103
    with isotopes, 148–149
    in natural products industry, 43
    quantification of, 103
Aflatoxins, 299
AFLP, *See* Amplified fragment length
        polymorphism (AFLP)
Agilent Technologies, 109, 111
Agrochemicals, 24–25
    indication, 99
Agro-ecosystem management, 17–19
    biodiversity in, 18–19
    smallholder farmers of tea agroforests, 22–29
Alkaloids, 60, 171
    in black cohosh, 78*f*
Aloe vera, 163–164, 164*f*–165*f*
Amaranth dye adulteration in bilberry, 9–10
AMDIS, *See* Automated mass spectral
        deconvolution and identification
        system (AMDIS)
American ginseng with Asian ginseng, 92
American Herbal Products Association, 9
American Society for Testing Materials (ASTM)
        method
    D6686, 135
    D6866, 136
Amino acids, 8, 133
Amplified fragment length polymorphism
    (AFLP), 47, 53, 73
AMS, *See* Accelerator mass
        spectrometry (AMS)
Analysis of variance (ANOVA), 92, 96, 99, 116
    for full chromatograms, 100*t*
    variable 5215, 101*t*
Anharmonic oscillator model, 217
Anomalous deuterium/hydrogen ratios, 138
ANOVA-PCA, 94–95
Anthocyanins, 9–10
Anthropogenic sources of radioactivity, 125
APT, *See* Attached proton test (APT)
Aroma profiling, 277
Artificial intelligence, 119
Asafoetida powder, 275*t*
Astragalus membranaceus, 35, 46
Astragalus penduliflorus, 35
ASTREE electronic tongue, 264*f*
Atmospheric ^{14}C activity, 127–130
Attached proton test (APT), 189, 191*f*
Attenuated total reflection-infrared
    (ATR-IR), 234
AuthenTechnologies LLC., 44, 46, 47, 62, 63
Authentication of botanicals, 31, 34
    botanical taxon circumscription, 32–34
    morphological
        limitations, 38
        materials and methods, 34–37
        sampling, 38
    vouchering, 39–40
Authentication of natural products, 50, 63
Automated mass spectral deconvolution and
        identification system (AMDIS),
        110–111

B

Banana root (*Musa acuminata*), 46
Bastard varieties, 6
Bayesian analysis, 207
Bayesian probability theory, 202

Benzaldehyde, 133
Benzoic acid, 133
Bilberry
    BRM for, 218–219
Bilberry, amaranth dye adulteration in, 9–10
    quantitative determinations, 10
Bilberry adulteration, 233
Bilberry spectrophotometric method, 11
Bioactive agents, 7
Bio-based material verification, 135–136
Bio-based products for fuel production, 139
Biochemical isotope effects, 149, *See also*
        Fractionations
Biochemical transformations, 146
Biodiversity, 18–19
    in agro-ecosystem management, 18–19
Biomarkers, 133
Biosynthesis, 131
Bitterness intensity, 272–275
Black cohosh, 36, 48, 166*f*, 219, 249
    alkaloids in, 78*f*
    authenticity of, 73–84
    commercial products, 84
    eclectics and, 69, 70
    and formononetin, 77, 79*t*–83*t*
    morphology, 70
    phylogeny, 48, 49*f*
    polyphenol markers in, 76*f*
    quality of, 84–86
    taxonomy, 70–72
    unadulterated, 77, 84, 85*f*
    for women's health, 70
Blueberry leaf, (*Vaccinium angustifolium*),
        162
Botanical identification, 287
Botanical identification, NIR, 228–229
    instrument performance, 229–231
    ruggedness, 229
    selectivity, 229
    traceability, 229
Botanical identity, 158
Botanical literature, 32, 35, 36
Botanical metabolomics, 170–172
Botanical quality, 16–18
    GAPs framework and sensory discernment,
        27–29
Botanical raw materials, chemical complexity,
        281–282
Botanical R&D, quality and authenticity, 1–2
Botanical reference materials (BRM), 36–37,
        218–219, 247, 280
Botanicals, 1
Botanicals, conformation of, 125–127
    atmospheric ^{14}C activity, 127–130
    ^{14}C determinations by AMS and LSC, 137
    isotopic cases of interest, 138–139
    stable isotope ratio analysis, 137–138

stable isotopes, 130–131
    bio-based material verification, 135–136
    increasing specificity, 131–135
Botanical taxon circumscription, 32–34
Bovine spongiform encephalopathy, 294
Breeding practices, 17
Brerton, Richard, 93
British Pharmacopeia, 9–10
BRM, *See* Botanical reference materials
        (BRM)
Broad targeting, 92
Bruker Avance-III spectrometers, 193

C

^{14}C activity levels, 128, 129*f*
Calibration, 137
California poppy herb, 223–224
CAM, *See* Crassulacean acid metabolism
        (CAM)
Camellia sinensis L., 227
Camo Software, 96, 98
Canyon Diablo Troilite (CDT), 144
Carbon dioxide (CO_2), 126
Carbon isotope ratios, 145–146
Carboxylic acid, 218
Catechin compounds, 25
$^{13}C/^{12}C$, 130
^{14}C decay, 126, 126*f*
^{13}C detection by NMR, 186–191
^{14}C determinations by AMS and LSC, 137
Celery (*Apium graveolens*), 46*t*
Center for Herbal Identity, 84
Certificate of Analysis (COA), 222
Cetyl trimethylammonium bromide, 60
21 CFR 111, 2, 43
Chemical and sensory fingerprinting, 262
    electronic nose, 262–264
    electronic tongue, 264–265
    follow-up of oils shelf life, 268–269
    olive oils ranking based on bitterness
        intensity, 272–275
    quality control of vegetal oils, 265–267
    sensory profiling and selection of spices
        based on origin, 275–278
    sensory profiling of olive oils, 270–272
    vegetal products, 265
Chemical ecology, 16
Chemical fingerprinting, 38
Chemical markers, 122, 247
Chemical profile of botanicals, industrial
        processes on, 282
    extraction process, 284–286
    post-harvest treatments, 282–284
    purification and enrichment, 286–287
Chemometrics, 93–94, 191, 221
    defined, 221

Chemotaxonomic classification, 170
Chenomx, 193
    NMR Suite, 193
C–H fundamental and overtone stretch Bands,
        217–218, 217t
Chinese Herbal Materia Medica, 233
ChromaDex
    bilberry botanical reference material, 219f
Chromatogram, 92, 102f, 104f
Chromatographic deconvolution, 111, 112f–113f
Chromatographic fingerprinting, 287
Chromatography, 112
Cimicifuga racemosa L., See Black cohosh
Cimifugin, 76–77
Cimiracemoside F, 73, 77
Cinnamic aldehyde in cassia oil, 135
Cinnamomum, 44
Circumscription of species boundaries, 32
Classification methods, 93
Climate and GAP, 21
Clover honey, 192
Cluster analysis, 117
    qC-NMR spectra, 194
13C-NMR spectrum, 189
Column chromatography, 242
Combustion reactor, 134f
Commercial samples of botanicals, 38
Complete reduction to amplitude frequency table
        (CRAFT), 202
    application to botanical sampels, 204–213
Complex natural mixtures, 184–185
    13C detection by NMR, 186–191
    component ID and bookkeeping, 193
    1H detection by NMR, 186
    NMR data acquisition and processing,
        193–194
    NMR spectroscopy, advantages, 185–186
    qC-NMR spectra, 194
        13C satellites, 194–196
    sample preparation, 193
    statistical analysis, 191–193
Component ID and bookkeeping, 194
Compound exchange format (CEF) files, 111
Compound-specific analysis, 135
Compound-specific isotope analysis, 147
Computer-assisted database building, 193
Condensation, 131, 146
Confusion matrix, 119, 120f
Conservation agriculture, 297
Convention on Biodiversity, 2
Convention on International Trade in Endangered
        Species, 2
Convention on International Trade in Endangered
        Species of Wild Fauna and Flora
        (CITES), 295–296
Conversion/pyrolysis reactions, 144
Cordyceps, 44

Coriander (Coriandrum sativum), 46
Corn's stable isotope signature, 136
Correlation optimized warping (COW), 96
Cosmetics
    cultivated plants, 296
        GAP, 296–299
    eco-responsible production, 302
        organic farming, 303
        sustainable farming, 302
    obtaining plants with an optimal content,
        299–301
    quality and safety guaranties, 294–295
    use of plants in, 294
    wild plants, 295–296
Cosmic radiation, 125
Cosmogenic radionuclides, 125
Coupled human and natural systems (CHANS)
        framework, 27–29, 28f
Coupling constant, 206
C-3 process, 130
C-4 process, 130
CRAFT, See Complete reduction to amplitude
        frequency table (CRAFT)
Cranberry juice project, 110–111, 114
    results, 121t
Crassulacean acid metabolism (CAM), 130
    photosynthesis, 145–146
Crop(s)
    management practices, 297
    production plans and GAP, 20
    rotations, 20
Cross breeding of plants, 300
Cryoprobes, 185, 189
    cryoprobe-assisted 13C-NMR, 190
Crypoprobes, 186
Crystallization, 147
13C satellites, 194–196
Cultivated plants, 296
    GAP, 296–299
Cultivation, 297
Current Good Manufacturing Practice (cGMP),
        6, 227–228
$\delta^{13}C$ value, 133f
Cyanuric acid, 8

D

Daidzin, 204
Data analysis, 52, 57, 60
Data filtering, 115
Data preprocessing, 220
Data processing, 110–111, 194
Data quality, 116, 121, 122
Decision tree (DT), 119
Deconvolution process, 208
Decoupling, 189–190
Degrees of freedom, 97

26-Deoxyactein, 166*f*
Deoxyribonucleic acid (DNA), 295
    barcoding, 43, 47, 61
    chromatograms, 61, 61*f*
    extraction, 60
    fingerprinting, 47, 73
    sequencing, 48, 50, 53, 55, 60–63
        DNA extraction, 60
        gene selection and amplification, 60–62
        and reference material comparison,
            62–63
        validation, 63
    species authentication tests, 45*t*
    testing, 51–53, 52*t*
    toolkit, 51–60
Department of Agriculture in Maryland, 215
Deuterium, 130, 143
Developing classification models, 119–121
Dietary supplement cGMPs, 227–228
Dietary Supplement Health and Education Act
    (1994) (DSHEA), 2, 204
Digitalis lanata, 1
Direct-detection ¹³C-NMR, 188, 196
Discriminant factorial analysis (DFA), 276
DNA, *See* Deoxyribonucleic acid (DNA)
DNA-authenticated Standard Reference
    Materials, 47
DNA-based species authentication, 47
DNA-based technologies, 46–47
1Dnoig-d2o-600, 167
DOSY, 200
Drug Importation Act (1848), 6
Drying technique, 298

E

Echinacea
    ChromaDex BRM for, 233
Echinacea angustifolia, 227, 234
Echinacea pallida, 35, 227, 234
Echinacea purpurea, 227, 234
Ecological cues, 17
Economically motivated adulteration (EMA),
    232–234
    polyphenols, 234
Eco-responsible production, 302
    organic farming, 303
    sustainable farming, 302
ED, *See* Energy drinks (ED)
Electron donating, 217
Electronic nose, 262–264
    FOX electronic nose, 263*f*, 266
    HERACLES electronic nose, 263*f*, 268
Electronic sensing instruments, 262
Electronic tongue, 264–265
    ASTREE electronic tongue, 264*f*
Electron ionization (EI) spectra, 110

EMA, *See* Economically motivated adulteration
    (EMA)
Endangered Species Act, 2
Energy and GAP, 21
Energy drinks (ED), 158, 173–177
Energy Independence and Security Act
    (2007), 136
Environmental contamination, 20
Ethanol, 127
    concentration, 114*f*
Ethylene tertiary-butyl ether (ETBE), 139
Evaporation, 130, 146
Evaporative light scattering detection
    (ELSD), 73
Excitation sculpting (ES) method, 193
Extractable lipids, 133
Extraction from fresh plants, 298
Extra virgin olive oil (EVOO), 109
    experimental, 111–112

F

FAO, *See* Food and Agriculture Organization
    (FAO)
FDA, *See* Food and Drug Administration (FDA)
Federal Biobased Products Preferred
        Procurement Program (FB4P), 135
Fingerprinting
    for botanicals, 287–288
    chemical, 38
    chemical and sensory, *See* Chemical and
        sensory fingerprinting
    chromatographic, 287
    deoxyribonucleic acid (DNA), 47, 73
    genetic, 185
    HPTLC fingerprints, 248–250
    profiling, 73, 267
Fission reactor, 126
Floral morphology, 33
Food and Agriculture Organization (FAO), 19
Food and Drug Administration (FDA), 6, 43, 85,
        91, 138, 204
Food products, adulterated, 5, 8–9
    nonprotein nitrogen-containing
        compounds, 8
Forest tea populations, 22–23
Formononetin, 77–78
    and Black cohosh, 79*t*–83*t*
Fossil fuels, 127
Fourier transformation, 209
Fourier transform near infrared (FT-NIR)
        method, 222
FOX electronic nose, 263*f*, 266
Fractionations, 130–131, 145–146
French Guiana, 295
Fruition, 135
Fusel alcohols, 118–119

G

Gaan, 25
Gaiwan, 26
GAP, *See* Good agricultural practices (GAP)
Gas chromatography, 133, 134*f*, 252
Gas chromatography–mass spectrometry
       (GC-MS), 158
    screening method, 8, 109
Gas chromatography/quadrupole time-of-flight
       (GC/QTOF), 109
Gas sensor arrays technology, 262
Gene specificity, 63
Genetically modified organisms (GMO), 55
Genetically modified (GM) products, 303
Genetic authentication, 48
Genetic fingerprint, 185
Genistin, 204
Germander (*Teucrium*), 38
Ginseng (*Panax quinquefolius*), 172–173
Ginsenosides in ginseng, 11–12
Global meteoric water line (GMWL), 147
Global spectrum deconvolution (GSD)
       algorithm, 194
Glucose, 164
Glycosylation, 205
Golini, Jeff, 226
Gongfu cha dao method, 26–27
Good agricultural practices (GAP), 16, 19,
       296–299
    climate, 21
    crop production plans, 20
    energy, 21
    harvest and on-farm processing and
       storage, 21
    seeds, 20
    site selection, 19–20
    social factors, 22
    soil, 21
    tea cultivation by smallholder farmers,
       23–29
    waste management, 22
    water irrigation and drainage, 21
Good manufacturing practices (GMP), 1–2,
       6, 227
Ground ginseng, 172
Gum oleoresin, 275

H

Half-life, 127
Harmonic oscillator model, 217
Harvest and on-farm processing and storage, 21
Hatch–Slack process, 130
Heavy isotopes, 147–148
HERACLES electronic nose, 263*f*, 268, 276*t*
Herbal dietary supplement, 1

Herbal personal care and cosmetics, 1
Herbal Reference Barcode (HERB™)
       Database, 62
Herbal sexual enhancement, 232
Herbarium specimens, 33, 36
$^2H/^1H$, 130
Hierarchical clustering, 117–118, 118*f*
High-performance liquid chromatography
       (HPLC), 73, 133
High-performance thin-layer chromatography
       (HPTLC), 9–10, 242
    apparatus, 243–244
    fingerprint, specification for, 251–255
    HPTLC fingerprints, 248–250
    identification of botanicals, 250–251
    for identification of botanicals, 245–248
    methods and their validation, 245–248
    SOP for, 255–257
    standardization of methodology, 244
    theoretical background, 242–243
    vs. thin layer chromatography, 242
High pressure liquid chromatography
       (HPLC), 158
High-vacuum technology, 137
^{1}H-NMR analysis, 186
Honey, 148, 184–185
    analysis of, 184
    carbohydrates in, 188
    clover honey, 192
    contamination, 184
    ^{1}H- and ^{13}C-NMR spectra, 187*f*
    ^{1}H-NMR of the sugar region, 188*f*
    hydrogen isotope ratios, 149
    Japanese Bamboo honey, 190*f*
    mono- and oligosaccharides in, 185
    NMR analysis of, 186
    statistical analysis of spectra, 191
Hooke's law, 217
Horwitz Trumpet, 11
Hotelling T^2 values, 93, 99, 100*f*
Human–environment interactions, 17
Hybridization, 17, 22, 33
    DNA, 52, 53
Hydrocarbon-based materials, 126
Hydrogen
    bonding, 217
    and oxygen isotope ratios, 146–147
Hydrogen-containing molecules, 135

I

IAEA, *See* International Atomic Energy Agency
       (IAEA)
Identification of compounds, 112
Ilex, 170
Inductively coupled plasma atomic emission
       spectroscopy (ICP-AES), 227

Infraspecific taxa, 33
Initial post-harvest processing, 298
Instrumental TLC, 242
Integrated pest management, 20
Intercropping schemes, 20
International Atomic Energy Agency (IAEA),
        138, 144
International Olive Council (IOC) sensory
        test, 111
Inter-simple sequence repeat (ISSR), 53
Isoflavones, 77, 204, 204f
Isoflavonoid glycosides, 204
Isoscapes, 148
    to assess geographic origin, 149–151
    continuous, 150
    discrete, 150
Isotope ratio mass spectrometer (IRMS),
        131, 144
    for compound-specific isotope
        measurement, 134f, 147
Isotopes
    defined, 143
    fractionations, 146
    hydrogen, 143
    hydrology, 151
Isotopic analysis, 125
Isotopic cases of interest, 138–139
Isotopic fractionation, 151
Isotopic signature, 133

J

Japanese Bamboo honey, 190f

K

Kabuli asafoetida, 277
Kavain, 166, 166f
Kava kava, 166f
Kjeldahl and Dumas methods, 8, 11
Kudzu (Pueraria lobata), 204

L

Linnaean system of taxonomy, 50
Liquid chromatography-mass spectrometry
        (LC-MS), 73
Liquid chromatography (LC)/photodiode array
        detection method, 9
Liquid chromatography-tandem mass
        spectrometry (LC-MS/MS)
        method, 8
Liquid oxidizer, 137
Liquid scintillation counting (LSC),
        135–136
Lorentzian decay function, 201

M

Madison Metabolomics Consortium Database
        (MMCD), 193
Mahalanobis, 216, 221
Malic acid, 164
Maltodextrin, 218–219
Marijuana, 151
Markers, See also Biomarkers; Chemical markers
    in Asian Actaea species, 76f
    compounds, Actaea species, 75t
    frequency filter and, 115
Markov chain Monte Carlo calculations, 202
MassHunter ProFinder, 115
MassHunter Unknowns Analysis, 111, 113
Masslynx© (Waters Corp.), 95
Mass profiler professional (MPP), 116
Mass spectrometry (MS), 9
Matricaria recutita L., 294
Median baselining, 115
Medicinal plants, TLC and HPTLC methods for
        identification, 246t
Melamine, protein products adulterated with, 8
    quantitative determinations, 10
Melissa, 254
Metabolomics, 163, 170–172
Methylxanthines, 25
MiSeq (Illumina), 56
Mixed crop systems, 22–23
Moisture variation, 220
Molecular vibrators, 217
Muira Puama (Ptychopetalum olacoides), 46
Multidimensional chemometric analysis, 135
Multiple prediction models, 121
Multiplicative scatter correction (MSC), 96
Multivariate statistical analysis, 287
Multivariate statistics, botanical quality and
        authenticity, 107–109
    alignment, 112–115
    cranberry juice experimental, 110–111
    data filtering, 115
    developing classification models, 119–121
    EVOO experimental, 111–112
    general workflow, 109–110
    identification of compounds, 112
    statistical analysis, 116–119

N

Naïve Bayes (NB), 119
National Center for Complementary and
        Alternative Medicine, 47, 61
National Institute of Standards and Technology
        (NIST), 47, 111, 137
National Institutes of Health, 47, 86, 228
Natural ingredients, 2

Natural products
  characterization by NIR, 234
Natural products industry, adulteration in, 43, 47
Natural sources of radioactivity, 125
Near-infrared methods, 8
Near-infrared (NIR) methods, 215–216
  analysis of botanical NIR data, 218–225
  characterization of natural products, 234
  dietary supplement cGMPs, 227–228
  economically motivated adulteration
      screening, 232–234
    polyphenols, 234
  for identification, 230f–231f
  identification and qualification of botanicals,
      226–227
  principles and theory of, 216–218
  validation for botanical identification,
      228–229
    instrument performance, 229–231
    ruggedness, 229
    selectivity, 229
    traceability, 229
Near infrared spectroscopy (NIR), 158
Neural network (NN), 119
Next-generation sequencing (NGS) test, 48,
      56–57, 58t–59t
NMR, See Nuclear magnetic resonance (NMR)
NMR-based quantification methods, 201
NMR SBASE, 164, 166–168, 172
  for kavain, 167f
  of Kava kava, 167f
NMR spectroscopy, 178, 199, 214
  advantages, complex natural mixtures,
      185–186
Noise analysis, 111
Nonprotein nitrogen compounds, 8
Nontargeted detection of adulterants, 91–92, 105
  chemometrics, 93–94
  classical statistical analysis, 94–95
  materials and methods
    data processing, 96–97
    instrumentation, 95
    sample analysis, 96
    sample preparation, 95
  sample complexity, 92
Nontargeted NMR approaches of qualitative
      assessment, 170
  botanical metabolomics, 170–172
Normalization, 96
Nuclear blasts, 128f
Nuclear magnetic resonance (NMR), 9, 73,
      157–158, 199–200
  automation in, 152–163
  ^{13}C detection by, 186–191
  data acquisition and processing, 193–194
  ^{1}H detection by, 186

highly complex mixture, 163
LC-NMR, 200
linearity of, 161f
NMR SBASE, 164, 166–168
nutraceuticals and botanicals
  energy drinks, 173–177
  ginseng, 173
  undeclared drug substance in
      regenerect, 177
qC-NMR spectra, 194
  ^{13}C satellites, 194–196
qualitative and quantitative assessment,
      163–166
quantitative results, 168–169
reproducibility of, 159f
screening of nutraceuticals, 178
spectrum of nutraceutical Regenerect, 177
strengths, 158–162
Null hypothesis, 117

O

Ochratoxins, 299
Odor map, 270, 276
Oils shelf life, 268–269
Olive oils, 149
  ranking based on bitterness intensity,
      272–275
  sensory profiling of, 270–272
Online mass spectrometric detection, 199
$^{18}O/^{16}O$, 130
Organic farming, 303
Organoleptic analysis, 37
Orthogonal partial least squares discriminant
      analysis (O-PLS-DA), 191, 192f
Oxidation/reduction reactions, 144

P

Paleothermometry, 130
Panax ginseng, 11–12, 46, 173, 250
Panax quinquefolius, 11–12
Parahydroxy-benzaldehyde, 133
Partial least square discriminant analysis
      (PLSDA), 93, 119
Partial least squares (PLS) regression, 234
PCA, See Principal components analysis (PCA)
PCR, See Polymerase chain reaction (PCR)
Pelletron accelerator, 137
Perkin–Elmer counters, 137
Pet foods, high protein, 8
Pharmaceutical industry, 216
Pharmacognostic literature, 35
Phosphoenolpyruvate (PEP) carboxylase, 145
Photodiode array (PDA), 9, 73, 95
Photon transmission, 137

Photosynthesis, 126, 130, 145–147
Photosynthetic enzymes, 145
Phytochemicals, 17, 158
Picrohiza, 46
Plantain (*Plantago* spp.), 1
Plant breeding, 300
Plant growers, 297
Plant identification terminology, 34
Planting of perennial polycultures, 20
Plants precursors, 130
Plant taxonomy, 54
Polar/nonpolar solvent, 92
Polycyclic aromatic hydrocarbon (PAH), 299
Polymerase chain reaction (PCR), 47, 295
    amplification of DNA loci, 53
    and CRAFTing, 213*f*
    quantitative (qPCR) techniques, 54
    site-targeted, 54–55
Polyphenols, 25, 73, 74, 76*f*, 84, 115, 234
Predicted relative standard deviation of
        reproducibility (pRSD$_r$), 11
Principal components analysis (PCA), 92–94,
        116, 116*f*, 170, 211, 222
    and authentic SMP, 98*f*
    of isoflavones, 212*f*
    and soybean dietary supplements, 213*f*
Product and ingredient characterization, 7
Product specificaitons, method precision,
        accuracy, and intended purpose,
        11–12
Protein products
    adulterated with melamine, limitations, 8
Proteins, 133
Protium, 143
Proton
    decoupling, 189
    spectra, 205
Pu-erh tea tasting, 26
Pure species, 34
Pyropheophytin, 107
Pyrosequencing, 56

Q

Q statistic values, 93, 100*f*
Quality assurance/quality control
        (QA/QC), 44
    of plant materials, 148–149
Quality control
    of vegetal oils, 265–267
Quality specifications, 7
Quantification, NMR acquisition, 168
    accurate phase correction, 169
    experiment choice, 169
    flat baselines, 168
    lactic acid in blueberry, 169*f*

    pulse width calibration, 168
    relaxation delay, 169
    signal-to-noice, 169
Quantitation, 161
Quantitative, cryoprobe-assisted 13C-NMR
        (qC-NMR), 185
Quantitative analysis, 188
Quantulus liquid scintillation counters, 137
Quaternary carbons, 189
Quinidine, 159*f*–160*f*
Quinine, 159*f*–160*f*

R

Radiation damping, 186
Radioactive decay, 125
Radiocarbon, 129*f*
Radiocarbon
    dating, 127
Radionuclides, 125–126
Random amplified polymorphic DNA
        (RAPD), 53
Recognition, pattern, 233
Recursion, 114
Red Dye No. 2, 9
Reference material comparison, 62–63
Regenerect, 177
Relative standard deviation of reproducibility
        (rRSD$_r$), 11
Renewable bio-fuel, 136
Restriction fragment length polymorphism
        (RFLP), 47, 53
Retention indexes (RI), 110
Retention time alignment, 113
Retention time locking (RTL), 109
Rhizomes of *Curcuma longa*, 234
Risk-Identity Pyramid, 44, 44*f*
Rosewood (*Aniba rosaeodora* Ducke), 295
Royal jelly, 149
Rubidium-87, 125
Rubisco, 145–146

S

Sample class prediction (SCP) models, 109, 121
Sample complexity, 92
Sanger DNA sequencing, 57
Secondary metabolites, 16–17
Seeds
    and GAP, 20
    and seedlings, 297
Sensory profiling of olive oils, 270–272
Sensory test, 108*t*
Sequencing by ligation (SOLiD), 56
Shandong Sinoglory Health Food Co., Ltd., 95
Shengma, See Black cohosh

Shimming, 201
Signal-to-noise spectra, 206
Silica columns, 60
Site selection and GAP, 19–20
Site-specific natural isotope fractionation
    nuclear magnetic resonance
    (SNIF-NMR), 184
Skim milk powder (SMP), 92
Skullcap (*Scutellaria*) species, 38
Smallholder farmers of tea agroforests, 22–29
Smallholder management practices, 22–29
Social factors and GAP, 22
Soft independent modeling of class analogy
    (SIMCA), 92–93, 99, 170
    application, 93, 94*f*
Soil and GAP, 21
Soybean dietary supplements, 205*f*, 208*f*
    Bivariate analysis of, 210*f*
    carbon spectrum of, 209*f*
    CRAFT stacked plot, 210*f*
    proton spectrum of, 208*f*
    PS1D spectrum of, 209*f*
Soybeans (*Glycine max*), 204
Soybean supplements, 204–205
Soy protein isolate (SPI), 95
Spatiotemporal isotope variation, 148
Specimen identification, 35
Spectral deconvolution/curve fitting
    methods, 188
Spectral patterns, 216
Spectrophotometric assay, 9
Spectrum, 92
    builder, 193
Stable isotopes, 130–131
    bio-based material verification, 135–136
    botanical authentication, 144*f*
    defined, 143–145
    increasing specificity, 131–135
    mass spectrometer measuring CO_2, 132*f*
    ratio analysis, 137–138
    use in assessing authenticity, 145
Standard deviation, 117
Standardization, 244
Standard operating procedures (SOP), 159, 244
Standard quantitation tools, 114–115
Statistical analysis, 94–95, 116–119
Statistical quality control (SQC), 277
Stressors to spur production, 17
Strontium, 147–148, 150
Suess, Hans, 128
Suess effect, 128
Sugar resonances, 186
Sulfoaildenafil, 177
Sulfur isotope ratios, 146
Summary plot, 113
Sunflower oils, 265–266

Support vector machines (SVM), 119
Sustainable farming, 302

T

Taxonomic scaffold, 50
Tea agroforests, 22–23
Tea astringency, 25
Tea farmers, smallholder, 22
    botanical quality through sensory
        discernment, 25–26
    gaps in tea agroforests, 23–25
    management effects on tea quality, 25
    tea management systems, 22–23
    tea preparation protocol for sensory
        evaluation, 26–27
Tea in Yunnan, China, 22–27
    quality, 25–26
Tea preparation protocol for sensory evaluation,
    26–27
Tea tasting, 25
Terrace teagardens, 22–23
Terroir, 16–17
Thermal decomposition, 111
Thin-layer chromatography (TLC), 158, 222,
    242, 244
Time-domain analysis, 202
Time-domain deconvolution, 202
Time-of-flight instruments, 111
Toxic plants, 297
Tracing isotope ratios, 184
Trifluoroacetic acid (TFA), 95
Trimethylsilyl propanoic acid (TSP), 167
Triterpene glycosides, 73, 77, 84
Triterpene saponin ginsenosides, 11–12

U

Ultra high performance liquid chromatography
    (UHPLC), 95
Ultraviolet radiation, 125
Umami (*brothy taste*), 25
Undeclared drug substance in regenerect, 177
Unintentional contamination, 1
United States Geological Survey
    (USGS), 138
Universal primers, 55
Unnatural chemical ingredients, 1
Uranium-238, 125
US Pharmacopeia, 95

V

Vaccinium angustifolium, 162
Vaccinium myrtillus L. (Bilberry), 9–10
Vaccinium spp., 162*f*

Vanillin derived from pine trees, 138
Vegetal oils, quality control, 265–267
    odor map of, based on PCA, 267*f*
Vegetal products, 265
Verbascum L. (*mulleins*), 170
Verpoorte, Robert, 170
Viagra®, 177
Vienna-PeeDee Belemnite (V-PDB), 131*f*, 144
Vienna-Standard Mean Ocean Water
    (V-SMOW), 132*f*, 144
Visual inspection, 221
Vouchering, 39–40

W

Waste management and GAP, 22
Water irrigation and drainage, 21
Water-soluble polyphenols, 115
Wild plants, 295–296
World Health Organization (WHO), 19, 38
Wrick, Kathie, 157

Z

Z-transform process, 115